ENGINEERING CELLS AND PATHWAYS

ENGINEERING CELLS AND PATHWAYS

ShivSanjeevi Sripathi

Delve PUBLISHING

www.delvepublishing.com

Engineering Cells and Pathways

ShivSanjeevi Sripathi

Delve Publishing
4164 Lakeshore Road
Burlington, ON L7L 1A4
Canada
www.delvepublishing.com

Email: orders@arclereducation.com

ISBN: 978-1-77956-351-4 (Hardcover)

ABOUT THE AUTHOR

ShivSanjeevi Sripathi completed his Master's in Biotechnology from Mumbai University in 2008. He was awarded for academic excellence in both his Bachelors degree and Masters of Science for securing the second rank at the university level and first rank in his college. This was followed by a short foray into research on a stem cell project and later patent writing and immunoassays. This was followed by his entry into the teaching field where he taught biology and chemistry to 11th and 12th-grade students along with English grammar and communication skills to people of all age groups. He is an author/co-author of books on biotechnology/life sciences (more than 50 published books), which are in several international libraries. These books cover several topics like "depression: insights", "quantum biology", "viral epidemics of the 21st century", "yoga: research insights", "remote sensing in biology", "bionics", artificial intelligence in biology", molecular wires", "mitochondria", "stem cells", "nanotechnology", "camouflage" and "gene editing", to name a few. In his free time, he researches and gives talks on ancient Indian practices, temples and systems. This research compiles the science of traditional practices and how to interpret them in today's context. He has taught and interpreted several mantras and shlokas free of cost to a global audience as a service. He has appeared in a vernacular TV show (in Hindi) on a series of talks covering different temples and customs titled "Bharat ek punah khoj" (translates as searching again for the secrets of India) and on "Simha talks". His core belief is the ancient and powerful prayer "Tamaso maa jyotir gamaya" or "Lead us from darkness to the light".

TABLE OF CONTENTS

Preface....................*xvii*

Chapter 1 **Why Synthetic Biology and Engineering: Our Association with Nature Spans Eons**.................... **1**

References.................... 68

Chapter 2 **Synthetic Biology: Glimpses of Concepts and Examples**.................... **69**

2.1. Introduction.................... 70

2.2. On-Off Switches in Synthetic Biology.................... 72

2.3. Synthetic Biology and Application in Vaccines.................... 76

2.4. Synthetic Biology to Target Pathogens.................... 77

2.5. Synthetic Biology and Drug Discovery.................... 77

2.6. Identification of Targets.................... 79

2.7. Synthetic Biology and Disease Models.................... 79

2.8. Production of Drugs by Metabolic Engineering.................... 80

2.9. Toxicology.................... 81

2.10. Using Cells for Screening.................... 81

2.11. Aptamers in Health.................... 81

2.12. Case Studies: Synthetic Biology in Human Health.................... 82

2.13. Conclusion.................... 84

References.................... 88

Chapter 3 **Few Platforms for Engineering and Production: The Quest for Fuels and Precursors**.................... **95**

3.1. Producing Fuels and Chemicals Using "Cell Factories".................... 96

3.2. Food Products and Nutraceuticals.................... 132

3.3. Final Words.................... 147

References.................... 153

Chapter 4 Machine Learning for Predicting Metabolic Pathways 155
4.1. Introduction 156
4.2. Case Study 1 162
4.3. Case Study 2 170
4.4. Final Words 195
References 196

Chapter 5 Engineering Cells for Drug Synthesis and Delivery 199
5.1. Introduction 199
5.2. Cell Surface Drug Modification 201
5.3. Cells as Delivery Devices 204
5.4. Anucleate Cells as Delivery Systems 206
5.5. Engineering RBCs for Drug Delivery 206
5.6. RBC Biomimetics 208
5.7. Platelets as Delivery Vehicles 208
5.8. Platelet Biomimetics 209
5.9. Mesenchymal Stem Cells as Delivery Systems 210
5.10. Neutrophils as Delivery Systems 211
5.11. Bacteria as Delivery Systems 213
5.12. Synthetic Biology Towards Improved Delivery Devices 214
5.13. Cell Engineering for Cancer Immunotherapy 215
5.14. Engineering Nanomaterials with Bio-Membranes 218
5.15. Red Blood Cell Engineering for Smart Insulin Delivery 222
5.16. β-Cells Integrated with Microneedle Patches 224
5.17. Advantages of Living Cell Delivery System 226
5.18. Challenges 228
5.19. Future Perspectives 229
5.20. Conclusion 230
References 231

Chapter 6 Engineering Microbial Cells 245
6.1. Introduction 245
6.2. Genetic Engineering to Modify Antibiotics 248
6.3. Genetic Engineering to Enhance Antibiotic Production 250
6.4. Engineered Phages to Fight Superbugs 252
6.5. Microbial Control Systems 253

6.6. Improving Catalytic Efficiency at Organelle Level 254
6.7. Increasing Strain Performance at Cell Level.................................. 259
6.8. Reinforcing Cell-Cell Communication at the Consortium Level....... 263
6.9. Other Synthetic Biology Tools to Engineer Microbial Communities . 268
6.10. Molecular Tools for the Introduction of Biosynthetic Pathways 269
6.11. Enzyme Engineering 271
6.12. Genome-Scale Metabolic Models (GEMs).................................... 273
6.13. Random Mutagenesis of Chassis Strain 273
6.14. Tolerance and Transporter Engineering...................................... 274
6.15. Lipid and Membrane Morphology Engineering 276
6.16. Conclusion 276
References 279

Index 299

PREFACE

Ever since advancements in the human brain, our pursuit of "something better" has led to the production and manufacturing of numerous products tailored to our requirements. Just as we have clothes, food items, and gadgets designed according to current trends, the concept of metabolic engineering emerges as a transformative innovation. Metabolic reactions involve breaking down or synthesizing molecules that sustain the cycle of life. Metabolic engineering refers to the deliberate modification of metabolic pathways. This can entail either tweaking existing pathways to enhance the production of a target molecule or reducing the production of an unwanted molecule (referred to as a metabolite in this context). Another approach involves constructing a new pathway by combining the genes encoding several enzymes (even from different organisms) to build novel metabolic systems. This aspect of engineering and synthetic biology paves the way for creating new systems—and perhaps even life!

"Nor can it be supposed that the diversity of chemical structure and process stops at the boundary of the species, and that within that boundary, which has no real finality, rigid uniformity reigns. Such a conception is at variance with any evolutionary conception of the nature and origin of species. The existence of chemical individuality follows of necessity from that of chemical specificity, but we should expect the differences between individuals to be still more subtle and difficult of detection. Indications of their existence are seen, even in man, in the various tints of skin, hair, and eyes, and in the quantitative differences in those portions of the end-products of metabolism which are endogenous and are not affected by diet, such as recent researches have revealed in increasing numbers. Even those idiosyncrasies with regard to drugs and articles of food which are summed up in the proverbial saying that what is one man's meat is another man's poison presumably have a chemical basis."

—Archibald Garrod, *Inborn Errors of Metabolism: The Croonian Lectures delivered before the Royal College of Physicians of London, in June, 1908* (1909), 2–3

The promise of metabolic engineering lies in its ability to produce a variety of molecules to overcome challenges associated with natural resources or nonrenewable sources. Specific enzymes or pathways have been transferred from rare organisms into microbial hosts to facilitate the production of numerous molecules. Synthetic biology integrates the data and materials of various pathways and molecules, while metabolic engineering compiles this information and optimizes the production of desired pathways or molecules. This field is founded on recombinant DNA technology, beginning with alterations or the design of genetic sequences, followed by their assembly into a vector or target organism. Techniques include PCR and its advancements, OMICS tools, CRISPR-Cas9, and recombinases, among others.

This book addresses the need for this technology by first exploring applications involving plants and microbes, followed by detailed methodologies. The use of artificial intelligence (AI) and machine learning (ML) in this field, along with relevant case studies, is also examined. This is particularly pertinent given the widespread integration of these technologies into almost every aspect of life. Numerous examples of metabolic engineering are presented to help readers understand the concepts and developments. While this field is undeniably exciting, several challenges must be addressed. These include prioritizing ETHICAL and RESPONSIBLE research (given the involvement of genetic modification), achieving uniformity in research, ensuring standardization, and maintaining clear communication.

In critical times such as these—marked by wars, pandemics, and emerging diseases—the promise of cells serving as factories to produce vaccines or drugs through ethical and just research and production offers a beacon of hope.

"The potential for synthetic biology and biotechnology is vast; we all have an opportunity to create the future together."

—Ryan Bethencourt

Chapter 1

WHY SYNTHETIC BIOLOGY AND ENGINEERING: OUR ASSOCIATION WITH NATURE SPANS EONS

Shiv Sanjeevi Sripathi

"Microbiology teaches us that size is no measure of significance; even the smallest microbe can wield immense power. Microbes are the silent warriors; these tiniest beings possess the strength to spark a revolution that can have monumental impacts in the world of science."

—A□oo Denish

Research on microbial cells has developed into a "microbiome market" where the past ten years have seen an investment of US$1.7 billion on the human microbiome and medicine. Furthermore, when we consider agriculture, the application of microbes in various sustainable products is experiencing a boost with a Compound Annual Growth Rate (CAGR) of 15–18% (Berg, Rybakova, and Fischer, 2020). Figure 1.1 by Berg et al. (2020) comprehensively summarizes the vital timelines of microbiology.

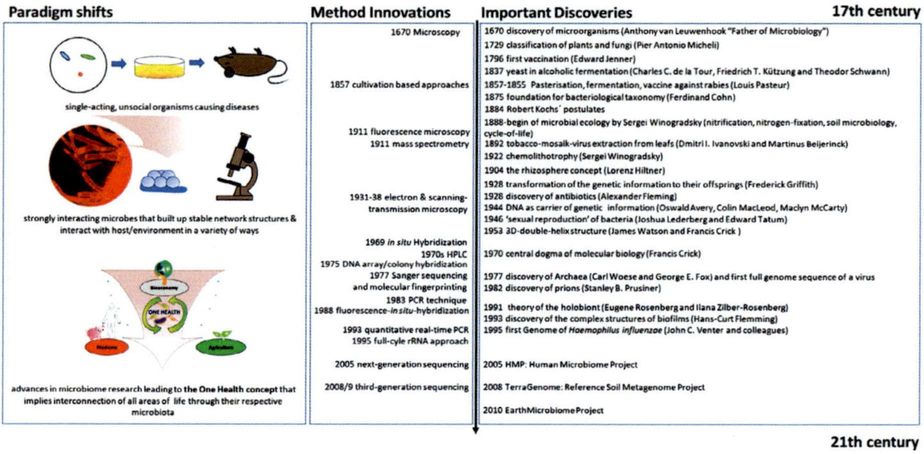

Figure 1.1

Source: Berg, G., Rybakova, D., Fischer, D. et al., 2020: *"Open Access. This article is licensed under a Creative Commons Attribution 4.0 International License, which permits use, sharing, adaptation, distribution, and reproduction in any medium or format, as long as you give appropriate credit to the original author(s) and the source, provide a link to the Creative Commons license, and indicate if changes were made. The images or other third-party material in this article are included in the article's Creative Commons license unless indicated otherwise in a credit line to the material. If material is not included in the article's Creative Commons license and your intended use is not permitted by statutory regulation or exceeds the permitted use, you will need to obtain permission directly from the copyright holder. To view a copy of this license, visit http://creativecommons.org/licenses/by/4.0/. The Creative Commons Public Domain Dedication waiver (http://creativecommons.org/publicdomain/zero/1.0/) applies to the data made available in this article unless otherwise stated in a credit line to the data."*

"We inherit every one of our genes, but we leave the womb without a single microbe. As we pass through our mother's birth canal, we begin to attract entire colonies of bacteria. By the time a child can crawl, he has been blanketed by an enormous, unseen cloud of microorganisms--a hundred trillion or more. They are bacteria, mostly, but also viruses and fungi (including a variety of

yeasts), and they come at us from all directions: other people, food, furniture, clothing, cars, buildings, trees, pets, even the air we breathe. They congregate in our digestive systems and our mouths, fill the space between our teeth, cover our skin, and line our throats. We are inhabited by as many as ten thousand bacterial species; those cells outnumber those which we consider our own by ten to one, and weigh, all told, about three pounds--the same as our brain. Together, they are referred to as our microbiome--and they play such a crucial role in our lives that scientists like [Martin J.] Blaser have begun to reconsider what it means to be human."

— Michae□Specter

Given this interest, our association with microbes can be traced back through history. For instance, the nutritional value of microbes as food is shown in the Figure 1.2.

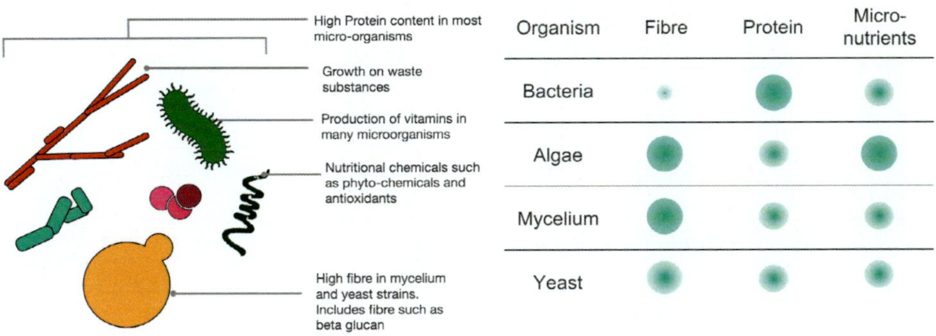

Figure 1.2.

If we consider the application of microbes in the food industry, the applications are only on the rise. Gholami-Shabani et al. (2023) have exhaustively summarized the applications as shown in Table 1.1.

Table 1.1.

Group	Genera/Species	Product/Application(s)
Bacteria	*Acetobacter aceti*	Vinegar fermentation
	Acetobacter pasteurianus	Vinegar fermentation, cocoa fermentation
	Bacillus coagulans	Cocoa, glucose isomerase (food additive), fermented soybeans
	Bacillus licheniformis	Protease (food additive)
	Bacillus subtilis	Fermented soybeans, protease, glycolipids, riboflavin-B2 (food additive)
	Bacillus paranthrasis	Food waste biodegradation
	Bacillus siamenis	
	Bacillus tequilensis	
	Bacillus cereus	
	Bacillus velezensis	
	Bifidobacterium animalis	Fermented milks with probiotic properties
	Bifidobacterium breve	Common in European fermented milks
	Brachybacterium alimentarium	Gruyere and beaufort cheese
	Brevibacterium flavum	Amino acid producer, malic acid, glutamic acid, lysine, monosodium glutamate (food additives)
	Corynebacterium ammoniagenes	Cheese ripening
	Enterobacter aerogenes	Bread fermentation
	Enterococcus durans	Cheese and sourdough fermentation
	Enterococcus faecium	Soybean, dairy, meat, vegetables
	Exiguobacterium acetylicum	Food waste biodegradation
	Burkholderia contaminans	
	Klebsiella pneumonia	Tempeh fermentation, production of vitamin B12

	Lactobacillus acetotolerans	Vinegar fermentation
	Lactobacillus acidophilus	Fermented milks, probiotics, vegetables
	Lactobacillus brevis	Bread fermentation, wine, dairy
	Lactobacillus buchneri	Malolactic fermentation in wine, sourdough
	Lactobacillus casei	Dairy starter, cheese ripening, green table olives
	Lactobacillus delbrueckii	Yogurt and other fermented milks, mozzarella
	Lactobacillus fermentum	Fermented milks, sourdough, urease (food additive)
	Lactobacillus ghanensis	Cocoa
	Lactobacillus helveticus	Starter for cheese, cheese ripening, vegetables
	Lactobacillus hilgardii	Wine malolactic fermentation
	Lactobacillus kefiri	Fermented milk (kefir), reduction of bitter taste in citrus juice
	Lactobacillus kimchii	Kimchi fermentation
	Lactobacillus oeni	Malolactic fermentation of wine
	Lactobacillus paracasei	Cheese fermentation, probiotic cheese, probiotics, wine, meat
	Lactobacillus plantarum	Fermentation of vegetables, malolactic fermentation, green table olives, dairy, meat
	Lactobacillus sakei	Fermentation of cheese and meat products, beverages
	Lactobacillus salivarius	Cheese fermentation
	Lactobacillus sanfranciscensis	Sourdough fermentation
	Lactobacillus versmoldensis	Raw fermented sausage
	Lactococcus lactis	Dairy starter, nisin (protective culture)

	Pediococcus acidilactici	Meat fermentation and biopreservation of meat, cheese starter
	Pediococcus pentosaceus	Meat fermentation and biopreservation of meat
	Propionibacterium acidipropionic	Meat fermentation and biopreservation of meat
	Propionibacterium freudenreichii	Cheese fermentation (emmental cheese starter), probiotics
	Streptococcus natalensis	Natamycin (food additive)
	Weissella genus	Cocoa fermentation, Fermented sausage, Croatian cheese fermented from raw milk, Chinese yogurt
	Zymomonas mobilis	Beverages fermentation
Yeast	*Debaryomyces hansenii (Candida famata)*	Fermentation of blue veined cheese, Fermentation white brined cheeses, ripening of smear cheeses, and sausages and dry-meat products
	Candida guilliermondii	Citric acid fermentation (fodd additive)
	Geotrichum candidum	Fermented dairy products (ripening of many soft and semi-hard cheeses and make a positive contribution to the development of taste and aroma or fermented milks), bioformation of flavor on glucose, peptone, maize oil and meat extract
	Kluyveromyces marxianus	Cheese ripening, lactase (food additive)
	Kluyveromyces lactis	Cheese ripening, lactase (food additive)
	Saccharomyces bayanus	Kefir fermentation, juice and wine fermentation
	Saccharomyces cerevisiae	Probiotic culture, bioethanol, cocoa fermentation, Sake fermentation, cheese-ripening, beer fermentation, bread, invertase (food additive)

	Saccharomyces pastorianus	Beer fermentation, removes and transforms Fusarium trichothecene mycotoxins during fermentation of brewer's wort
	Saccharomyces unisporus	Fermented milk products such as kefir and koumiss
	Schizosaccharomyces pombe	Wine fermentation
	Zygosaccharomyces rouxii	Soy sauce
Filamentous molds	*Aspergillus flavus*	Laccase, lipases, pectinase, protease, α-amylases (food additive)
	Aspergillus fumigatus	Amylases, cellulases, chitosanases
	Aspergillus awamori	Amylases
	Aspergillus sydowii	Amylases
	Aspergillus nidulans	Invertases, laccases
	Aspergillus niger	Beverages, industrial production of citric acid, α-amylases, cellulases, chitosanases, galactosidases, invertases, lipases, naringinases, phytases, tannases, amyloglucosidases, pectinase, celluloses, glucose oxidase, protease (food additives)
	Aspergillus terreus	Amylases, celluloses, phytases
	Aspergillus tamarii	Amylases
	Aspergillus oryzae	Soy sauce, beverages, α-amylases, chitosanases, amyloglucosidase, naringinases, proteases, lipases (food additives)
	Aspergillus sojae	Soy sauce, beverages, α-amylases, amyloglucosidase, lipase (food additives)
	Penicillium camemberti	White mold cheeses (maturation of soft cheeses, such as camembert and brie)
	Penicillium notatum	Glucose oxidase (food additive)
	Penicillium roqueforti	Blue mold cheeses
	penicillium candidum	Bioformation of flavor on glucose, peptone, maize oil and meat extract

	Penicillium nalgiovense	Bioformation of flavor on glucose, peptone, maize oil and meat extract
	Rhizopus oligosporus	Tempe-type fermentation
	Rhizopus oryzae	Tempeh fermentation, Soy sauce, koji
	Fusarium venenatum	Mycoprotein (meat-like)

Source: *Gholami-Shabani, M., Shams-Ghahfarokhi, M., & Razzaghi-Abyaneh, M, 2023: © 2023 The Author(s). Licensee Intech Open. This chapter is distributed under the terms of the Creative Commons Attribution 3.0 License, which permits unrestricted use, distribution, and reproduction in any medium, provided the original work is properly cited.*

On the same lines, several microbes have shown promise in various research studies for other uses in the food industry in terms of detoxification and transformations that have been presented as a Table 1.2 by Gholami-Shabani et al., (2023).

Table 1.2.

Group	**Genera/Species**	**Mycotoxin Decontamination**	**Mechanism of Inhibition**
Bacteria	*Alcaligenes fecalis*	Ochratoxin	Biodegradation to ochratoxin-α due to carboxypeptidase activity
	Bacillus licheniformis	Zearalenone	Biocontrol due to cell-wall adsorption
	Bacillus megaterium	Ochratoxin	Biocontrol due to cell-wall adsorption
	Bacillus pumilus	Ochratoxin	Biocontrol due to cell-wall adsorption
		Zearalenone	Biotransformation due to esterase activity
	Bacillus subtilis	Aflatoxin-B_1	Biotransformation into less toxic products due to laccase activity
		Ochratoxin	Biocontrol due to cell-wall adsorption
		Zearalenone	Biotransformation into Zearalenone-14-phosphate

	Bacillus velezensis	Aflatoxin-B_1	Biotransformation into less cytotoxic products
	Bifidobacterium animalis	Patulin	Biocontrol due to cell-wall adsorption
	Devosia insulae	Deoxynivalenol	Biotransformation into 3-keto- deoxynivalenol
	Escherichia coli	Aflatoxin-B_1	Biotransformation into less toxic products
		Zearalenone	Biocontrol due to cell-wall adsorption
	Gluconobacter oxydans	Aflatoxin-B_1	biotransformation of mycotoxins (Physical binding to bacterial cell wall proteins and polysaccharides)
		Ochratoxin	
		Citrinin	
		Patulin	
	Lactiplantibacillus plantarum	Aflatoxin-B_1	Production of antimicrobial peptides/ involves non-covalent bonds between the cell wall components and the mycotoxin dissolved in the liquid medium
		Ochratoxin	
		Zearalenone	
	Levilactobacillus brevis	Ochratoxin	Biocontrol due to cell-wall adsorption
		Aflatoxin-B_1	
	Lactiplantibacillus pentosus	Ochratoxin	
		Aflatoxin-B_1	
	Lacticaseibacillus paracasei	Ochratoxin	
		Aflatoxin-B_1	
	Lacticaseibacillus casei	Ochratoxin	Biodegradation and/or adsorption

Yeast	*Candida guilliermondii*	Patulin	Biotransformation into E-ascladiol with short-chain dehydrogenase
	Candida parapsilosis	Zearalenone	Glucozyltransferase activities toward main derivatives of Zearalenone
	Candida utilis	Zearalenone	Biodegradation by the combination of probiotics with fungal cell-free extracts
		Ochratoxin	Biocontrol due to cell-wall adsorption
	Kodameae ohmeri	Patulin	Biotransformation into hydroascladiol
	Komagataella phaffi	Fumonisin B_1	Biotransformation due to fumonisin esterase
	Pichia caribbica	Patulin	Enzymatic biodegradation
	Rhodosporidium spp.	Patulin	Biotransformation into desoxypatulinic acid
	Rhodotorula mucilaginosa	Patulin	Enzymatic biotransformation/ Cell-wall adsorption
	Rhodotorula glutinis	Patulin	Biocontrol agent
	Saccharomyces cerevisiae	Patulin	Adsorption to proteins and polysaccharides in the cell-walls
		Ochratoxin	Biological degradation due to dechlorination, hydrolysis, hydroxylation, and conjugation
		Aflatoxin M_1	Biological Biodegradation and/or adsorption
	Saccharomyces pastorianus	Deoxynivalenol;	Biodegradation and/or adsorption
		Zearalenone	

Filamentous molds	*Aspergillus niger*	Ochratoxin	Biodegradation into ochratoxin-α by extracellular ochratoxinase/ Biotransformation pathway
	Byssochlamys nivea	Patulin	Biodegradation
	Clonostachys rosea	Zearalenone	Detoxify Zearalenone through the enzyme zearalenone lactonohydrolase
	Rhizopus oryzae	Aflatoxin	Enzymatic biodegradation
	Trichoderma reesei		
	Cladosporium uredinicola	Aflatoxin-B_1	Biotransformation and detoxification

Source: *Gholami-Shabani, M., Shams-Ghahfarokhi, M., & Razzaghi-Abyaneh, M, 2023:*

Further, as mentioned earlier, microbes emerge as vital in agriculture where they play several roles in facilitating plant growth and providing an edge to crops under stress. Suman et al. (2022) summarized key roles of microbes in agriculture, especially in the context of "sustainable agriculture" (Table 1.4).

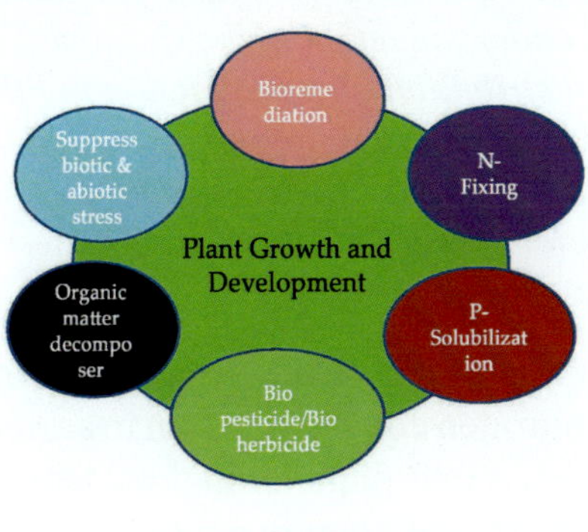

Figure 1.3.

Source: Suman, J., Rakshit, A., Ogireddy, S. D., Singh, S., Gupta, C., & Chandrakala, J. (2022). Microbiome as a Key Player in Sustainable Agriculture and Human Health. Front. Soil Sci. 2:821589. doi: 10.3389/fsoil.2022.821589: *Copyright ©*

Table 1.3.

Microbiome	Crop	Mechanism	References
Bacillus amyloliquefaciens 5113 and *Azospirillum brasilense NO 40*	Wheat	Increase enzyme activity in drought conditions to protect wheat from water stress	(69)
Mesorhizobium sp.	Bengal gram	Enhance the acquisition of nutrients and increase the yield	(70)
Bacillus licheniformis *K11*	Pepper	Auxin and ACC deaminase production and reduce drought stress	(71)
Erwinia species and *P. chlororaphis* sp.	Coffee	Production of organic acids and increase the availability of insoluble phosphate to soluble phosphate in soil	(72)
Bacillus thuringiensis AZP2	Wheat	Stress tolerance enhancement	(73)
Pseudomonas putida H-2-3 sp.	Soybean	Induced tolerance against abiotic stress	(74)
Burkholderia gladioli and *Penicilliumaculeatum* sp.	Oil palm	Increase the availability of P in peat soil	(75)
Pseudomonas aeruginosa FP6 sp.	Chili	Siderophore production to increase the availability of iron	(76)
Metarhizium sp.	Bean (*Phaseolus vulgaris*)	Increases the diversity of plant growth by suppressing the activity of the *Fusarium solani* pathogen	(77)
Streptomyces, *Luteimonas*, and *Enterobacter* sp.	Tomato	Inoculants affect the bacterial structure and promote plant growth	(78)
Plasmodiophora brassicae sp.	Chinese cabbage (*Brassica rapa*)	Changes the bacterial and fungal community concerning healthy plants	(79)
Pseudomonas syringae sp.	Kiwifruit	Affects phyllosphere microbiome and reduces the diversity	(80)
Ralstonia solanacearum sp.	Tomato	Alters the root exudation profile	(81)
Methylobacterium sp.	Potato	Alteration of endophytic microbial community	(82)
Rhizoctonia solani sp.	Lettuce	Increases the diversity of Gammaproteobacteria	(83)

Source: Suman, J., Rakshit, A., Ogireddy, S. D., Singh, S., Gupta, C., & Chandrakala, J. (2022). Microbiome as a key player in sustainable agriculture and human health. Front. Soil Sci. 2:821589. doi: 10.3389/fsoil.2022.821589:

The idea of "microbial protein" (MP) is garnering much interest given the challenge of incomplete nitrogen conversion seen in agriculture where not all the applied nitrogen is converted into plant/animal protein. The production of MP in bioreactors has been reported to circumvent the application of agrochemicals and allow for increased harvest of the biomass (Ciani et al., 2021). Table 1.4, compiled by Ciani et al. (2021), summarizes the nutrient content, costs, and the GHG (greenhouse gas) emission profile of various plant, animal, and microbial sources to serve as an illustration of the future of MP (Figure 1.3).

Table 1.4.

	Protein Content (%) [a]	Land Productivity [b] (kg prot ha^{-1} y^{-1})	Use of Resources			GHG (kg CO_2-eq kg $prot^{-1}$)	Bulk Price (EUR Kg^{-1})	Price per kg of Protein (EUR kg $prot^{-1}$)
			Land Use (m^2 kg $prot^{-1}$)	Energy Use (MJ kg $prot^{-1}$)	Water Use (L kg $prot^{-1}$)			
Animal Proteins								
Beef	25 [15]	2.4–132 10–20 [19]	76–166 [20] 200–4160 [21]	26.9–210 [21] 179–283 [20]	112,000 [22]	48–839 [21]	4.04 [23]	16.16
Pork	20 [15]	29–250 70–80 [19]	40–76 [20] 69–341 [21]	77.5–329 [21]	57,000 [22]	18.5–64.2 [21]	1.06 [23]	5.30
Poultry	31 [15]	75–333	30–134 [21]	83.3–247 [21]	34,000 [22]	12–60 [21]	1.54 [23]	4.97
Fish	17–20 [24]	322–833 70–210 [19]	12–31 [20]	67–630 [20]	-	4–173 [21]	3.70–28 [25]	18.50–164.70
Eggs	13 [24]	74–385	26–135 [20,21]	10.4–210 [21]	29,000 [22]	11–56 [21]	1.38 [26]	10.61
Milk	3.5 [24]	91–1000 60–70 [19]	1–110 [21]	17.5–383 [21]	31,000 [27]	13–73 [21]	0.35 [28]	10.0
Insects (mealworms or superworms)	45–53 [29]	588–714 180–420 [19]	14–17 [29]	156–186 [29]	23,000 [27]	13–17 [29]	3.70 [30]	6.98–8.22
Cultured meat	19 [31]	435–10,000 70–350 [19]	0–23 [20]	164–555 [20]	1734–4420 [31]	10–40 [20]	54–91 [32]	284.21–478.94
Vegetable Proteins								
Soybeans	35 [15]	909–1124	8.9–11 [21]	2.1–20.5 [21]	6034–7428 [22]	0.2–1 [21]	0.33 [23]	0.94

Wheat	12 [15]	38–1492	6.7–260 [21]	5.2–51.4 [21]	15,042–15,567 [22]	0.9–20.6 [21]	0.20 [23]	1.67
Microbial Proteins								
Mycoprotein (fresh)	11.25 [33]	662.25	15.1 [33]	-	6000 [33]	14.2 [33]	-	-
Quorn ™ mince and pieces [c]	13.8–14.5 [33]	370–454 1644–1834	22–27 [33] 5.45–6.08 [34]	147–154 [34]	12,000–14,000 [33]	25–40 [33] 38.27–44.56 [34]	5.86 d	36–38
Arthrospira platensis (spirulina)	46–65 [17,35]	5000–18,181	0.55–2 [20]	15–45 [13,20]	800–3000 [13]	1–5 [13,20,36]	5 [37]	7.7
HOB Solein®	65–75 [38]	60,000 [38]	0.17	-	200 [38]	0.4 [38]	5 [39]	7.6
Metanotrophs Feedkind®	71 [40]	192,307	0–0.052 [40]	-	10–29 [40]	2.23–2.65 [40]	1.52 [40]	3

Source: Ciani M, Lippolis A, Fava F, Rodolfi L, Niccolai A, Tredici MR. Microbes: Food for the Future. *Foods*. 2021; 10(5):971. https://doi.org/10.3390/foods10050971:

Note: According to the authors, "[a]The percentage of protein is referred to dry or fresh weight depending on the way in which the product is available on the market. [b] Land productivity was calculated from land use values or as reported by [19]. [c] Quorn™ products refer to mince and pieces with a content of mycoprotein of about 95%. [d] Quorn™ retail price.

Given the importance of proteins in the human and animal diet, the quest for new sources of this nutrient has ushered in the idea of SCP or single cell protein. This refers to proteins sourced from microbes like bacteria and multicellular organisms like algae and fungi. Further, research is also exploring the use of various waste materials for use as raw materials for producing SCPs. This can help address the issue of waste management and the feeding of the ever-growing population.

(Ritala et al., 2017; Bratosin, Darjan, and Vodnar, 2021)

A scheme of the production of SCP from microbes has been in Figure 1.4 by Bratosin, Darjan, and Vodnar (2021) to comprehend the utility of this field of research.

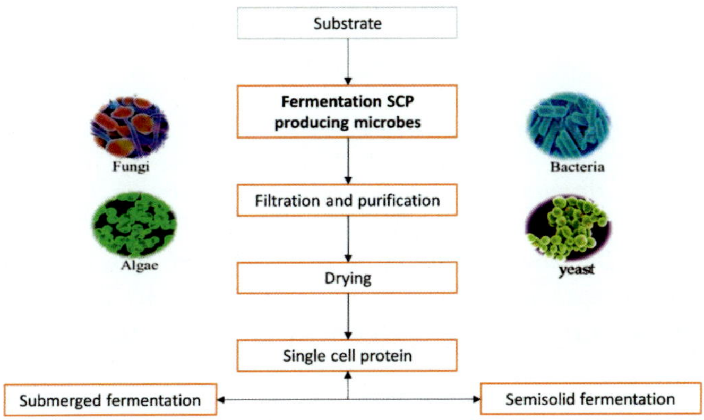

Figure 1.4.

Source: Bratosin BC, Darjan S, Vodnar DC. Single Cell Protein: A potential substitute in human and hnimal nutrition. sustainability. 2021; 13(16):9284. https://doi.org/10.3390/su13169284: *© 2021 by the authors. Licensee MDPI, Basel, Switzerland. This article is an open access article distributed under the terms and conditions of the Creative Commons Attribution (CC BY) license (https://creativecommons.org/licenses/by/4.0/).*

The combined approach of using waste and cultivating microbes to produce MP/SCP has been depicted by Ciani et al., (2021) in Figure 1.5.

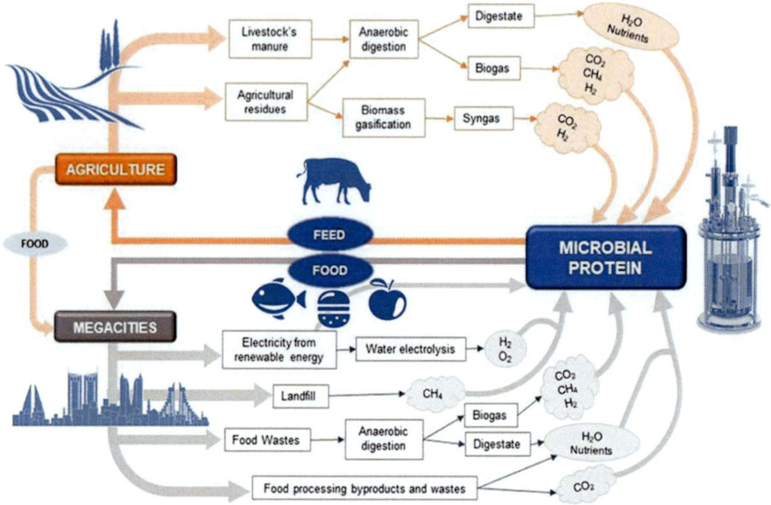

Figure 1.5

Source: Ciani M, Lippolis A, Fava F, Rodolfi L, Niccolai A, Tredici MR. Microbes: Food for the Future. Foods. 2021; 10(5):971. https://doi.org/10.3390/foods10050971: *© 2021 by the authors. Licensee MDPI, Basel, Switzerland. This article is an open access article distributed under the terms and conditions of the Creative Commons Attribution (CC BY) license (https://creativecommons.org/licenses/by/4.0/).*

Table 1.5 summarizes the SCP obtained from algal species cultivated on various sources including waste.

Table 1.5.

Algae	Substrate	SCP (%)
Chaetomorpha antennina	Soda ash effluent	14.0–18.2%
Ulva fasciata	Soda ash effluent	13.7–18.6%
Chlorella sp.	Tofu waste	52.32% ± 3.31
	Tempeh waste	52.00% ± 1.80
	Cheese waste	15.43% ± 2.55
Chlorella salina	Saline sewage effluents	51%
Gracilaria domingensis	Natural habit	6.2%

Gracilaria birdiae	Natural habit	7.1%
Laurencia filiformis	Natural habit	18.3%
Laurencia intricate	Natural habit	4.6%
Palmaria palmate	Natural habit	8–35%
	Natural habit	8.0–35.0%
	Natural habit	11.9–21.9%
	Natural habit	13.5%
	Natural habit	12–21%
Chondrus crispus	Natural habit	20.1%
Porphyra umbilicalis	Natural habit	15–37%
Gracilaria verrucosa	Natural habit	7.0–23.0%
Chlorella sorokiniana	Wastewater	45%
Scenedesmus obliquus	Wastewater	52%
Spirulina	Salinated water	48.59%
	Desalinated wastewater	56.17%

Source: Bratosin BC, Darjan S, Vodnar DC. Single Cell Protein: A potential substitute in human and animal nutrition. Sustainability. 2021; 13(16):9284. https://doi.org/10.3390/su13169284: *© 2021 by the authors. Licensee MDPI, Basel, Switzerland. This article is an open access article distributed under the terms and conditions of the Creative Commons Attribution (CC BY) license (https://creativecommons.org/licenses/by/4.0/).*

Table 1.6 summarizes the SCP obtained from bacterial species cultivated on various sources including waste.

Table 1.6.

Bacteria	**Substrate**	**SCP (%)**
Afifella marina STW181	Commercial shrimp feed	>46
Bacillus cereus	Ram horn	68
Bacillus licheniformis	Potato starch processing waste	38

Bacillus pumilis	Potato starch processing waste	46
Bacillus subtilis	Ram horn	71
Bacillus subtilis sp	Soybean hull	26
Corynebacterium ammoniagenes	Glucose + fructose	61
Cupriavidus necator	Synthetic growth medium	40–46
Escherichia coli	Ram horn	66
Haloarcula sp. IRU1	Petrochemical wastewater	76
Methylococcus capsulatus, Ralstonia sp., Brevibacillus agri, Aneurinibacillus sp.	Methane (natural gas)	67–73
Methylomonas and Methylophilus spp. Metilococ capsulatus	Gas and liquid products of sewage	<41
Methylophilus methylotrophus	Methanol	81
Methylococcus capsulatus (bath)	Methane	53
Methylophilus methylotrophus	Methanol	<70
Methylophilus sp	Supernatant and biogas	24
Methylocapsa acidiphila	Methane	59
Methilomonas.sp	Natural gas	69.3
Methilomonas.sp	Biogas and supernatant of sewage sludge	56
Rhizospheric diazotrophs (whole microbial community	Brewery wastewater	>55
Rhodopseudomonas blastica	Wastewater from a latex rubber	66.7
Rhodopseudomonas palustris	Sludge and sago starch processing	72–74
Rhodocyclus gelatinosus	Poultry slaughterhouse wastewater	67.6

Rhodobacter sphaeroides P47	Pineapple waste	66.6
Rhodopseudomon as palustris P1	Fermented pineapple extract	65
Rhodobacter sphaeroides Z08	Soyabean wastewater	52
Rhodovulvum sulphidophilum	Glutamate malate medium	15,6
Rhodocyclus gelatinosus	Pig farm waste	50.6
Rhodopseudomonas and *R. fulvum*	Sugar refinery wastewater	58
Rhodocyclus gelatinosus	Miso-like effluent medium	63
Rhodobacter sphaeroides P47	Dehydrated medium from pineapple peel waste	66.6
Rhodocyclus gelatinosus	Seafood processing wastewater	50
Rhodopseudomonas sp.	Synthetic medium	11
Rhodopseudomonas palustris Rhodopseudomonas sp.	Wastewater from noodle production	50
Rhodopseudomonas sp. CSK01	Municipal wastewater	60.1
Rhodocyclusg elatinosus R7	Tuna condensate	56
Rhodobacter capsulatus	Synthetic medium	45
Rhodopseudomonas acidophila	Synthetic medium	23
Rhodocyclus gelatinosus	Cassava waste	56
Rhodopseudomonas palustris	Simulated wastewater	45
Rhodopseudomonas palustris	Photosynthetic sludge	74

Source: Bratosin BC, Darjan S, Vodnar DC. Single Cell Protein: A potential substitute in human and animal nutrition. Sustainability. 2021; 13(16):9284. https://doi.org/10.3390/su13169284:

Table 1.7 summarizes the SCP obtained from fungal species cultivated on various sources including waste.

Table 1.7

Fungi	Substrate	SCP (%)
Cladosporium cladosporioides	Rice bran	10%
Penicillium citrinum	Rice bran	10%
Pleurotus florida	Wheat straw	63%
Chrysonilia sitophilia	Lignin	39%
Aspergillus flavus	Rice bran	10%
Aspergillus niger	Apple pomace	17–20%
Aspergillus niger	Banana waste	18%
Aspergillus niger	Rice bran	11%
Aspergillus niger	Citrus pulp	25.6%
Aspergillus niger	Potato starch processing waste	38%
Aspergillus niger	Waste liquor	50%
Aspergillus ochraceus	Rice bran	10%
Aspergillus oryzae	Rice bran (deoiled)	24%
Fusarium semitectum and sp1 and sp2	Rice bran	10%
Fusarium venenatum	Glucose (Product: Quorn™)	44%
Monascus ruber	Rice bran	9%
Trichoderma harzianum	Cheese whey filtrate	34%
Trichoderma virideae	Citrus pulp	32

Source: Bratosin BC, Darjan S, Vodnar DC. Single Cell Protein: A potential substitute in human and animal nutrition. sustainability. 2021; 13(16):9284. https://doi.org/10.3390/su13169284: © *2021 by the authors. Licensee MDPI, Basel, Switzerland. This article is an open access article distributed under the terms and conditions of the Creative Commons Attribution (CC BY) license (https://creativecommons.org/licenses/by/4.0/).*

To highlight the importance of this research field, Table 1.8 summarized by Ritala et al., (2017) outline the patents and companies filed from 2001–2016 for SCPs isolated from various organisms (Tables 1.9–1.11).

Table 1.8. Industrial establishments involved in SCP production.

Company, Country	Microorganism	Substrate	Patent, web site/ references
Algaeon[b]	*Euglena gracilis*	CO_2	http://algaeon-inc.com/
Amoco (BP), USA[a]	*Candida utilis*	Ethanol	Rudravaram et al., 2009
Bega Cheese Ltd [b]	*Saccharomyces*	Wheat	http://www.smh.com.au/business/bega-snaps-up-vegemite-as-part-of-460m-deal-20170118-gtu7wk.html
Bellyyeast, FR[c]	*Kluveromyces*	Whey	Rudravaram et al., 2009
BlueBio Tech Int. GmbH, DE[b]	*Spirulina, Chlorella* sp.	CO_2	www.bluebiotech.de
Blue Green Foods[b]	*Aphanizomenon flos-aquae*	CO_2	https://bluegreenfoods.com/
Calysta Inc. UK[b]	Soil microbes	Methane	www.calysta.com
Cangzhou Tianyu Feed Additive Co. Ltd., CN[b]	Yeast powder (n.a.)	n.a.	www.cztymy.com
CBH Qingdao Co., Ltd, CN[b]	Bacterial fermentation (n.a.)	n.a.	www.cbhcn.com
Cyanotech, USA[b]	*Spirulina platensis*	Sodium bicarbonate, CO_2	www.cyanotech.com/ US19920959649, US19960641159
E3Live[b]	*Aphanizomenon flos-aquae*	CO_2	https://www.e3live.com/
Earthrise, USA[b]	*Spirulina* sp.	CO_2	www.earthrise.com/
E.I.D Parry Ltd., Parry Nutraceuticals Division, IN[b]	*Arthrospira platensis, Chlorella vulgaris*	CO_2	www.parrynutraceuticals.com/#
Euglena Co. Ltd.[b]	*Euglena*	CO_2	http://www.euglena.jp/en/
FEBICO[b]	Spirulina, *Chlorella*	CO_2	http://www.febico.com/
Flint Hills Resources[b]	*Saccharomyces cerevisiae*	Corn	https://www.fhr.com/newsroom
Hainan Simai Enterprising Ltd.[c]	Spirulina	CO_2	

Imperial Chemical Industries, UK (now: AkzoNobel, NL)[a]	*Methylotrophus*	Methanol	Rudravaram et al., 2009; Johnson, 2013
IFP, FR[c]	*Candida tropicalis*	n-alkanes	Rudravaram et al., 2009
Kanegafuichi, JP[c]	n.a.	n.a.	Rudravaram et al., 2009
Klamath Valley Botanicals LLC[b]	*Aphanizomenon flos-aquae*	CO_2	http://klamathvalley.com/
KnipBio[b]	*Methylobacterium extorquens*	methanol	www.knipbio.com
Lallemand Inc., CA[b]	Yeast and bacteria	n.a.	www.bio-lallemand.com
LeSaffre, FR[b]	Yeast	n.a.	http://www.lesaffre.com/
Liquichemica, IT[b]	*Candida maltosa*	n-alkanes	Rudravaram et al., 2009
Marlow Foods Ltd. UK[b]	*Fusarium venenatum*	Glucose syrup	www.quorn.co.uk
Mayanmar Spriulina Factory[c]	Spirulina	CO_2	
Mondelez Int.[b]	Yeast	brewer's spent grain	http://www.mondelez-international.com/
Nucelis[b]	*Yarrowia lypolitica*	n.a.	https://www.nucelis.com/
Nutrinsic[b]	*Bacteria*	starch, brewing, other waste waters	http://nutrinsic.com/ ZA201003590
Phillips Petroleum Company USA (Ohly)[b]	*Pichia* sp. *Torula* sp.	Sugar feed stock	Rudravaram et al., 2009
Qingdao Zhongtai Poultry Ind. Professional Cooperatives, CN[c]	Lactic acid bacteria	Whey, Whey & soybean meal, wheat & rice bran, beer lees, jujube, urea	CN102894183 CN102987056
Roquette Klötze GmbH & Co[b]	*Chlorella*	CO_2	http://www.algomed.de/en/homepage/
Shanghai Gentech Ind. Group Co. Ltd. CN[d]	n.a.	n.a.	CN103843971

Shanghai Tramy Green Food Co, CN[b]	*Aspergillus oryzae, Saccharomyces cerevisiae, Tricoderma* sp.	Bean dregs and soybean processing water: Soybean dregs, Bean waste water	CN103098979 CN103156051
Skotan S.A., PL[b]	*Yarrowia lipolytica*	n.a.	http://www.skotansa.pl/
Skystone Feed Yixing Co, CN[b]	*Aspergillus niger*	Blue-green algae	CN103749957
Tangshan Top Bio-Technology Co., Ltd., CN[b]	*Saccharomyces* sp.	n.a.	www.tuopobio.com
Taiwan Chlorella Manufacturing Co.[c]	*Chlorella*	CO_2	
TerraVia, USA[b]	Alga	n.a.	http://algavia.com/
UniBio A/S, DK[b]	Methanotrophic bacteria	Natural gas	www.unibio.dk PA199900690
Unilever[b]	Yeast	brewer's spent grain	www.unilever.com
Vega Pharma Ltd., CN[b]	Bacterial	n.a.	www.vegapharma.com

n.a. not available.

[a] *Company inactive or has merged or been taken over by another company, with a new name.*

[b] *Active in SCP production.*

[c] *Current activities in SCP unknown.*

Active through partners.

Source: Ritala A, Häkkinen ST, Toivari M and Wiebe MG (2017) Single cell protein—state-of-the-art, industrial landscape and patents 2001–2016. front. Microbiol. 8:2009. doi: 10.3389/fmicb.2017.02009:

Table 1.9. *Patents on SCPs from Algal Sources*

Patent number	Title	Publication date	Assignees and Inventors
WO2017050917 (A1)	Novel method for the culture of unicellular red algae	2017-03-30	**Fermentalg**; Cagnac, O., Richard, L., Labro, J.
CN105861379 (A)	Method for preparing algae powder by drying peculiar smell-free blue algae	2016-08-17	**Jiangnan U**.; Yang H., Ping S., Wang T., Wang W., Zhang L.
WO2016092583 (A2)	Method for growing microalgae, and device for implementing said method	2016-06-16	**Lavanga V**.; Lavanga V. and Farne S.
CN105483014 (A)	Production technology for high-density culture of Chlorella by utilizing fermentation method	2016-04-13	**Qingdao Kehai Biological Co., Ltd**.; Li Y., Xia X., Xu J., Cui J., Li X., Wang D., Chi S., Sun H., Zhao S., Zhao S., Wang B., Cui X.
US2016021923 (A1)	High-Protein Gelled Food Products Made Using High-Protein Microalgae	2016-01-28	**Solazyme Inc**.; Paulsen S., Klamczynska B., Plasse K., Bowman C.
AU2015271929 (A1)	Novel microalgal food compositions	2016-01-21	**Solazyme Inc**.; Brooks, G., Franklin, S., Avila, J., Decker, S. M., Baliu, E., Rakitsky, W., Piechocki, J., Zdanis, D., Norris, L. M.
US2015374012 (A1)	High-Protein Food Products Made Using High-Protein Microalgae	2015-12-31	**Solazyme Inc**.; Klamczynska B., Echaniz A., Zhu R.

US2015320086 (A1)	Microalgae Meal	2015-11-12	**Solazyme Inc.**; Piechocki J., Hansen S. L., Loy D. D., Genther-Schroeder O., Stokes R., Van Emon M.
CN104946536 (A)	Isochrysis zhangjiangensis culture method	2015-09-30	**Dalian U.**; Yang H., Wang X., Xu P., Pan Y., Yu Y.
WO 2013086302 (A1)	Fractionation of proteins and lipids from microalgae	2013-06-13	**Old Dominion U. Research Foundation**; Kumar, S., Hatcher, P. G.
US 20100303990 (A1)	High Protein and High Fiber Algal Food Materials	2010-12-02	**Solazyme, Inc.**; Brooks, G., Franklin, S., Avila, J., Decker, S.M., Baliu, E., Rakitsky, W., Piechocki, J., Zdanis, D., Norris, L.M.
TW20070136127	SCP producing method	2009-04-01	**Zen U Biotechnology Co Ltd**; Chen C.H., Lu C.Y.
KR20040069402	Production of rice cake and noodles, which contain well balanced nutrients and have improved product quality, using fresh-water chlorella	2004-09-21	**Kim Soon Bok**; Kim S. B.
WO 2001005414 (A1)	Algae protein polysaccharide extraction and use thereof	2001-01-25	**Qing Qi, Jian Ding**; Qi, Q., Ding, J.

Source: Ritala A, Häkkinen ST, Toivari M and Wiebe MG (2017) Single cell protein—state-of-the-art, industrial landscape and patents 2001–2016. Front. Microbiol. 8:2009. doi: 10.3389/fmicb.2017.02009:

The search was carried out in Espacenet and focused on patents from companies with known activities with SCP from algae, in addition to use of keywords (alga AND "single cell protein").

Table 1.10. *Patents on SCPs from Bacterial Sources*

Patent number	Title	Publication date	Assignees & Inventors
US201519779A	Microorganisms for the enhanced production of amino acids and related methods	2015-07-16	**Calysta Inc;** Doss, B.D., Giver. L.J., Luning. E.G., Regitsky. D.D., Saville. R.M., Resnick, S.M., Silverman. J.A.
CN104489281A	Method for preparing single-cell protein feed additive by processing wastes	2015-04-08	**Jiangsu Qianyaotang Traditional Chinese Medicine Res. Inst. Co. Ltd.**; Zhang K., Zhang Z.
CN104472867A	Method for preparing single-cell protein feedstuff by utilizing waste liquid in production of ginkgo leaf extracts and application	2015-04-01	**Jiangsu Qianyaotang State Medical Res. Inst. Co. Ltd.**; Zhang, K., Zhang, Z.
CN104450513A	Full-automatic factory full-wave band closed circulating water real-time monitoring breeding device	2015-03-25	**Zhu Zuyang**; Zhu Z.
US2015044327A	Methylotrophs for aquaculture and animal feed	2015-02-12	**Knipbio;** Marx, C. J., Feinberg, L.F.

US2016333384A	Carbohydrate-enriched recombinant microorganisms	2015-01-16	**Calysta Energy Llc; Calysta Inc;** Giver, L.J., Mueller, J., Regitsky, D.D., Saville, R.M., Silverman, J.A.
CN103918874A	Method for improving quality of manioc wastes by using mixed fermentation technology	2014-07-16	**U. of Sichuan Agricultural**; Chen, X., Jia, G., Liu, G., Tang, J., Tang, X., Zhao, H.
CN103484395A	Bacterial strain used for preparing single-cell protein from methanol, and applications of bacterial strain	2014-01-01	**Henan Coal Chemical Ind. Group Inst. Co**.; Chao, Y., Jiang, Y., Jiao, Z., Li, N., Li, W., Li, Z., Song, C., Su, M., Wang, Y., Wei, L., Zhang, X.
CN103156051A	Method using composite bacterium to ferment bean dregs to manufacture protein feed	2013-06-19	**Shanghai Tramy Green Food Co. Ltd**.; Li, L., Shen, J., Yuan, H.
US2014323694A	Multiphase porous flow reactors and methods of using same	2013-05-23	**Calysta Energy Inc;** Gosse, J.L., Harwood, T., Thust, S., von Keitz, M.G.
CN102987054A	Production process of fermented feed	2013-03-27	**Qingdao Tianrui Ecological Technology Co. Ltd**.; Qu T.
CN102987056A	Fermentation method for whey fermentation liquor for fermenting feed	2013-03-27	**Qingdao Zhongtai Poultry Ind. Professional Cooperatives**; Qu T.

CN102978271A	Method for producing carotinoid and single-cell protein via transforming cellulose pyrolytical liquid and levo-glucosan through photosynthetic bacteria	2013-03-20	**Nanjing Normal U.**; Chen, Y., Liu, J., Sun, H., Wei, M., Zhao, Y.
CN102972622A	Preparation method of whey fermented forage	2013-03-20	**Qingdao Tianrui Ecological Technology Co. Ltd.**; Qu T.
CN102960538A	Unicellular protein feed prepared from fermented dragon fruit peel and production method of unicellular protein feed	2013-03-13	**Guangxi Zhuang Autonomous Region Ct for analysis and test research**; Lu, A., Mo, J.
CN102894183A	Preparation method of whey fermented feed	2013-01-30	**Qingdao Zhongtai Poultry Ind Professional Cooperatives**; Qu T.
CN101507469A	Preparation method of composite zymoprotein	2009-08-19	**Jiangxi Purun Mechanical Co. Ltd.**; Yang, T.,
US2009114602A	Biosolids-based food additive for animal feed, methods of production, and business application thereof	2009-05-07	**Nutrinsic Corp; Oberon Fmr Inc; Procell Investments Ltd**; Logan, A.J., Swenson, R.P. Jr., Seth, S.T.
US2004241790A	Method of fermentation	2004-07-09	**Cockbain Julian; Norferm Da; Statoil Asa; Statoilhydro Asa; Bioprotein As; Calysta As;** Eriksen, H., Joergensen, L., Strand, K.

US2003138878A	Method	2003-07-24	**Golding L., Johannessen A., Kleppe G., Larsen J., Moen E., Norferm Da Stavanger, Statoil ASA**; Moen, E., Larsen, J., Kleppe, G., Johannessen, A.

Source: Ritala A, Häkkinen ST, Toivari M and Wiebe MG (2017) Single cell protein—state-of-the-art, industrial landscape and patents 2001–2016. Front. Microbiol. 8:2009. doi: 10.3389/fmicb.2017.02009: Copyright © 2017 Ritala, Häkkinen, Toivari and Wiebe. This is an open-access article distributed under the terms of the Creative Commons Attribution License (CC BY). The use, distribution or reproduction in other forums is permitted, provided the original author(s) or licensor are credited and that the original publication in this journal is cited, in accordance with accepted academic practice. No use, distribution or reproduction is permitted which does not comply with these terms).

Table 1.11. *Patents on SCPs from Fungal Sources*

Patent number	Title	Publication date	Assignees; Inventors
CN106011198A	A prepared from lignocellulosic biomass in agroforestry synthetic starch, cellulosic ethanol and the method of single cell protein	2016-10-12	**Tonglu Tang Won Biotechnology Co. Ltd.**; Ma, H., Ma, Y., Xu, X., You, C., Zhang, W., Zhang, Y.
CN105884104A	A membrane distillation—nanofiltration / reverse osmosis process as a combination of a high concentration of useful components of fermentation waste recycling and waste water purification method	2016-08-24	**Beijing Forestry U.**; Zhang, L., Qu, D., Kang, Y., Feng, L., Cheng, X.
CN105861343A	A preparation of high-lysine with methanol Single Cell Protein H. polymorpha yeast and its application	2016-08-17	**Nanjing U. of Technology**; Dai, Z., Dong, W., Jiang, M., Ma, J., Wu, R., Zhang, W.
CN105505803A	Saccharomycete and application thereof	2016-04-20	**Huayan Medical Res. Inst. Co. Ltd.**; Wang X.

CN104054907A	Preparation method for protein feed	2014-09-24	**Hunan Agricultural Products Proc. Inst**.; Fu, F., He, J., Li, G., Liu, W., Shan, Y., Su, D.
CN103881932A	Method for producing single-cell protein from water-chestnuts by virtue of fermentation	2014-06-25	**U. of Hezhou**; Pan, B., Su, L.
CN103849575A	Production method of single-cell protein	2014-06-11	**Nanchang University**; Gao, Z., Huang, H., Liu, Y., Ma, X., Ruan, R., Wan, Y., Wang, Y., Wu, X., Zheng, H.,
CN103844269A	Method for producing rice and noodle food through single-cell protein nutrient solution prepared by fruit and vegetable fermentation	2014-06-11	**Lan Jingmo**; Lan J.
CN103749957A	Preparation method for blue-green algae single-cell protein feed	2014-04-30	**Skystone Feed Yixing Co. Ltd**.; Liu, F., Xue, H., Yu, J., Zou, S., Zou, Y.
CN103627645A	Method for preparing carotenoid-enriched yeast single-cell protein by using bean curd yellow water for fermentation	2014-03-12	**Shenyang Chemical Technology U**.; Chu, C., Li, H., Liu, G., Yang, H.
CN103361279A	Method for producing single-cell protein by using pulping waste liquor	2013-10-23	**Beijing Yingli Shengke New Material Technology Co. Ltd**.; Zhuang, Y., Yin, Y., Huang, Y., Fan, H., Chen, T.
CN103275875A	*Trichoderma koningii*, and compound microbial agent composition and application thereof	2013-09-04	**An Jianping; Ma Weichao**; An, J., Ma, W.,
CN103271233A	Method for producing single-cell protein feed by means of fermentation of gallnut residues	2013-09-04	**Jishou U**.; Tian, H.
CN103103137A	*Saccharomycete* for hydrolyzing starch, producing lactic acid and fixing nitrogen, as well as application thereof	2013-05-15	**U. Xi An Jiaotong**; Huang, J., Lin, J., Liu, J., Long, J., Wang, L., Wu, Y., Yang, S.

CN103045494A	*Pichia pastoris* for efficiently converting methanol to produce single cell protein and application of *Pichia pastoris*	2013-04-17	**Coal Biochenmical High Tech Engineering Co. Ltd. Yima Coal Group**; Cao, M., Fan, C., Gou, W., Hu Yuansen; Qiao, G., Tian, Y., Wang, L., Wang, X., Wei, H., Xu, Z., Zhang, G., Zhang, Y., Zhu, G.
CN103027183A	Feed production method	2013-04-10	**Anhui Normal U.**; Liu, A., Lu, C., Peng, P., Zhang, Y.
WO13021138A2	Yeast flakes enriched with vitamin D2, compositions containing same, method for preparing same, uses thereof, and device for implementing the method	2013-02-14	**Lesaffre and Cie**; Fuentes, J.L., Knobloch, C., Mouly, I.
CN102911882A	*Rhodosporidium kratochvilovae* and application of same in preparation of carotenoid and single-cell protein	2013-02-06	**Nanjing Normal U.**; Chen, Y., Liu, J., Sun, H., Wei, M., Zhao, Y.
CN102815795A	Method for processing starch wastewater as well as product and application thereof	2012-12-12	**Guangxi U. for Nationalities; Nanning Suboante Biochemical and Scient. Co. Ltd**; Huang, J., Lan, L., Lan, P., Li, M., Liao, A., Shi, M., Wu, R., Xie, T.
US2010303778A	Composition for human and/or animal nutrition, uses thereof and yeasts	2010-12-02	**D'Auvergne Clermont 1 U., Du Droit et de la Sante Lille 2 U., Lesaffre and Cie**; Vandekerckove, P., Sivignon, A., Poulain, D., Desreumaux, P., Darfeuille, M.A., Simon, J.L., Pignede, G. Pierre uille Michaud Arlette

US2010092611A	Coated dried active yeasts and food products containing the same	2010-04-15	**Danstar Ferment Ag, Lallemand Sas**; Brouzes, J., Degre, R.
US2009232942A	Yeast preparations with improved antioxidant properties and uses thereof	2009-09-17	**Danstar Ferment Ag, Lallemand Sas**; Baulez, M., Degre, R., Forbes, W., Zhang, Z.
CN101386817A	Method for producing yeast single cell protein by blue algae fermentation	2009-03-18	**U. Jiangnan**; Yang, H., Li; K., Zhang, L., Wang, W.
US2004185162A	Edible fungi	2004-09-23	**Marlow Foods Ltd.**; Blanchard, R., Finnigan, T.J.A.

Source: Ritala A, Häkkinen ST, Toivari M and Wiebe MG (2017) Single cell protein—state-of-the-art, industrial landscape and patents 2001–2016. Front. Microbiol. 8:2009. doi: 10.3389/fmicb.2017.02009: Copyright © 2017 Ritala, Häkkinen, Toivari and Wiebe. This is an open-access article distributed under the terms of the Creative Commons Attribution License (CC BY). The use, distribution, or reproduction in other forums is permitted, provided the original author(s) or licensor are credited and that the original publication in this journal is cited, in accordance with accepted academic practice. No use, distribution, or reproduction is permitted which does not comply with these terms.

Table 1.12.

Species	Ile §	Leu	Val	Lys	Phe	Tyr	Met	Cys	Trp	Thr	Ala	Arg	Asp	Glu	Gly	His	Pro	Ser	Ref.
Arthrospira platensis *	6.7	9.8	7.1	4.8	5.3	5.3	2.5	0.9	0.3	6.2	9.5	7.3	11.8	10.3	5.7	2.2	4.2	5.1	[
Spirulina maxima	6.0	8.0	6.5	4.6	4.9	3.9	1.4	0.4	1.4	4.6	6.8	6.5	8.6	12.6	4.8	1.8	3.9	4.2	[
Spirulina sp. *LEB 18* *	4.4	8.0	4.6	2.9	5.7	3.2	1.6	0.47	2.5	4.9	6.5	4.9	9.2	10.7	5.2	2.7	4.0	4.3	[
Aphanizomenon sp	2.9	5.2	3.2	3.5	2.5	-	0.7	0.2	0.7	3.3	4.7	3.8	4.7	7.8	2.9	0.9	2.9	2.9	[
Heterochlorella luteoviridis *	1.8	8.1	2.9	8.7	5.4	2.7	1.8	0.4	0.6	5.2	11.1	5.6	0.3	1.3	9.6	1.8	5.5	6.8	[
Chlorella pyrenoidosa *	6.2	3.4	5.2	8.1	3.8	1.2	3.3	2.8	n.d.	3.4	5.1	5.9	8.1	7.8	9.8	1.6	n.d.	2.8	[
Chlorella vulgaris *	3.8	8.8	5.5	8.4	5.0	3.4	2.2	1.4	2.1	4.8	7.9	6.4	9.0	11.6	5.8	2.0	4.8	4.1	[
Chlamydomonas reinhardtii *	1.7	6.9	3.0	8.1	5.0	3.1	2.0	0.3	0.3	4.2	10.5	9.2	0.4	0.7	8.0	2.1	4.6	6.2	[
Tetraselmis chuii *	3.5	7.5	4.9	5.7	4.8	3.1	2.5	2.9	2.4	4.1	6.1	9.6	14.4	12.3	6.7	1.6	3.7	4.3	[
Tetraselmis CTP4	1.1	2.2	1.5	1.7	1.4	0.8	0.6	0.3	0.4	1.3	2.0	1.7	2.9	3.6	1.6	0.1	1.3	1.2	[
Dunaliella salina	4.0	9.6	7.2	6.0	6.9	4.9	2.8	1.6	0.2	5.2	11.0	8.2	9.6	12.4	8.7	1.7	5.2	4.8	[
Haematococcus pluvialis	0.5	1.2	0.6	0.7	0.6	0.4	0.1	–	n.d.	0.6	1.3	0.7	1.4	1.9	0.9	–	–	0.9	[
Scenedesmus obliquus	3.6	7.3	6.0	5.6	4.8	3.2	1.5	0.6	0.3	5.1	9.0	7.1	8.4	10.7	7.1	2.1	3.9	3.8	[
Soybean	5.3	7.7	5.3	6.4	5.0	3.7	1.3	1.9	1.4	4.0	5.0	7.4	1.3	19.0	4.5	2.6	5.3	5.8	[
FAO/WHO	2.8	6.6	3.5	5.8	–	–	–	–	1.1	3.4	–	–	–	–	–	–	1.9	–	

Source: Molfetta M, Morais EG, Barreira L, Bruno GL, Porcelli F, Dugat-Bony E, Bonnarme P, Minervini F. Protein sources alternative to meat: state of the art and involvement of fermentation. Foods. 2022; 11(14):2065. https://doi.org/10.3390/foods11142065: © 2022 by the authors. Licensee MDPI, Basel, Switzerland. This article is an open access article distributed under the terms and conditions of the

Key: § Ile—isoleucine, Leu—leucine, Val—valine, Lys—lysine, Phe—phenylalanine, Tyr—tyrosine, Met—methionine, Cys—cysteine, Trp—tryptophan, Thr—threonine, Ala—alanine, Arg—arginine, Asp—aspartate, Glu—glutamate, Gly—glycine, His—histidine, Pro—proline, Ser—serine.

In fact, the concept of microbial meat as an alternative to meat has emerged as an interesting option for vegans. For instance, Molfatta et al. (2022) summarized and compared the amino acid content of microalgae as opposed to the content of soybean as presented in Table 1.12 with the standards for essentia amino acids for chi dren aged two to five years (FAO/WHO).

This has led to a "microbial food market" where a slew of companies are actively on the quest to bring up SCPs as alternatives to meat. Figure 1.6 summarize the approaches and several players in this market.

Looking at the fie d of medicine, we are fami iar with how microbes have served as a source of "antibiotics" from the famous penicillin to the list of antibiotics that are only increasing. For instance, Table 1.13 documented by Danquah et al., (2022) summarizes a few such antimicrobials.

Name	Focus	Country	GMO	Organism	Product
Shiru	Ingredients	USA	✓	Yeast and fungi	Flourish™; protein discovery platform for food ingredients (currently: gelling agents.)
Oobli (formerly Joywell)	Ingredients	USA	✓	Yeast: *Pichia pastoris*	Sweet proteins
EVERY company (formerly Clara Foods)	Ingredients	USA	✓	Undisclosed	Eggs: EVERY EggWhite™, EVERY Protein™, Pepsin: EVERY Pepsin
Geltor	Ingredients	USA	✓	E.coli	Collagen: PrimaColl®
Melibio	Ingredients	USA	✓	Undisclosed	Honey
Impossible foods	Ingredients	USA	✓	*Pichia pastoris*	Animal alternative processed meats
Perfect day	Ingredients	USA	✓	Yeast	Dairy: Whey
Better Dairy	ingredients	UK	✓	Undisclosed	Dairy: Casein
Change Foods	Ingredients	Australia/ USA	✓	Undisclosed	Dairy: Casein
New Culture Food	Ingredients	USA	✓	Undisclosed	Dairy: Casein
Nutropy	Ingredients	France	✓	Yeast	Dairy: Casein
Standing ovation	Ingredients	France	✓	Undisclosed	Dairy: Casein
Formo (formerly LegenDairy)	Ingredients	Germany	✓	Undisclosed	Dairy :'bioidentical milk protein'
Remilk	Ingredients	Israel	✓	Yeast	Dairy: Milk proteins
Mycotechnology	Ingredients	USA	✗	Filamentous Fungi: multiple	Fermented plant proteins: FermentIQ™, Bitter blocking and flavour modulating: ClearIQ™
Motif foodworks	Ingredients	USA	✓	Yeast	Animal free ingredients: Haem: HEMAMI™,
Planet A Foods	Ingredients	Germany		Yeast	Cocoa butter
C16 Bioscience	Ingredients	USA	✓	Coculture: oleaginous yeast +	Palm oil
Melt&Marble	Ingredients	Sweden	✓	Yeast	Animal Fats: beef fat
Nourish ingredients	Ingredients	Australia	✓	Yeast	Animal Fats
Corbion	Ingredients	Netherlands	✓	Microalgae: *Chlorella* spp., *Prototheca*	Lactic Acid and its derivatives/ Algal oil SCO and SCP
Manus Bio	Ingredients	USA	✓	Multiple	Food ingredients: diverse
Skotan	Feedstock/aquaculture / Ingredients	Poland	✗	*Yarrowia lipolytica*	Animal Feed: Equinox Horse food, Caniflow dog and cat food,
Lallemand	Feedstock/aquaculture / Ingredients	Canada	✓	Multiple	Bscteria and yeast speciality products in many areas.
Unibio	Feedstock/aquaculture	Denmark	✗	*Methylococcus capsulatus*	Animal Feed: Uniprotein®
Novus International	Feedstock/aquaculture	USA	✗	B.licheniformis	Animal Feed Supplement: Cibenza®, protease from B.licheniformis
Kiverdi	Feedstock/aquaculture	USA	✓	Undisclosed	Fish feed: CO2 AQUAFEED , Palm oil
Novonutrients	Feedstock/aquaculture	USA	✗	Bacterial	Fish feed
Nutrinsic (acquired by iCell)	Feedstock/aquaculture	USA	✗	Bacterial	Fish feed
KnipBio	Feedstock/aquaculture	USA	✓	*Methylobacterium extorquens*	Fish feed: KnipBio Meal (Taurine and Carotenoid enriched)
Deep branch	Feedstock/aquaculture	UK	✗	Undisclosed	Fish Feed: Proton™
Calysta	Feedstock/aquaculture / SCP	UK	✗	Methane oxidizing bacteria	Animal and Fish feed: FeedKind®. SCP: Positive Protein
String Bio	Feedstock/aquaculture/ SCP	India	✓	Bacterial methanotroph	Animal and Fish feed: aquaculture and poultry. SCP: human food
Enifer	Feedstock/aquaculture / SCP	Finland	✗	Filamentous Fungi : *Paecilomyces*	Mycoprotein SCP and feed: PEKILO®
Solar foods	SCP for humans	Finland	✗	Chemoautotrophic bacteria	Bacterial SCP food
Quorn (Marlow Foods)	SCP for humans	UK	✗	*Fusarium venenatum*	Quorn® meat-alternative products: various
Air protein	SCP for humans	USA		Undisclosed	Air Protein: protein powder
MyForest Foods (formerly Atlast Food Co)	SCP for humans	USA	✗	Filamentous Fungi	MyBacon® Strips
The protein brewery	SCP for humans	Netherlands		Filamentous Fungi	Mycoprotein SCP
Natures Fynd	SCP for humans	USA	✗	*Fusarium yellowstonensis*	Fy Protein™: Meat Patties and cream cheese substitute
Marmite (Unilever)	SCP for humans	UK	✗	Brewers yeast	Marmite
Meati™ Foods	SCP for humans	USA	✗	Undisclosed	Mycoprotein SCP: Meati™, Whole Cuts substitute
Prime roots	SCP for humans	USA	✗	*Aspergillus oryzae*	Mycoprotein SCP: charcuterie, ham, bacon substitute
Brevel	SCP for humans	Israel	✗	Microalgae	Microalgae protein supplement
Superbrewed Food	SCP for humans	USA	✗	Gut Bacterium	Fortified protein milk and cream
Bosque Foods™	SCP for humans	Germany	✗	Undisclosed	Mycoprotein SCP: Whole cuts
Mycorena	SCP for humans	Sweden	✗	Filamentous Fungi	Mycoprotein B2B SCP: Promyc®
Enough	SCP for humans	UK	✗	*Fusarium venenatum*	ABUNDA B2B mycoprotein seafood alternative ingredient
The Mediterranean Food Lab	Fermented animal product alternative	Israel	✗	Multiple	Meat alternatives: Fermented Plant protein
Miyoko Creamery	Fermented animal product alternative	USA	✗	Multiple	Fermented plant milk cheese products
MATR Foods	Fermented animal product alternative	Denmark	✗	Fungi	Meat alternatives: Fermented Plant protein

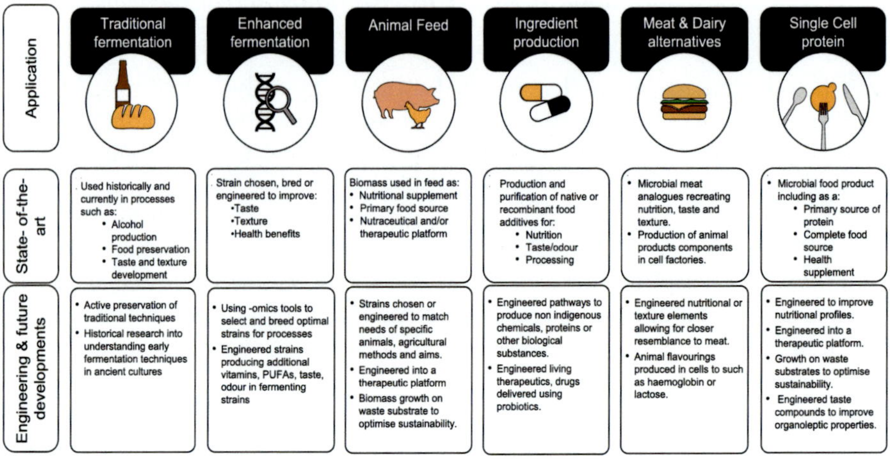

Figure 1.6.

Source: Graham and Ledesma-Amaro, 2023: "*Open access this article is licensed under a creative commons attribution 4.0 international license, which permits use, sharing, adaptation, distribution and reproduction in any medium or format, as long as you give appropriate credit to the original author(s) and the source, provide a link to the Creative Commons license, and indicate if changes were made. The images or other third party material in this article are included in the article's Creative Commons license, unless indicated otherwise in a credit line to the material. If material is not included in the article's Creative Commons license and your intended use is not permitted by statutory regulation or exceeds the permitted use, you will need to obtain permission directly from the copyright holder. To view a copy of this license, visit http://creativecommons.org/licenses/by/4.0/.*"

Table 1.13

Bacteriocin	Producer of Bacteriocin	Susceptible Microorganisms
Nisin A	*Lactococcus lactic subsp. lactis*	*E. fecalis ssp. Liquefaciens, Streptococcus equinus, Staphylococcus epidermidis (S. epidermidis), S. aureus, Streptococcus uberis (S. uberis), Streptococcus dysgalactiae (S. dysgalactiae), Streptococcus agalactiae (S. agalactiae), Streptococcus suis (S. suis) Mycobacterium avium subsp. Paratuberculosis*

Nisin ANisin V	*L. lactis* NZ9700L. lactis NZ9800nisA:M21V	*Listeria monocytogenes*
Pediocin A	*Pediococcus pentosaceus* FBB61	*Clostridium perfringens*
Enterocin M	*Enterococcus faecium AL41*	*Campylobacter spp., Clostridium spp.*
Enterocin CLE34	*Enterococcus faecium CLE34*	*Salmonella pullorum*
Enterocin E-760	*Enterococcus durans, Enterococcus faecium, Enterococcus hirae*	*Salmonella enterica serovar Enteritidis, S. enterica serovar Choleraesuis, S. enterica serovar Typhimurium, S. enterica serovar Gallinarum, E. coli O157:H7, Yersinia enterocolitica, S. aureus, Campylobacter jejuni*
Lacticin 3147	*Lactococcus lactis DPC3147.*	*S. dysgalactiae, S. agalactiae, S. aureus, S. uberis, Mycobacterium avium subsp. paratuberculosis*
Macedocin ST91KM	*Streptococcus gallolyticus subsp. macedonicus* ST91KM	S. agalactiae, S. dysgalactiae, S. uberis, S. aureus

Source: Amaning Danquah C, Minkah PAB, Osei Duah Junior I, Amankwah KB, Somuah SO. Antimicrobial compounds from microorganisms. *antibiotics*. 2022; 11(3):285. https://doi.org/10.3390/antibiotics11030285: *© 2022 by the authors. Licensee MDPI, Basel, Switzerland. This article is an open access article distributed under the terms and conditions of the Creative Commons Attribution (CC BY) license (https://creativecommons.org/licenses/by/4.0/).*

The scope of natural products, which are naturally derived molecules sourced from microbes, plants, and animals, is being probed for their ability to heal and prevent diseases. These include (but are not limited to) anti-cancer, anti-inflammatory, anti-microbial, and immunosuppressive functions. Pharm et al., (2019) reviewed and tabulated such activities that are presented in Table 1.14 to enable a reader to understand the potential of this field.

Table 1.14.

Name	Origin	Biological activity	References
Antibiotic			
Erythromycin A (1)	Saccharopolyspora erythraea	Antibacterial	McGuire et al., 1952; Zhang et al., 2010; Cobb et al., 2013
Tetracycline (2)	Streptomyces rimosus	Antibacterial	Chopra and Roberts, 2001; Demain, 2009
Vancomycin (3)	Amycolatopsis orientalis	Antibacterial	Geraci et al., 1956; Dasgupta, 2012
Streptomycin (4)	Streptomyces griseus	Antibacterial	Schatz et al., 1944; Waksman et al., 1946
Nisin A (5)	Lactococcus lactis	Antimicrobial	Li and Vederas, 2009; Gyawali and Ibrahim, 2014
Reuterin (6)	Lactobacillus reuteri	Antimicrobial	Talarico et al., 1988; Gyawali and Ibrahim, 2014
Antifungal Agents			
Amphotericin B (7)	Streptomyces nodosus	Antifungal	Abu-Salah, 1996; Tevyashova et al., 2013
Ieodoglucomide C (8)	Bacillus licheniformis	Antifungal	Tareq et al., 2015
Anticancer and Antitumor			
Bleomycin (9)	Streptoalloteichus hindustanus, Streptomyces verticillus	Squamous cell carcinomas, Hodgkin's lymphomas and testis tumors	Einhorn and Donohue, 2002; Demain and Vaishnav, 2011
Ddaunorubicin (10)	Streptomyces peucetius and various related strains	Acute lymphoblastic or myeloblastic lymphoma	Di Marco et al., 1981; Giddings and Newman, 2013
Immunosuppressant/Anti-inflammatory Agents			
Rapamycin (11)	Streptomyces rapamycinicus (formerly, Streptomyces hygroscopicus ATCC 29253), Streptomyces iranensis, and Actinoplanes sp. N902-109	Immunosu-ppressive, antifungal, antitumor, neuroprotective, neuroreg-enerative, and lifespan extension activities, growth inhibitory activity against several fungi	Vezina et al., 1975; Mann, 2001; Law, 2005; Pan et al., 2008; Anisimov et al., 2011; Song et al., 2015; Yoo et al., 2017

FK506 (12)	Streptomyces tsukubaensis and several Streptomyces species	Immuno-suppressive, antifungal, anti-inflammatory, neuroprotective and neuroregenerative activities, rheumatoid arthritis treatment	Tanaka et al., 1987; Mann, 2001; Migita and Eguchi, 2003; Demain, 2014; Ban et al., 2016; Yoo et al., 2017
Biofilm-Inhibitory Agents			
Cahuitamycins (13)	Streptomyces gandocaensis	Inhibitors of Acinetobacter baumannii biofilms	Park et al., 2016
Others			
Avermectins (14)	Streptomyces avermitilis	Onchocerciasis and lymphatic filariasis	Shen, 2015
Mollemycin A 20 (15)	Streptomyces sp. (CMB-M0244)	Gram-positive and Gram-negative bacteria, antimalarial activity	Blunt et al., 2016
Lipstatin (16)	Streptomyces toxytricini	Pancreatic lipase inhibitor for obesity and diabetes	Weibel et al., 1987; Sanchez et al., 2012

Source: Pham JV, Yilma MA, Feliz A, Majid MT, Maffetone N, Walker JR, Kim E, Cho HJ, Reynolds JM, Song MC, Park SR and Yoon YJ (2019) A review of the microbial production of bioactive natural products and Biologics. Front. Microbiol. 10:1404. doi: 10.3389/fmicb.2019.01404:

"Plants are our food, oxygen, and medicine. Some even say they are one of the most pleasurable experiences on earth! From the flowers to the trees and the seas filled with coral dreams; the earth's natural flora has inspired and enhanced humans for as long as time can tell. That's why the power of plants is the key to unlocking our enjoyment of life."

— *Natasha Potter*

Apart from serving as a source of food, clothing, and shelter, plants have been extensively researched for their ability to produce molecules that heal and prevent countless ailments. For instance, Ferdosh (2023) reviewed the recorded activities of select plants from Guam that have been presented in Table 1.15.

Table 1.15.

Botanical Name	Local Name	Chemical Constituents	Reported Pharmacological Activities in Different Plant Parts
Euphorbia hirta	Golundrina	Leaves: tannins, saponins, steroids, terpenoids, flavonoids, triterpenoids, polyphenol, glycoside, anthocyanins, and coumarins [18].	Leaves: antimicrobial [19–22]; anti-inflammatory [23]; anticancer [24]; antioxidant and antifungal [25].
Phyllanthus amarus	Maigo' Lålo'	Leaves and roots: saponins, tannins, phenolics, anthocyanins, alkaloids, triterpenoids, and glycosides, steroids, flavonoids, and coumarins [26].	Leaves: antibacterial [27–29]; antiviral, antibacterial, antiplasmodial, anti-inflammatory, antimalarial, antimicrobial, anticancer, antidiabetic, hypolipidemic, antioxidant, hepatoprotective nephroprotective, and diurectic properties [30].
Premna obtusifolia	Åhgao	Leaves: flavonoids, saponins, tannins, and triterpenoids/ steroids [31].	Antibacterial [32], Anticancer [33]; Antioxidant, antiarthritic, anticholesterol, anti-inflammatory, antimicrobial, antihelmintic, antidiabetic, and anticancer/cytotoxic [31].

Psidium guajava	Åbas	Leaves: carbohydrates, proteins and amino acids, carotenoids, flavonoids, alkaloids, polyphenols, saponins, tannins, glycosides, and sterols [34].	Leaves: antidiabetic [35, 36]; antioxidant [37, 38]; anti diarrheal [39,40]; antibacterial [38, 41]; anticancer [42, 43].
Urena lobata	Dadangse Ahgaga	Leaves: flavonoids, alkaloids, steroids, saponins, and tannins [44].	Leaves: antimicrobial [44]; antiarthritic activity [45]; analgestic and anti-inflammatory activity [46]; antidiabetic [47].

Source: Ferdosh S. Ethnobotanical review of selected medicinal plants in guam for the treatment of urinary tract ailments and their pharmacological properties. *Scientia pharmaceutica.* 2023; 91(3):43. https://doi.org/10.3390/scipharm91030043: *© 2023 by the author. Licensee MDPI, Basel, Switzerland. This article is an open access article distributed under the terms and conditions of the Creative Commons Attribution (CC BY) license (https://creativecommons.org/licenses/by/4.0/).*

A slew of medicinal herbs that documented anti-allergic and anti-viral activities for use against the COVID-19 virus was summarized by Ahmad et al., (2021), which has been presented in Tables 1.16–1.17.

Table 1.16.

Category. Sl no	Botanical Name/ Common Name/ Family/Part	Immunomodulatory Activity	Anti-Viral Activity	Anti-Allergic/Anti Asthmatic/Anti-Inflammatory/ Respiratory Disorders
C1.1	*Acacia catechu* (L.f.) Willd./ Khadira/ Fabaceae/Leaves, bark, heartwood	Aqueous and alcoholic extract increased phagocytic response showed by peritoneal macrophages. The extracts inhibited TNF-α and the production of NO, IL-10. Dose: 100 and 200 mg/kg Sunil et al. (2019)	Aqueous, hydroalcoholic and n-butanol extract showed anti-HIV-1 activity by inhibiting viral protein and Tat IC_{50}: 1.8 μg/m□Nutan et al. (2013)	Aqueous extract of leaves showed inhibitory effects on histamine synthesis in rat peritoneal as well as mast cells. Dose: 100 mg/ kg Prasad et al. (2009), Negi and Dave (2010)

C1.2	*Adhatoda vasica* Nees/Adusa/ Acanthaceae/ Leaves	Methanolic extract of leaves inhibit DTH reactiveness, increased the percentage neutrophil adhesion, promoting increased phagocytic activity vis-à-vis increased concentration of lytic enzymes for more effective killing. Dose: 400 mg/kg Vinothapooshan and Sundar (2011)	Ethanolic leaf extract inhibit the activity of HIV-Protease. HIV-protease plays a significant part in the replication cycle. Singh et al. (2010)	Alcoholic extract inhibited IgE-dependent basophil mediator release. Dose: 20 mg/kg Gibbs (2009), Hossain and Hoq (2016)
C1.3	*Aegle marmelos (*L.) Correa/Bael/ Rutaceae/Root, stem bark, fruits	Alcoholic extract stimulates immune system by acting through cellular and humoral immunity. Dose: 100 and 500 mg/kg Patel and Asdaq (2010)	Purified seselin showed inhibitory potential over multiple SARS-COV-2 targets and holds a high potential to work effectively as a novel drug for COVID-19. Nivetha et al. (2020)	Aqueous extract inhibit production of nitric oxide (NO) by rat peritoneal cells, anti-histamine effect, and membrane stabilization activity. Dose: 200 mg/kg Kumari et al. (2014)
C1.4	*Anacyclus pyrethrum* (L.) Lag. / Akkal kadha/ Asteraaceae/Root	Petroleum ether extract showed cellular and humoral immunity. Dose: 50–100 mg/kg Sharma et al. (2010)	Pyrethrin act as ligands to bind with viral proteins to prevent the binding of host receptors preventing the fusion lead viral replication in COVID 19 Kumar et al. (2019)	-
C1.5	*Andrographis paniculata* (Burm.f.) Nees/ Kalmegh/ Acanthaceae/ Leaves	Isolated compound of andrographolide modulate immune responses by regulating macrophage phenotypic polarization and MAPK and PI3K signaling pathways regulate macrophage polarization. Dose: 10 µg/m□(In vitro) and 1 mg/kg (In vivo) Wang et al. (2010)	Alcoholic extract inhibits the viral titer in A549 cells transfected with SRV. They showed the activity through p38 MAPK/Nrf2 pathway. Dose: 50 **µg/ml** Churiyah et al. (2015), Wintachai et al. (2015)	Andrographolide attenuate allergic asthma by inhibition of the NF-kappaB signaling pathway. Dose: 0.1, 0.5, and 1 mg/kg Bao et al. (2009)

C1.6	*Carica papaya* L./Papaya/ Caricaceae/ Leaves, fruits	Alcoholic extract of fruit pulp and seed enhanced phagocytic activity of peritoneal macrophages is correlated with T helper 1 cytokine response. Interferon-gamma increases the phagocytosis process. Dose: 0.11 g/ml extract every day using a gastric cannula Amin et al. (2019)	Aqueous extract of the C. papaya leaves increases the expression of the envelope and NS1 proteins in DENV-infected THP-1 cells. IC $_{50}$: 100 µg/m□ Sharma N. et al. (2019)	Alcoholic extract of leaves in mouse model of ovalbumin- (OVA) induced allergic asthma down regulates IL-4, IL-5, eotaxin, TNF-α, NF-κB, and iNOS levels thus exhibits anti-inflammatory effect. Dose: 100 mg/kg Inam et al. (2017)
C1.7	*Cassia occidentalis* L./Kasunda/ Fabaceae/Aerial part, seeds	Isolated rhein suppresses the functional responses of the T- and B-lymphocytes and also suppresses lymphoproliferation in splenocytes.Dose: 10 µM Panigrahi et al. (2016)	Alcoholic extract showed that the plant possessed an anti-HIV property through inhibition of viral reverse transcriptase activity.IC $_{50}$: >100 mg/ml Estari et al. (2012)	Isolated anthraquinone showed anti-asthmatic potential by decreasing mRNA expression of Th1/Th2 cytokine in lung tissue. Dose: 250, 500 and 2000 mg/kg Xu et al. (2018)
C1.8	*Cocculus hirustus* (L.) Diels/ Patalagarudi/ Menispermaceae/ Whole plant	Methanolic extract showed significantly enhanced specific and non-specific activity on various immune paradigm in cyclophosphamide induced immunosuppresed animals.Dose: 200 mg/kg Mallik and nayak (2014)	Found effective against all strains of dengue virus and SARS CoV 2 in *in vitro* studies, hence under phase 2 clinical trial as phytopharmaceutical drug against COVID 19 at 12 centers. (https://www.clinicaltrialsarena.com/news/sun-pharma-covid-19-trial/)	The methanolic leaf extract showed significant analgesic activity in mice as well as significant anti- inflammatory activity using *in vitro* and *in vivo* rat models. Dose: 100 mg/kg Sengottuvelu et al. (2012)
C1.9	*Cordia myxa* L./Sapistan/ Boraginaceae/ Fruits	Aqueous extract of *C. myxa* fruits significantly increased the delayed type hypersensitivity (DTH), mitotic index (MI) of bone marrow and spleen cells Ali et al. (2015)	Dichloromethane, ethyl acetate, and methanol stem extracts showed anti-viral potential against HIV-1 using the syncytia formation assay. IC 50: 21.8 µg/ml Rashed (2014)	Hydroalcoholic extract inhibit the oxidant stress factors that lead to progression of colitis.Dose: 100 mg/kg Ranjbar et al. (2013)

C1.10	*Curcuma longa* L./Haldi/ Zingiberaceae/ Rhizome	Lyophilized turmeric was found to decrease spleen weight, decrease the proportion of $CD4^-$, $CD8^+$ T cells, and decrease phagocytic activity. Dose: 1 and 2% (w/w) Kim et al. (2014). Polysaccharide fraction of aqueous extract of *C. longa* inhibiting the secretion of IL-12 and PGE2. Dose: 0.8–500 µg/m□ Chandrasekaran et al. (2013)	Aqueous extract of *C. longa* suppressed the HBV replication and the transcription of HBV genes in HepG2 cells which produce HBV particles. Dose: 200 mg/L and 500 mg/L Kim et al. (2009). Isolated curcuminoids from aqueous extract of curcuma longa exhibited significant inhibitory activity against the neuraminidases from novel influenza H1N1 (WT) and oseltamivir-resistant novel H1N1 (H274Y mutant) expressed in 293 T cells. IC 50: 6.18 ± 0.64 to 40.17 ± 0.79 µg/m□ Dao et al. (2012). Virtual screening of curcumin and its analog found its activity SARS CoV 2 surface proteins and is under clinical trial. (https://chemrxiv.org/articles/Virtual screening of curcumin and its analogs against the spike surface glycoprotein of SARS-cov-2 and SARS-cov/12142383)	Alcoholic extract of C. longa ameliorates food allergy by maintaining Th1/ Th2 immune balance in ovalbumin challenged mice. Dose: 100 mg/kg Shin et al. (2015)
C1.11	*Cynodon dactylon* (L.) Pers./Doorva/ Poaceae/Whole plants	Fresh juice of the grass increased humoral antibody response upon antigen challenge, significant increase in antibody titer in the haemagglutination antibody assay and plaque forming cell assay. Dose: 250 and 500 mg/kg Mangathayaru et al. (2009)	Alcoholic dried extract showed virustatic and virucidal activity against porcine reproductive and respiratory syndrome virus (PRRSV) and also significantly inhibits replication of PRRSV. Dose: 0.78 mg/ml Pringproa et al. (2014)	Chloroform extract of whole plant produces a bronchodilation via antimuscarinic calcium channel blocking activators and phosph-odiesterase inhibition activity. Dose: 5, 10, 50 and 100 mg/kg Patel et al. (2013)

C1.12	*Jatropha curcas* L./ Euphorbiaceae/ Leaves, roots	Phytoconstituents of hydroalcoholic extract ameliorated both cellular and humoral antibody response. Dose: 0.25, 0.5, 1 mg/ kg Abd-Alla et al. (2009)	Successive extract of *J. curcas* was evaluated by inhibition of HIV replication as determined by HIV p24 antigen ELISA showed 100% inhibition by methanolic and 97.19% inhibition by aqueous extract. IC_{50}: 0.0255–0.4137 mg/ml (aqueous) and 0.00073–0.1278 mg/ml (Methanolic) Dahake et al. (2013)	Isolated jatrophacine showed anti-inflammatory potential by inhibiting production of nitric oxide in LPS-induced RAW 264.7 macrophages. IC_{50}: 0.53 µM Yang et al. (2019)
C1.13	*Mollugo cerviana* (L.) Ser./ Grishmasundara/ Molluginaceae/ Whole plant	Alcoholic extracts increase NO release by peritoneal cells. Dose: 25 µg/m□ Ferreira et al. (2003)	Alcoholic extract exhibits antiviral properties for both chikungunya virus and dengue virus. Dose: 1.8 mg/ml Jain et al. (2019)	Hydroalcoholic extract inhibit the levels of lipid peroxides, acid phosphatase, and gamma-glutamyl transpeptidase activity. Dose: 1 mg/g Sadique et al. (1987)
C1.14	*Nigella sativa* L./Kalonji/ Ranunculaceae/ Seeds	Aqueous extract of *N. sativa* enhance the proliferative capacity of splenocytes and T lymphocytes, suppression of IFNγ secretion from splenocytes. Dose: 10, 50, and 100 g/ ml Majdalawieh et al. (2010)	Nige□idine and α-hederin found to have the best potential to act as COVID-19 treatment in docking studies. (https://chemrxiv.org/articles/Identification of compounds from nigella sativa as new potential inhibitors of 2019 novel corona virus Covid-19 molecular docking study/12055716/1) *N. sativa* seeds oil possesses a striking antiviral effect against MCMV infection. Dose: 100 mg/100 ml/ mouse Umar et al. (2016)	Aqueous extract of seed showed sensory receptors mediating reflex bronchoco-nstriction and tachykinin receptor antagonists. Dose: 3.3% w/w extract Boskabady et al. (2003)

C1.15	*Ocimum sanctum* L./ Tulsi/Lamiaceae/ Leaves	Aqueous extract of leaves showed regulation of IL-2 production and exhibited leukocytosis and augmentation of T& B cells. Dose: 250 mg/kg Goel et al. (2010)	Hydroalcoholic extract showed promising antiviral properties against H9N2 virus by inhibition of a stage in viral intracellular multiplication and non-specific interference with virus-cell interactions. Dose: 135, 67, 33 mg/0.1 ml Ghoke et al. (2018)Tulsinol and dihydroeugenol have been found effective against SARS CoV 2 in molecular docking studies. (https://papers.ssrn.com/sol3/papers.cfm abstract id = 3554371)	Alcoholic extracts showed anti-asthmatic potential through inflammatory mechanism by inhibiting LTC4, LTA4 and COX-2 in HL-60 cell lines and reduction in inflammation in asthma mice model. IC_{50}: 1–10 µg/m□Dose: 100 mg/kg Soni et al. (2015)
C1.16	*Phyllanthus emblica* L./Amla/ Phyllanthaceae/ Fruits	Alcoholic extract of fruits stimulate B and T lymphocyte and restored the interleukin production considerably. Dose: 10 mg to 1 mg/ ml Sai Ram et al. (2002)	Fractionated alcoholic extract inhibit HIV reverse transcriptase activity. IC 50: >100 mg/ml Estari et al. (2012)	Alcoholic extract exhibits anti-inflammatory and anti-oxidant activity by protecting RAW264.7 cells from oxidative damage by increasing glutathione content and total superoxide dismutase activity, suppressing MDA content and decreasing release of pro-inflammatory mediators. IC 50: 0.677 ± 0.029 mg/ ml Li W. et al. (2020)

C1.17	*Solanum nigrum* L./Makoi/ Solanaceae/ Seeds, barriers	Isolated polysaccharides of significant increment in the percentage of $CD4^+$ T lymphocyte and a decrease in the percentage of $CD8^+$ T lymphocyte of tumor-bearing mice peripheral blood. Dose: 90, 180, 360 mg/kg Li et al. (2009)	Chloroform extract decreased the expression or function of HCV NS3 protease in a dose dependent manner and GAPDH remained constant. Dose: 100 μg/μL Javed et al. (2011)	Petroleum ether extract of berries inhibits asthma by inhibiting increase in leukocyte and eosinophil count, protection against mast cell degranulation and resisting contraction due to presence of β-sitostero□Dose: 50, 100 and 200 mg/kg Nirmal et al. (2012)
C1.18	*Valeriana wallichii* DC./ Valerianaceae/ Roots	Alcoholic root extract inhibited HCV by binding with HCV NS5B protein. Dose: 250 μg/m□Ganta et a□ (2017)	Alcoholic extract and its fraction inhibit HCV by binding with HCV NS5B protein. Dose: 200 μg/m□Ganta et a□ (2017)	Crude extract showed protection against airway disorders through relax ion of the low K^+ (25 mM)-induced contractions with a mild effect on the contractions induced by high K+ (80 mM). Dose: 0.03–3.0 mg/ml Khan and Gilani (2012)
C1.19	*Vitex negundo* L./Renuka/ Verbanaceae/ Leaves	Hydroalcoholic extract of leaves of *V. negundo* activate the phagocytic cells such as macrophages and neutrophils. Dose: 200 mg/kg Lad et al. (2016)	Alcoholic extract of leaves inhibits HIV-1 reverse transcriptase activity in *in vitro* assay thus exhibits anti-HIV activity. Dose: 200 μg/ ml Kannan et al. (2012)	*V. negundo* leaf oil inhibits COX-2 without much interfering COX-1 pathways. Dose: 500 μL/kg Chattopadhyay et al. (2012)
C1.20	*Withania somnifera (*L.) Dunal/Asgand/ Solanaceae/Roots	Aqueous suspension of root showed potent inhibitory activity toward mitogen induced proliferative response of T-lymphocyte and delayed-type hypersensitivity reaction. Dose: 1000 mg/kg Rasool and Varalakshmi (2006)	Hydro-alcoholic root extract of *W. somnifera* showed antiviral properties against IBD virus by cytopathic effect reduction assay. Dose: 25 μg/m□Pant et al. (2012). Withanone and withaferin a have been found effective against SARS CoV 2 in bioinformatics studies and asgandh	Aqueous extract of withania root inhibit histamine and 5-HT in early phase and prostaglandins in delayed phase of inflammatory reaction. Dose: 1000 mg/kg Sahni and Srivastava (1993)

			extract is under clinical trial. (https://www.researchsquare.com/article/rs-17806/v1), (http://www.bioinformation.net/016/973206-30016411.pdf)	
C1.21	*Zingiber officinale* Roscoe/Sunthi/ Zingiberaceae/ Rhizome	Volatile oil of ginger influences both cell-mediated immune response and nonspecific proliferation of T lymphocyte. Dose: 0.125, 0.25, and 0.5 g/kg Zhou et al. (2006)	Aqueous extract effective against HRSV-induced plaque formation on airway epithelium by blocking viral attachment and internalization. IC_{50}: >150 µg/m□Chang et al. (2013)	Aqueous and alcoholic extract showed anti-asthmatic effect by reducung inflammation through suppression of Th2-mediated immune response. Dose: 500 mg/kg (alcoholic extract)720 mg/kg (aqueous extract) Khan et al. (2015)
C2.1	*Abutilon indicum* (L.) Sweet/Tuthi/ Malvaceae/Aerial parts	Alcoholic extract showed stimulatory effect on T lymphocytes. Increasing doses of showed higher HA titer value, restoration of WBC count. It also increased lymphocyte and E-rosette formation. Dose: 200 and 400 mg/kg Gaikwad and Krishna Mohan (2012)	Alcoholic extract of leaves showed anti-MCV and anti-HSV activities. Dose: 0.4 µg/ml Vimalanathan et al. (2009)	Methanolic extract of aerial part showed mast cell stabilizing and anti-inflammatory activity. Dose: 250 and 500 mg/kg Mehta and Paranjape (2008)
C2.2	*Achyranthes aspera* L./ Apamarga/ Amaranthaceae/ Root	Polyphenolic compounds of hydroalcoholic extract showed cytokine based immunomodulatory role. Dose: 100 mg/kg Narayan and Kumar (2014)	Alcoholic extract showed potential activity against herpes simplex virus type-1 and type-2 by inhibiting the early stage of multiplication in vero cells. Mukherjee et al. (2013)	Ethyl acetate fraction from methanolic extract showed *in vitro* anti-inflammatory activity. IC 50: 50 = 76 ± 0.14 Khuda et al. (2013)

C2.3	*Aloe vera* (L.) Burm.f./ Ghrit kumari/ Asphodelaceae/ Roots, leaves	Aloe vera gel administration did not increase ovalbumin (OVA)-specific cytotoxic T lymphocyte (CTL) generation in normal mice. Dose: 100 mg/kg Im et al. (2010)	Isolated anthraquinone showed anti-viral activity by inhibiting virus replication. IC_{50}: 13.70 ± 3.80 to 62.31 ± 3.05 Borges-Argáez et al. (2019)	Polysaccharide isolated from gel showed anti-allergy potential by inhibition of type 2 helper T cell (Th_2) immune response, increase in IL-10 production and stimulating type 1 regulatory T (Tr1) cells activation. Dose: 50 and 100 mg/kg Lee D. et al. (2018)
C2.4	*Alstonia scholaris* (L.) R.Br./Saptaparni/ Apocynaceae/ Bark	Aqueous extract enhanced phagocytic activity. Dose: 50 mg/kg Iwo et al. (2000)	Aqueous and alcoholic plant extract showed anti-viral potential against coxsackie B2, polio virus and herpes simplex virus. Dose: 2.8 mg/kg Antony et al. (2014)	Alcoholic extract inhibited inflammatory response by through reduction in ovalbumin-provoked airways allergic inflammatory stress. Dose: 10, 25, and 50 mg/kg Zhao et al. (2017)
C2.5	*Azadirachta indica* A.Juss./ Neem/Meliaceae/ Leaves	Dried powdered leaves significantly enhanced the antibody titers against new castle disease virus (NCDV) antigen. Dose: 2 g/kg Sadekar et al. (1998)	Isolated polysaccharides from aqueous extract of the leaf virucidal aginst Poliovirus-1 (inhibiting initial stage of viral replication). IC_{50}: 80 μg/m□and 77.5 μg/m□ Faccin-Galhardi et al. (2012)	Aqueous leaves extract showed anti-inflammatory and analgesic activity by in chemical and thermal induced pain models in albino rats. Dose: 500 mg/kg Buchineni et al. (2014)
C2.6	*Berberis aristata* DC. / Daruharidra/ Berberidaceae/ Bark	Isolated berberine inhibited the suppressed viral infection-induced up-regulation of TLR7 signaling pathway. Dose: 20 mg/kg Yan et al. (2018)	Isolated compound of berberine inhibited EV71 replication by down regulating autophagy and MEK/ ERK signaling pathway. IC_{50}: 7.43 to 10.25 μM Wang et al. (2017)	Hydroalcoholic extract showed anti-inflammatory potential, which may be attributed to its inhibitory activity on macrophage-derived cytokine and mediators. Dose: 50, 100, and 200 mg/kg Kumar et al. (2016)

C2.7	*Bergenia ciliata* (Haw.) Sternb./ Pashanbheda/ Saxifragaceae/ Stem	Alcoholic extract stimulated the expression of CD69 on lymphocytes. Dose: 3.13 and 6.25 mg/ml Tumova et al. (2018)	Alcoholic extract showed potent anti-viral activity against both influenza virus a and HSV-1. IC $_{50}$: >6.25 µg/m□Rajbhandari et a□ (2009)	Alcoholic extract exhibited significant anti-inflammatory activity in carrageenan-induced rat paw edema manner. Dose: 300 mg/kg Sinha et al. (2001)
C2.8	*Camellia sinensis* (L.) Kuntze/ Chary/Thecae/ Leaves	Aqueous extract of *C. sinensis* changes hematological profile, immuno potentiating cells, cellular response in splenectomised mice. Dose: 250 and 500 mg/kg Gomes et al. (2014)	Hydroalcoholic extract of *C. sinensis* inhibited ADV replication in post-adsorption stage. IC 50: 6.62 µg/m□ Karimi et al. (2016)	Aqueous extract showed anti-asthmatic potential by increasing expression of Th1 cell-specific anti-asthmatic biomarkers (tumor necrosis factor-ß and
				interferon-γ) and decreasing the expression of anti-asthmatic cytokines in the lungs. Dose: 25 µg/m□Heo et a□ (2008)
C2.9	*Cannabis sativa* L./Vijaya/ Cannabaceae/ Leaves	Cannabinoids indicating mainly immunosuppressive effects on macrophages, NK cells, T lymphocytes and their ability to produce cytokines. Dose: 5 mg/day Killestein et al. (2003)	*C. sativa* inhibited viral DNA synthesis. It inhibits the replication cycle of various types of DNA or RNA viruses Jassim and Naji (2003) Hot water extract *C. sativa* reduced the plaque forming ability. Dose: 300–500 µg/m□ Kurokawa et al. (1993)	Oil extract of *C. sativa* protective effect against COPD through affecting the expression of specific airway epithelial cell genes that modulate Th1 processes using *in vitro* assay. Dose: 2.4, 1.2, and 0.6 µg/m□Mamber et al. (2020)

C2.10	*Cassia fistula* L./Amaltas/ Fabaceae/Bark	Hydro alcoholic extract of *C. fistula* increased antibody titer against salmonella typhimurium 'O' antigen and significant enhancement in skin thickness in DNCB sensitized albino rats. Dose: 125 mg/ kg, 250 mg/kg and 500 mg/kg Laxmi (2015)	Hydroalcoholic extract of *C. fistula* suppressed extracellular HBV DNA production. Dose: 100 µg/m□Indrasetiawan et al. (2019)	Isolated rhein showed anti-inflammatory activity by modulating levels carrageenan-induced hind paw edema, croton oil-induced ear edema, cotton pellet-induced granuloma and acetic acid-induced vascular permeability models. Dose: 10 mg/kg Antonisamy et al. (2019)
C2.11	*Cinnamomum verum* J.Presl./ Daarchini/ Lauracea/Stem, bark	Bark suspension increased the phagocytic index in carbon clearance test, neutrophil adhesion and serum immunoglobulin levels and antibody titer values. Dose: 10 and 100 mg/ kg Niphade et al. (2009)	Aqueous extract provide treatment against influenza virus infections in vero cells transfected with H7N3 influenza. Brochot et al. (2017)	Isolated procyanidine showed reduction in the elevated levels of total protein, albumin, goblet cell hyperplasia and inflammatory cell infiltration in lung tissue. Dose: 10, 30, and 100 mg/ kg Kandhare et al. (2013)
C2.12	*Cissampelos pareira* L./ Akamai/ Menispermaceae/ Aerial parts, roots	Isolated alkaloid fraction of alcoholic extract modulates both T and B cell mediated immune response. Dose: 100 mg/kg Bafna and Mishra (2010)	Alcoholic extract of aerial part of *C. pareira* inhibits the viral replication and ability to down-regulate the production of TNF-α, a cytokine implicated in severe dengue disease. IC_{50}: ≥125 µg/m□Sood et a□ (2015)	Alkaloids fraction suppressed the production of nitric oxide, a critical mediator in inflammation. Dose: 100 mg/kg Bafna and Mishra (2010)
C2.13	*Cyperus rotundus* L./Musta/ Cyperaceae/ Rhizome	Aqueous, alcoholic, ethyl acetate and total oligomer flavonoids (TOF) extracts of *C. rotundus* influence humoral-mediated immunity by stimulating B and T cell proliferation. Dose: 1–1000 µg/ ml Soumaya et al. (2013)	Aqueous, alcoholic and ethyl acetate extract of *C. rotundus* inhibited the HBV DNA replication in HepG2.2.15 cell line. IC 50: 29.0, 21.5, 263.4 Xu et al. (2015)	Isolated sesquiterpenes from alcoholic extract showed anti-allergic potential against immediate-type as well as delayed-type hypersensi-tivityDose: 300 µg/ ml (in vitro) and 50–300 mg/kg (in vivo) Jin et al. (2011)

C2.14	*Daphne gnidium* L./Lota/ Thymelaeaceae/ Aerial part	Dichloromethane extract of the aerial exhibited strong antiretroviral activity by interference with HIV co-receptors, CCR5 and CXCR4. Vidal et al. (2012)	Dichloromethane extract of the aerial parts exhibited strong antiretroviral activity and absence of cytotoxicity and pure compounds were active against multidrug-resistant viruses irrespective of their cellular tropism. Dose: 10 μg/ml Vidal et al. (2012)	Ethyl acetate extract showed anti-inflammatory effects by inhibiting macrophage proinflammatory function by reducing LPS-induced production of IL-1β, TNF-α, COX-2-derived PGE2 and iNOS-II-synthesized NO. Dose: 1–100 **μg/ml** Harizi et al. (2011)
C2.15	*Ficus carica* L./Anjeer/ Moraceae/ Leaves, latex	Administration of extract ameliorated both cellular and humoral antibody response Patil et al. (2010)	Resuspension of latex in DMEM containing 1% ethanol able to interfere with the replication of CpHV-1. IC 50: 100 μg/ml Camero et al. (2014)	Tea infusion of leaves showed anti-allergy potential through promotion of dissociation of IgE from FcεRI receptors. Dose: 10 ml/kg Abe (2020)
C2.16	*Glycyrrhiza glabra* L./ Mulethi/ Fabaceae/Roots, rhizome and leaves	Aqueous root extract showed leukocyte count and phagocytic index increased as well as cellular immune response study, an enhancement in foot pad thickness was observed. Dose: 1.5 g/kg Mitra Mazumder et al. (2012)	Aqueous and alcoholic extracts of G. glabra verified hemagglutination (HA) test data through which amount of virus is quantified from the allantoic fluid of chicken embryos. Dose: 300 μg/ml Ashraf et al. (2017)	Saponin fraction showed anti-asthmatic potential in triple antigen sensitized rats by inhibition of mast cell degranulation. Dose: 100 mg/kg Patel et al. (2009)
C2.17	*Illicium verum* Hook.f./Takkola/ Magnoliaceae/ Fruit	Isolated lectins from *I. verum* showed immunomodulatory action by stimulating phagocytic function. Dose: 30 and 50 mg/kg Bouadi et al. (2015)	Aqueous, alcoholic and hydroalcoholic extracts exhibited inhibitory effects against NDV and avian reovirus. Dose: 0.24–3.9 mg/ml Alhajj et al. (2020)	70% alcoholic extract exert antiasthmatic effects through upregulation of Foxp3+ regulatory T cells and inhibition of Th2 cytokines. Dose: 50, 100, and 200 mg/kg Sung et al. (2017)

C2.18	*Mentha × piperita* L./ Peppermint/ Lamiaceae/ Leaves	Hydrodistilate fractions of *M. piperita* affect the functional responses of human PMNs and PBMCs. Dose: 2 mM and 12 μL Cosentino et al. (2009)	Alcoholic extract showed antiviral activity against RSV with a high selectivity index, and significantly decreased the production of NO, TNF-α, IL-6, and PGE2 in lipopolysaccharide-stimulated RAW 264.7 cells. IC_{50}: 10.41 μg/m□ Li et al. (2017)	Essential oil showed antispasmodic activity by regulating prostaglandins and nitric oxide synthase on rat trachea. Dose: 1–300 μg/m□de Sousa et al. (2010)
C2.19	*Mentha spicata* L./Spearmint/ Lamiaceae/ Leaves	Essential oil from *M. spicata* proliferate T-cells, IL-2 and potently inhibit the production of pro-inflammatory cytokine TNF-α production Orhan et al. (2016)	Aqueous extract exhibits anti-viral potential against porcine parvovirus (PPV) *in vitro* by efficiently killing them and control their multiplication in cells. IC_{50}: 0.0340 mg/ml Weili et al. (2011)	Ethyl acetate soluble fraction of leaves by inhibit antigen stimulated rat basophile. Prasad et al. (2009)
C2.20	*Momordica charantia* L./ Bitter guard/ Cucurbitaceae/ Leaves, fruits and seed	Alcohol and diethyl ether extract has been found that the exposure of neutrophils and macrophages stimulates both their capacity to ingest foreign particles and their intracellular killing activities. Dose: 250, 500, 1000 mg/ kg Mahamat et al. (2020)	Crude protein fraction of *M. charantia* strongly inhibited H1N1, H3N2 and H5N1 subtypes. IC_{50}: 40–200 g/m□ Pongthanapisith et al. (2013)	Alcoholic extract showed the highest reduction of LPS-induced NO, iNOS and prostaglandin E2 production and down regulates pro-interleukin-1β and NF- κB activation expression in RAW 264.7 macrophages Lii et al. (2009)
C2.21	*Morus alba* L./Sahatoot/ Moraceae/ Leaves, fruits	Isolated water-soluble polysaccharides stimulate murine RAW264.7 macrophage cells to release chemokines and proinflammatory cytokines. Lee et al. (2013). Alcoholic extract of leaves significant increase in the phagocytic	*M. alba* fruits juice and its fractions inhibit internalization and replication of MNV-1, whereas it may influence adherence or internalization of FCV-F9 virions. EC_{50}: 0.005 (MNV-1) and 0.25–0.30 (FCV-F9) Lee et al. (2014)	Juice of *M. alba* fruits inhibit production of NO and proinflammatory cytokines (TNF-α, IL-6), as well as the expression of NOS2 and PTGS2 in LPS-stimulated RAW264.7 macrophages. Dose: 0.1, 0.5, and 1 μg/m□Jung et a□ (2019)

		index and adhesion of neutrophils. Dose: 100 mg/kg and 1 g/kg Bharani et al. (2010)		
C2.22	*Nyctanthes arbor*-tristis L./Parijata/ Oleaceae/Leaves, flowers and seeds	Immunostimulant activity of NAFE seems to be mediated through splenocytes proliferation and increased production of cytokines, especially IL-2 and IL-6 of aqueous extract of *Nyctanthes arbor*-tristis. Dose: 400 and 800 mg/ kg Bharshiv et al. (2016)	n-Butanol fraction of alcoholic extract of protected encephalomyocarditis virus (EMCV) infected mice against semliki forest virus (SFV). Dose: 125 mg/kg Gupta et al. (2005)	Alcoholic extract showed anti-asthmatic and anti-tussive activity against histamine and acetylcholine cocktail induced asthma and citric acid induce cough in Guinea pig. Dose: 100, 200, and 300 mg/ kg Mathur et al. (2016). Extracted polysaccharide from leaves aqueous extract reduce the number of cough efforts without influencing the specific airway resistance, it triggers cough reflex provocation. Dose: 25 and 50 mg/kg Ghosh et al. (2015)
C2.23	*Ocimum basilicum* L./ Basil/Lamiaceae/ Leaves	Hydroalcoholic extract of leaves increased the IFN-γ/ IL-4 ratio and decreasing BALF levels of IgE, PLA2 and TP. Dose: 50,300, 600 mg/ kg Eftekhar et al. (2019b)	Alcoholic extract inhibits ZIKV replication in vero E6 cells. The extract seems to inhibit the virus at the step of attachment and entry into the host cell. IC_{50}: 1:134 Singh et al. (2019)	Hydroalcoholic extract showed therapeutic effect on asthma by reducing eosinophil's, monocytes, neutrophils percentage and increase in percentage of lymphocytes and antioxidant biomarkers levels. Dose: 0.75, 1.50, and 3.00 mg/ml Eftekhar et al. (2019a)

C2.24	*Oleo europea* L./Zaitoon/ Oleaceae/Leaves	Isolated oleuropein from hydroalcoholic extract showed lymphocyte activation and proliferation properties. Oleuropein exhibited a high degree of lymphocyte aggregation. Dose: 540 μg/m□Randon & Attard (2007)	Aqueous leaves extract showed anti-viral potential against Newcastle disease virus by restricting replication. Concentration: 1000 μg/m□Sa□ih et a□ (2017)	Essential oil from leaves inhibit NFB activation in monocytes and monocyte derived macrophages. Lucas et al. (2011)
C2.25	*Panax ginseng* C.A. Mey./ Ginseng/ Araliaceae/Roots	Ginsenosides increased the number of spleen plaque-forming cells, the titers of sera hemagglutinins as well as the number of antigen-reactive T-cells and splenocyte natural killer activity. Dose: 10 mg/kg Buriana et al. (1990)	Fermented extract improved the survival of human lung epithelial cells, inhibits RSV replication, suppressed the expression of RSV-induced inflammatory cytokine genes and the formation of ROS in epithelial cell cultures. Dose: 25 mg/kg Wang et al. (2018)	*P. ginseng* extract showed anti-asthmatic activity by restoring EMBP(eosinophil major basic protein), Muc5ac, CD40, and CD40L expression and mRNA and protein levels of IL-1, IL-4, IL-5, and TNF-α. Dose: 20 mg/kg Kim & Yang (2011)
C2.26	*Peganum harmala* L./ Harmal/ Nitrariaceae/ Aerial parts and seeds	Alcoholic extract (80%) of seed showed effects on zymosan-A activated neutrophils (PMNs). Dose: 25, 50, and 100 μg/m□Koko et al. (2008)	Alcoholic extract inhibits viral RNA replication and viral polymerase activity. IC 50: 9.87 μg/m□Moradi et al. (2017)	Alkaloid fraction of alcoholic extract showed potent antitussive, expectorant and bronchodilating activities in cough models of mice and Guinea pigs. Dose: Total extract (1650 mg/kg) and alkaloid fraction (90 mg/kg) Liu et al. (2015)
C2.27	*Phyllanthus amarus* Schumach. and Thonn. / Bhui amla/ Phyllanthaceae/ Whole plant	Alcoholic extract of aerial parts exhibited potent inhibitory action on both phagocytic and CD18 expression of phagocytes. Dose: 6.25–100 μg/m□ Jantan et al. (2014)	Aqueous extract inhibited cellular proliferation and suppressed HBsAg production in human hepatoma cells. Dose: 1 mg/ml Yeh et al. (1993)	Alcoholic extract attenuates asthma by modulating oxido-nitrosative stress SOD, GSH, MDA, and NO), immune-inflammatory

				makers (HO-1, TNF-α, IL-1β, and TGF-β1), and Th2 cytokines. Dose: 100 and 200 mg/kg Wu et al. (2019)
C2.28	*Picrorhiza kurroa* Royle ex Benth./Kutki/ Plantaginaceae/ Rhizome, leaves	Hydroalcoholic extract stimulate cell-mediated and humoral immunities, along with complement activity and phagocytic function. Dose: 25, 50, 100 mg/kg Sharma et al. (1994)	Isolated iridoids from chloroform fractionated alcoholic extract of inhibit expression of vpr in TREx-HeLa-vpr cells and these iridoid are naturally occurring vpr inhibitors. Dose: 5 and 10 µg/m□Win et a□ (2019)	Alcoholic extract showed anti-asthmatic potential by exhibiting relaxation effect against histamine and acetylcholine induced contraction model in Guinea pigs. Dose: 25 mg/kg (*in vivo)1*, 10 and 100 mg/ml (*in vitro*) Sehgal et al. (2013)
C2.29	*Piper longum* L./ Pipli/Piperaceae/ Fruits	Aqueous extract possessed a demonstrable immunostimulatory activity, both specific and nonspecific, as evident from the standard test parameters such as haemagglutination titer, macrophage migration index and phagocytic index. Dose: 225 mg/ kg Tripathi et al. (1999)	Butanol fraction of alcoholic extract possessed remarkable inhibitory HBV activity, against the secretion of hepatitis B virus surface antigen (HBsAg) and hepatitis B virus e antigen (HBeAg). IC_{50}: 0.15 mM for HBsAg and 0.14 mM for HBeAg Jiang et al. (2013)	Aqueous and pet ether extract showed anti-asthmatic potential by protecting against histamine induced bronchospasm, haloperidol induced catalepsy and passive paw anaphylaxis and by decreasing number of leukocytes in milk-induce leukocytes model. Dose: 50, 100, and 200 mg/kg Kaushik et al. (2012)
C2.30	*Piper nigrum* L./Marica/ Piperaceae/Fruits	Aqueous extract of *P. nigrum* capable of promoting the proliferative signaling pathways in splenocytes and enhance murine splenocyte proliferation. Dose: 50 and 100 µg/ml Majdalawieh and Carr (2010)	Isolated piperamides from *P. nigrum* inhibit coxsackie virus type B3 (CVB3). It inhibits the proliferation of VSMCs. IC_{50}: 21.6µM to10.6 µg/m□Mair et a□ (2016)	Isolated piperine acted partially through stimulation of pituitary adrenal axis. Dose: 5, 10, 20, and 40 mg/ kg Tasleem et al. (2014)

C2.31	*Pongamia pinnata* (L.) Pierre/Karanj/ Fabaceae/Seeds	Isolated oil impact on immune cell signaling events needed for continued recruitment of neutrophils/other cells. Dose: 0.3 or 0.5 g/kg Muniandy et al. (2018)	Aqueous extract interfered with HBsAg and thus probably may prevent HBV entry. Dose: 5 mg for 0.18 pg/ ml concentrations of the virus Mathayan et al. (2019)	Isolated isoflavone and showed inhibitory effects against NO production in LPS-stimulated BV-2 microglial cell thus anti-inflammatory effects. IC 50: 9.0 μM Wen et al. (2018)
C2.32	*Punica granatum* L./Anar/ Punicaceae/Fruit, peel	Aqueous extract showed, a significant decrease in nitric oxide levels and TNF-α levels. A significant diminution of iNOS, TNF-α and NF-κB expression was also observed. Dose: 0.65 g/kg Labsi et al. (2016)	Alcoholic extract inhibited influenza a PR8 virus replication in the MDCK cell line, it could suppress the amplification of the infectious influenza viruses. IC $_{50:}$ 6.45 μg/ ml Moradi et al. (2019)	Isolated galloyl-hexahydroxy-diphenoyl-glucose showed protective effect against acute lung injury and anti-inflammatory activity by inhibiting LPS-induced JNK and NF-κB activation and reduction in expression of the TNF-α, IL-6, and IL-1β genes in lungs. Dose: 5, 50, and 100 mg/kg Pinheiro et al. (2019)
C2.33	*Plantago major* L./Lahuriya/ Plantaginaceae/ Whole plants, seeds	Aqueous extract increased lymphocyte proliferation and secretion of interferon-γ at low concentrations (<50 μg/ml), but at high concentrations, it can inhibit this property (<50 μg/ ml). Dose: 50 μg/ml Chiang et al. (2003)	Isolated compound of caffeic acid from aqueous extract possesses interesting anti-HSV-1, anti-HSV-2 and anti-ADV-3 activities. Caffeic acid was found to inhibit HSV-1 replication. EC $_{50}$: 15.3 μg/ml Chiang et al. (2002)	Hydroalcoholic extract showed amelioration of asthma by increasing mean mast cells, alveolar epithelium thickness and glycoprotein accumulation. Dose: 100 mg/ kg Farokhi & Khaneshi (2013)
C2.34	*Psoralea corylifolia* L./ Babchi/Fabaceae/ Seeds	Hydroalcoholic extract stimulate natural killer cell activity. A positive response was also observed in the ADCC activity of spleen cells. Dose: 100 and 200 mg/kg Latha et al. (2000)	Aqueous extract found more effective in suppressing the virosis and reduced the mortality against virosis cellular and biochemical changes. Kiran Kumar et al. (2012)	Extract showed novel agent for asthma by inhibiting eosinophils accumulation into airways and modulating Th1/ Th2 cytokine balance. Dose: 200 and 400 mg/ kg Lee & Kim (2008), Wen et al. (2018)

C2.35	*Rhodiola rosea* L./Rhodora/ Crassulaceae/ Whole plant	Isolated compound of could promote the activation of T lymphocytes, differentiate them into CD4+ cell or CD8+ cell, and implement their functions. Dose: 12.5, 25, 50 µg Guan et al. (2011)	Alcoholic extract inhibits the entry and infection of ebola and marburg viruses. IC_{50}: 0.25 µg/m (ebo a virus)4.0 µg/m (marburg virus) Cui et al. (2018)	Isolated salidroside showed protective effect in acute lung injury by decrease in the W/D ratio, myeloperoxidase activity of lung, reducing protein concentration, macrophages in the bronchoalveolar lavage fluid and regulating inflammatory cytokines and NF-κB. Dose: 120 mg/kg Guan et al. (2012)
C2.36	*Santalum album* L./Sandalwood/ Santalaceae/Stem	Aqueous extract inhibited cell proliferation, nitric oxide production and CD14 monocyte. Dose: 30 mg/ml Gupta & Chaphalkar (2016)	β-Santa o from hexane extract exhibits anti-influenza A/HK (H3N2) virus by inhibition of viral mRNA synthesis. Dose: 100 µg/m Paulpandi et al. (2012)	Alcoholic extract showed *in vitro* anti-inflammatory activity as compared to Diclofenac. Dose: 500 mg/ml Saneja et al. (2009)
C2.37	*Saussurea lappa* (Decne.) C.B.Clarke/ Kutha/ Compositae/ Roots	Isolated compound of costunolide and dehydrocostus lactone showed suppressive effect on the expression of the hepatitis B surface antigen (HBsAg) in Hep3B cells. IC_{50}: 1.0–2.0 µM Chen et a (1995)	Hexane fraction of alcoholic extract suppress the HBsAg production by Hep3B cells. $IC_{50:}$ 1.0–2.0 µM Chen et al. (1995)	SML0417, epiligulyl oxide and elecampane camphor isolated from roots ameliorates allergic asthma in murine model by inhibiting antigen-induced degranulation, reduction in inflammatory signs and mucin production and expression and secretion of Th2 cytokines. Lee B. K. et al. (2018)

C2.38	*Sphaeranthus indicus* L./ Mundi/Asteriae/ Leaves, flowers	Petroleum ether extract from the flower heads of *S. indicus* increasing phagocytic activity, hemagglutination antibody titer and delayed type hypersensitivity. Dose: 200 mg/kg Bafna & Mishra (2007)	Alcoholic extract exhibits anti-virus potential against herpes simplex virus (HSV) and mouse corona. Dose: 0.4 µg/ ml Vimalanathan et al. (2009)	Alcoholic leaves extract inhibits prostaglandin synthesis. Dose: 100,200, and 400 mg/kg Meher et al. (2011)
C2.39	*Syzygium aromaticum* (L.) Merr. and L.M. Perry/Lavang/ Myrtaceae/Fruits	Aqueous and alcoholic suppressive effects on mouse macrophages and inhibit IL-1*β*, IL-6, and IL-10. Dose: 1000 µg/ ml Dibazar et al. (2015). Essential oil increased the WBC count and enhanced DTH response in mice. Carrasco et al. (2009)	Hydroalcoholic extract exhibits anti- viral activity against herpes simplex virus-1 evaluated on vero cell line using MTT assay. IC_{50}: 8.4 µg/m□Moradi et al. (2018)	Aqueous extract decreases neutrophil count and proteins leakage into bronchoalveolar lavage fluid. Dose: 200 mg/ kg Chniguir et al. (2019)
C2.40	*Terminalia chebula* Retz. /Halala/ Combretaceae/ Fruits	Aqueous extract increase in humoral antibody titer and delayed-type hypersensitivity in mice. Dose: 100–500 mg/kg Shivaprasad et al. (2006)	Hydroalcoholic extract of prevents the attachment as well as penetration of the HSV-2 to vero cells and efficacy to inhibit virus attachment and penetration to the host cells. IC_{50}: 0.01 ± 0.0002 µg/ ml Kesharwani et al. (2017)	Carbohydrate polymer from aqueous extract of dried ripe fruit showed antitussive efficacy in citric acid-induced cough efforts. Dose: 50 mg/ kg Nosalova et al. (2013). Ethyl acetate fraction showed antitussive efficacy on sulfur dioxide gas induced cough partially through modulation of opioid receptors. Dose: 500 mg/kg Haq et al. (2013)

C2.41	*Tinospora cordifolia (Willd.) Miers/Giloe/ Menispermaceae/ Stem*	Aqueous and alcoholic extract reduced bacterial load as compared to untreated macrophages. Dose: 100, 200, and 500 µg/m□A□suhaibani & Khan (2017)	Methanol and ethyl acetate mixture extract inhibits the growth of HSV. Dose: 50–100 µg/m□Pruthvish & Gopinatha (2018)	Hydroalcoholic extract ameliorates asthma through decreasing oxidative stress and inflammation through modulating glutathione homeostasis and regulation of NFκB and pro-inflammatory genes. Dose: 100 mg/kg Tiwari et al. (2014)
C2.42	*Tribulus terrestis* L./Gokhru/ Zygophyllaceae/ Whole plant	Saponin fraction increased phagocytic activity in dose dependent manner. Dose: 50, 100, 200 µg/m□ Tilwari et al. (2011)	Alcoholic extract showed antiviral potential against newcastle disease virus evaluated by titering *in vivo* vero cell line cu□ture. Dose: 80 µg/m□ Malik et al. (2018)	Hydroalcoholic fruit extract activate mast cell. EC 50: 1% extract with 0.1% HC Kang et al. (2017)
C2.43	*Ziziphus jujuba* Mill. /Unnab/ Rhamnaceae/ Fruits	Aqueous extract increase thymus and spleen indices as well as enhance the T-lymphocyte proliferation, hemolytic activity and NK cell activity. Dose: 1.3, 2.6, and 5.2 g/kg Yu et al. (2016)	Isolated betulinic acid showed antiviral activity on influenza virus by attenuating pulmonary pathology and down-regulation of IFN-γ □eve□ Concentration: 50 µM Hong et al. (2015)	Alcoholic extract showed inhibition of expression and activity of COX-2. Dose: 200, 400, and 600 mg/ kg Mesaik et al. (2018)
C2.44	*Zataria multiflora* Boiss./ Satar/Lamiaceae/ Whole plant, leaves	Obtained essential oils from hydrodistilation increase in the secretion of TNF-α, IFN-γ, IL-2 and decrease in IL-4. Dose: 10mg/one BALB/c and 7mg/ one C57BL/6 Jamali et al. (2020)	*Z. multiflora* destruction of virus infectivity or inhibition of early phases of viral proliferation cycle. Arabzadeh et al. (2013)	Hydro-alcoholic extract ameliorates allergic asthma by decreasing pro-inflammatory cytokines, increasing expression of anti-inflammatory cytokines gene and number of treg (FOXP3) in splenocytes. Dose: 200, 400, and 800 µg/m□Kianmehr et al. (2017)

C3.1	*Artemisia absinthium* L./ Vilayati afsantin/ Asteraceae/Roots	Alcoholic extract modulates the percentage expression and fluorescent intensity of CD86, CD40 and MHC II molecules on DCs. Dose: 100 µg/m□Azegu□ et a□ (2018)	Decoction effectively suppressed HBV DNA, HBeAg, and HBsAg. Dose: 15 ml (containing 1 g of dried extract) Ansari et al. (2018)	-
C3.2	*Datura metel* L./ Safed dhatura/ Solanaceae/ Leaves, fruits and seeds	-	Aqueous and alcoholic extract performed in vero cell line using MTT assay showed good antiviral activity. IC $_{50:}$ 2.5 mg/ml Roy et al. (2016)	Aqueous extract in ovalbumin challenged mice ameliorates asthma through promotion of naïve T cell development and reducing activated T cells. Dose: 0.56 mg/kg Rifa'i et al. (2014)
C3.3	*Elettaria cardamomum* (L.) Maton/ Chotielaichi/ Zingiberaceae/ Fruits	Essential oil overlapped with that of various canonical signaling pathways which support its immunomodulator activity. Han & Parker (2017)	-	The extract obtained from supercritical fluid extraction with carbon dioxide inhibit NF-kappa signaling pathway. Dose: 0.03% Souissi et al. (2020)
C3.4	*Embelia ribes* Burm.f./ Baberang/ Myrsinacea/ Fruits	-	Ethyl acetate extract exhibits antiviral activity MDCK cells infected with influenza virus A/Puerto rico/8/34 (H1N1). IC $_{50:}$ 0.2 µg/ ml Hossan et al. (2018)	Isolated embelin attenuates anti-inflammatory activity against carrageenan induced paw edema in rats. Dose: 20 mg/kg Mahendran et al. (2011)
C3.5	*Hedychium spicatum* Sm./ Kapurkachri/ Zingiberaceae/ Rhizome	Alcoholic extract increased phagocytosis, WBC and neutrophils count. Dose: 200–500 mg/kg Uttara & Mishra (2009)	-	Aqueous extract attenuates anti-histaminic action against histamine-induced bronchospasm in Guinea pig. Dose: 200 mg/ kg Ghildiyal et al. (2012)

C3.6	*Hyssopus officinalis* L./Zoofa/ Lamiaceae/ Flowers, leaves	Alcoholic extract of leaves inhibits plaque formation of both of the two strains of HSV-1 in vero E6 cells. Dose: 125 mg/kg Behbahani (2009)	Aqueous extract of flowers affects the levels of some cytokines (such as IL-4, IL-6, IL-17, and IFN-γ) in asthmatic mice. By detection of the expressions of MMP-9 and TIMP-1 and the morphological changes. Dose: 0.04 g/10 g Ma et al. (2014)	
C3.7	*Inula racemosa* Hook.f./Puskara/ Asteraceae/Root	Polysaccharide fraction of water extract showed immunomodulatory action by stimulating phagocytic function. Dose: 100–200 mg/kg Mishra et al. (2016)	-	Pet. Ether extract shows anti-asthmatic potential by mast cell degranulation. Dose: 50 and 100 mg/kg Vadnere et al. (2009)
C3.8	*Lepidium sativum* L./Chansur/ Cruciferae/Whole plants	Protein extract of *lepidium sativum* alter the proliferation induced by Con-A. Daoudi et al. (2013)	-	Isolated fractions from ethanol extract of whole plant inhibit bronchospasm induced by histamine and acetylcholine. Rehman et al. (2012), Prasad et al. (2009)
C3.9	*Leptadenia reticulata* (Retz.) Wight and Arn./ Meethi dodi)/ Apocynaceae/ Root, stem bark	Alcoholic extract increased hematological profile, GSH, SOD, CAT activity and decreased LPO levels in cyclophosphamide-induced rats. Dose: 100–200 mg/kg Pravansha et al. (2012)	-	Ethyl acetate fraction inhibit pro-inflammatory cytokines (IL-2, IL-6, TNF-α) and release of prostaglandin to prevent inflammation. Dose: 600 mg/kg Mohanty et al. (2015)

C3.10	*Magnolia officinalis* var. officinalis/ Himchampa/ Magnoliaceae/ Bark	-	Isolated compound magnolol and honokiol from petroleum ether extract of bark provoked IRF7 transcripts (magnolol) and reinforcing the host antiviral response via NF-κB pathways (Honokiol). Dose: 35 mg/L Chen et al. (2017)	Aqueous extract exhibits anti-allergic actions through inhibition of local immunoglobulin E, histamine re[illegible]ease and TNF-α production in 48/80 induced systemic anaphylaxis in rats. Dose: 0.001–1 g/kg Shin et al. (2001). Polyphenolic rich extract of *Magnolia officinalis* suppressed the production of inflammatory mediators, NO, pro-inflammatory cytokines, TNF-α and IL-6, and inhibition of TLR3 and NF-κB activation. Dose: 10 and 200 mg/kg Fang et al. (2015)
C3.11	*Mucuna pruriens* (L.) DC./ Kaunchbeej/ Fabaceae/Seeds	*M. pruriens* modulate the immune components like TNF-a, IL-6, IFN-l, IL-1b, iNOS and IL-2. Rai et al. (2017) Alcoholic extract of root influenced both humoral and cell mediated immunity. Dose: 100, 200 and 400 mg/kg Murthy & Mishra (2016)	-	Alcoholic extract of seeds of *M. pruriens* act on opoid receptor that located on airway passage and produce inhibitory effect. Dose: 500 mg/kg Nuzhat et al. (2013)

C3.12	*Piper betle* L./ Paan/Piperaceae/ Leaves	Alcoholic extract of *P. betle* leaves showed lymphocyte proliferation, interferon- γ receptors and the pro-duction of nitric oxide. It suppressed phytohaema-gglutinin stimulated peripheral blood lymphocyte proliferation. Dose: 500 mg/ kg Kanjwani et al. (2008)	-	Alcoholic extract of leaves decreased histamine and GM-CSF produced by an IgE-mediated hypersensitive reaction, and inhibited eotaxin and IL-8 secretion in a TNF-α and IL-4-induced allergic reaction. Dose: 10 mg/ml Wirotesangthong et al. (2008)
C3.13	*Sesamum indicum* L./Tila/ Pedaliaceae/Seed	Essential oil suppresses cellular immunity with the domination of Th2 responses and also modulate macrophages, dendritic cells proinflammatory functions. Dose:100 µg/m□Khorrami et al. (2018)	-	Aqueous extract reduces LPS induced inflammatory gene expression. EC_{50}: 100 ng/ml Deme et al. (2018)
C3.14	*Sida cordifolia* L./Beejband/ Malvaceae/Seeds	*S. cordifolia* increased production of T-cell precursor and passive influences on the production of cytokines. Dose: 2 gm/kg Tekade et al. (2008)	-	Alcoholic extract of seed inhibits paw edema and granuloma formation. Dose: 200 and 400 mg/ kg Singh S. et al. (2011)
C3.15	*Swertia chirayita* (Roxb.) H.Karst./ Chirayata/ Gentianacea/ Whole plant	-	Chloroform extract inhibit expression of viral protein R in hela cells harboring the TREx plasmid encoding full-length vpr (TREx-HeLa-vpr cells. Dose: 10 µM Woo et al. (2019)	Chloroform fraction exhibits bronchodilator effect by Ca^{2+} channel blockade. Dose: 0.1–3.0 mg/ ml Khan et al. (2012)

C3.16	*Tachyspermum ammi* (L.) sprague/Ajwain/ Umbelliferone/ Seed	Isolated glycoprotein from aqueous extract of seed proliferate B-cell enriched murine splenocytes and activated macrophages in releasing NO and promoted phagocytosis. Dose: 1 µg/m□Shruthi et al. (2017)	Seed oil neutralize antibody for Japanese encephalitis virus. Dose: 0.5 mg/ml Roy et al. (2015)	-
C3.17	*Tylophora indica* (Burm.f.) Merr./ Antamool/ Asclepiadaceae/ Roots, leaves	Alkaloidal fraction inhibit proliferation of splenocytes and both macrophages and T cells were found to be vulnerable to tylophora alkaloids. Ganguly et al. (2001)	-	Alcoholic extract of aerial part showed anti-inflammatory effect against carrageenan induced paw edema and cotton pellet induced granuloma. Dose: 100, 200 and 400 mg/kg Raj et al. (2006)
C3.18	*Viola odorata* L./Banafsha/ Violaceae/ Flowers	-	*V. odorata* effective against multiple protease inhibitors Gerlach et al.	

Source: Ahmad S, Zahiruddin S, Parveen B, Basist P, Parveen A, Gaurav, Parveen R and Ahmad M (2021) Indian Medicinal Plants and Formulations and Their Potential Against COVID-19–Preclinical and Clinical Research. Front. Pharmacol. 11:578970. doi: 10.3389/fphar.2020.578970: Copyright © 2021 Ahmad, Zahiruddin, Parveen, Basist, Parveen, Gaurav, Parveen and Ahmad. This is an open-access article distributed under the terms of the Creative Commons Attribution License (CC BY). The use, distribution, or reproduction in other forums is permitted, provided the original author(s) and the copyright owner(s) are credited and that the original publication in this journal is cited, in accordance with accepted academic practice. No use, distribution, or reproduction is permitted which does not comply with these terms.

"Eat your veggies" – on the lines of this saying most of us have heard, the physiological effects of select phytonutrients from a few vegetables have been presented below as compiled by Minich (2019).

Table 1.17.

Color	Food	Some Food Formats Researched	Basic (Phyto) Nutrient Profile	Researched Health Benefits
Red	Tomatoes	Juice, powder, raw, sauce (prepared with and without oil)	Carotenoids (e.g., lycopene), flavonoids, vitamin C	(i) Reduction in inflammatory markers (ii) Reduction in postprandial inflammation (iii) Improvement in lipid markers
Red	Strawberries	Freeze-dried as beverage	Polyphenols (flavonoids, phenolic acids, tannins), vitamin C	(i) Reduction in postprandial inflammation (ii) Reduction in pain due to osteoarthritis
Red	Beets	Cooked, raw juice	Betalains	(i) Reduction in inflammatory markers
Orange	Wild yam	Cooked	Phytestrogens	(i) Increase in estrogen and estrogen metabolites (ii) Phytestrogenic activity
Orange	Carrots	Not specified	Alpha- and beta-carotene	(i) Decrease in rate of breast and prostate cancer

				(ii) Phytestrogenic activity (iii) Association with estrogen metabolism
Orange	Orange fruits	Not specified	Bioflavonoids, beta-carotene, beta-cryptoxanthin	(i) Delay in ovarian senescence (ii) Lower risk for endometriosis
Yellow	Ginger	Standardized extract	Gingerols, shojails	(i) Decrease in nausea (ii) Increase in gastric emptying
Yellow	Citrus (lemons)	Juice, raw	Hesperidin, nobiletin, rutin, vitamin C	(i) Protective against gastric ulcer (ii) Antidiabetic (iii) Reduction in glycemic impact
Yellow	Pineapple	Juice	Bromelain, serotonin	(i) Enzymatic activity
Yellow	Bananas	Raw	Prebiotic fiber, serotonin	(i) Increase in bifidobacteria (ii) Reduction in bloating
Green	Leafy greens	Raw, spinach	Chlorophyll, folate, nitrates, phylloquinone	(i) Reduction in blood pressure (ii) Increase in nitric oxide (iii) Increase in blood flow

Green	Cruciferous vegetables	Not specified	Glucosinolates, isothiocyanates, sulforaphane	(i) Antioxidant action (ii) Reduction in platelet aggregation (iii) Reduction in thrombus formation
Blue-purple	Concord grape juice	Juice	Phenolic acids, stilbenes, anthocyanins, proanthocyanins	(i) Improvement in spatial memory and performance (ii) Improvement in reaction time on attention (iii) Increase in calm ratings
Blue-purple	Blueberries	Beverage, freeze-dried, raw	Flavonoids, procyanidins (monomeric and oligomeric form), flavonols (i.e., kaempferol, quercetin, myricetin), phenolic acids (mainly hydroxycinnamic acids), derivatives of stilbenes	(i) Improvement in measures of cognition (ii) Benefit to mood (iii) Improvement in neuroplas

Source: Minich DM. A review of the science of colorful, plant-based food and practical strategies for "eating the rainbow" [published correction appears in J nutr metab. 2020 Nov 28; 2020:5631762]. J Nutr Metab. 2019; 2019:2125070. Published 2019 Jun 2. doi:10.1155/2019/2125070: © 2019 Deanna M. Minich. This is an open access article distributed under the Creative Commons Attribution License, which permits unrestricted use, distribution, and reproduction in any medium, provided the original work is properly cited.

With this background of what is going on in this field (which is actually the tip of the iceberg), let us cover various aspects of how the metabolism of cells is tweaked by researchers to fuel production and biotechnological applications.

"Every person is a collective, a vast and complex gathering of interdependent life. Any description of 'human' must acknowledge these intimate strangers. Our bond with microbes is such that they are not so much riders, parasites, and assistants as part human. And we, it's becoming increasingly clear, may need to begin thinking of ourselves as part microbe."

— Guy P. Harrison, At Least Know This:
Essential Science to Enhance Your Life

REFERENCES

1. Berg, G., Rybakova, D., Fischer, D., et al. (2020). Microbiome definition re-visited: Old concepts and new challenges. *Microbiome, 8*, 103. https://doi.sorg/10.1186/s40168-020-00875-0.
2. Bratosin, B. C., Darjan, S., & Vodnar, D. C. (2021). Single cell protein: A potential substitute in human and animal nutrition. *Sustainability, 13*(16), 9284. https://doi.org/10.3390/su13169284.
3. Ciani, M., Lippolis, A., Fava, F., Rodolfi, L., Niccolai, A., & Tredici, M. R. (2021). Microbes: Food for the future. *Foods, 10*(5), 971. https://doi.org/10.3390/foods10050971.
4. Gholami-Shabani, M., Shams-Ghahfarokhi, M., & Razzaghi-Abyaneh, M. (2023). Food Microbiology: Application of Microorganisms in Food Industry. IntechOpen. https://doi.org/10.5772/intechopen.109729.
5. Graham, A. E., & Ledesma-Amaro, R. (2023). The microbial food revolution. *Nature Communications, 14*, 2231. https://doi.org/10.1038/s41467-023-37891-1.
6. Ritala, A., Häkkinen, S. T., Toivari, M., & Wiebe, M. G. (2017). Single cell protein—State-of-the-art, industrial landscape and patents 2001–2016. *Frontiers in Microbiology, 8*, Article 2009. https://doi.org/10.3389/fmicb.2017.02009.
7. Suman, J., Rakshit, A., Ogireddy, S. D., Singh, S., Gupta, C., & Chandrakala, J. (2022). Microbiome as a key player in sustainable agriculture and human health. *Frontiers in Soil Science, 2*, Article 821589. https://doi.org/10.3389/fsoil.2022.821589.

Chapter 2

SYNTHETIC BIOLOGY: GLIMPSES OF CONCEPTS AND EXAMPLES

Shiv Sanjeevi Sripathi

"The potential for synthetic biology and biotechnology is vast; we all have an opportunity to create the future together."

—Ryan Bethencourt

2.1. INTRODUCTION

Synthetic biology refers to an application of synthetic chemistry to life sciences and involves the engineering of biological entities or processes to possess functions that were previously not present. It aims at compiling data, manipulating chemicals and molecules, and enhancing health (Chopra and Kamma, 2006). It involves the introduction of an engineering concept into the arena of a cell to convert a cell into an industry to produce molecules (Trosset & Carbonell, 2015). The notion of a drug to target a particular and specific part of the body to enable recovery from disease was proposed by Paul Ehrlich a century ago (Mulinari, 2014). The extracts of microbes and plants offered sources of compounds that were pharmacologically active to find utility as such drugs (Maruta, 2009). The techniques in combinatorial chemistry to develop scaffolds of requisite characteristics, coupled with the manufacturing of solid and liquid phases, aided in the development of such molecules. Furthermore, instead of using natural products, there was a preference for high-throughput screening (HTS) to end up with drugs. However, this methodology is dependent on the nature of the libraries used as the starting stage of the analysis. The alterations in properties of such molecules and their diversity were applied as the major elements to facilitate success in the technique. A major challenge that arises is the cross-reactivity or non-reactivity: the binding of such molecules generated from chemical libraries could bind to non-intentional targets and cause adverse side effects (Harris et al., 2011).

Proteomics highlighted that molecules designed with high selectivity could also target unintended targets (Hu et al., 2014).

These issues can be resolved by the application of synthetic biology: using systems based on biology such as cell-free systems or synthesized cells. The first discovery was that the production of secondary metabolites by several organisms such as microbes or plants using units of massive proportions aided in the creation of a genetic design. It involved the manipulation of these pathways depending on enzymes for the production of novel natural products (Trosset & Carbonell, 2015) (Table 2.1).

Initial studies in this field involved the standardization of various networks to increase the success of such systems in prokaryotes. Eukaryotic cells are compartmentalized apart from the arrangement of DNA within the nucleus subject to extensive regulation. Despite these odds, synthetic biology yielded much success in yeast and cell lines. The approach of synthetic biology involves alterations of signal pathways apart from modifications in genes or new cells (Kis et al., 2015). A slew of approaches are emerging for use in the field of metabolic engineering. A metabolic pathway can be constructed by assembling different genetic constituents like the promoter and coding sequences. This is called pathway construction. Further control of translation can entail the use of synthetic RNAs and genome editing tools like CRISPR and TALENs. The enzymes can also be attached to a scaffold made of DNA, RNA, and proteins in what is called spatial organization. A dynamic pathway with a metabolic circuit with optimized flux can be designed.

Table 2.1. A Few Terminologies Are Given Below As Compiled by Lee, Vo, & Na (2018)

TAL effectors	Transcription activator-like effectors secreted by *Xanthomonas* bacteria, which recognize plant DNA sequences through a central repeat domain of ~34 amino acid repeats, and activate the expression of individual genes to aid bacterial infection
Bio-oscillators	Synthetic genetic circuits that mimic the natural genetic clocks of organisms, which induces the periodical behavior of a system
Toggle switch	Synthetic genetic circuits that display a stepwise function: ON or OFF
Logic gates	Synthetic genetic circuits consisting of multiple cellular sensors and actuators to perform precise logical operation (e.g., AND-, OR-, NOT-, and NAND-gates)
Riboswitches	RNA sequences that resize in 5' UTR and regulate transcription and/or translation by conformational changes upon ligand binding
Small regulatory RNAs (sRNAs)	Small non-coding RNAs that regulate translation by base-pairing with target mRNAs
Protospacer-adjacent motif (PAM)	5′-NGG-3′ sequence immediately following the target DNA sequence in the CRISPR-Cas system

Source: Lee H-M, Vo PNL, Na D. Advancement of metabolic engineering assisted by synthetic biology. *Catalysts*. 2018; 8(12):619. https://doi.org/10.3390/catal8120619:

Figure 2.1 shows the approaches of metabolic engineering. In this example, the product is n-butanol, which is produced by the assembly of different enzymes from three organisms: *R. eutrophus* (blue), *C. acetobutylicum* (red), *and S. collinus* (green).

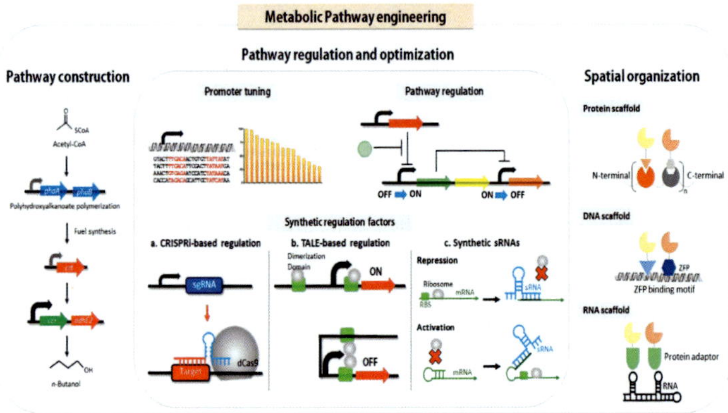

Figure 2.1.

Source: Lee H-M, Vo PNL, Na D. Advancement of Metabolic Engineering Assisted by Synthetic Biology. *Catalysts*. 2018; 8(12):619. https://doi.org/10.3390/catal8120619: *© 2018 by the authors. Licensee MDPI, Basel, Switzerland. This article is an open access article distributed under the terms and conditions of the Creative Commons Attribution (CC BY) license (http://creativecommons.org/licenses/by/4.0/).*

2.2. ON-OFF SWITCHES IN SYNTHETIC BIOLOGY

The reorganization of biological systems into switches or operators that can be controlled is a tool of synthetic biology. Akin to electrical switches, these elements have an "on" and an "off" state. While the former use current, the latter use a biological signal. These biological switches have been reported to be sensitive to the context and involve more complexities.

2.2.1. Transcriptional Switches

2.2.1.1. Tet-Off System

The first controller of transcription in mammals was developed as tetracycline-controlled transactivator (tTA). A combination of tetracycline (Tet) repressor, TetR, along with elements of operator (tetO) sequences, with the C-terminal domain of virion protein 16 (VP16) from herpes simplex virus. The use of an analog of tetracycline: Doxycycline, binds with tTA to cause it to enter an OFF state (Kistner et al., 1996).

2.2.1.2. Tet-On System

Subsequently, rtTA or a reverse tetracycline-controlled transactivator was derived by a mutation of the above to function with the analogs setting the system in an ON position (Gossen et al., 1995).

The switches can be controlled by the use of several molecules such as hormones or antibiotics that developed into transcriptional switches of the second generation that are controlled by light or temperature or amino acids have been reported.

2.2.2. Synthetic Transcription Factors

Transcription activators (effectors), synthetic TFs are designed to latch onto specific genome regions to alter gene expression (Keung, et al., 2015). TFs like doxycycline, uric acid macrolides, etc., can be engineered to detect drugs with small molecular weights or small metabolites. They can also be fused with reporter genes, which are fluorescent (e.g., GFP), serving as an assay for drug screening; a method invented by Fussenegger's group for the identification of antibiotics of the streptogramin group. The system involved the use of pristinamycin-induced protein (PIP) produced by an operon of *Streptomyces coelicolor* that confers resistance to streptogramin. The system could be operated in both "on" as well as "off" states and was reported to be controlled by antibiotics belonging to the group of streptogramins (Fussenegger, et al., 2000).

Table 2.2 summarizes several transcriptional switches in mammals and their application in medicine (Kis et al., 2015).

Table 2.2. *Transcriptional Switches in Mammals*

Sr. No.	Name of Switch	Type Of Switch	Effector	Potential
1.	ART from *Chlamydia pneumoniae*	On	L-arginine	Gene therapy
2.	Electro-genetic from *Aspergillus nidulans*	On	Acetaldehyde and electricity	A mammalian cell with electric circuits. The transcription can be affected with current.

3.	GAL4-ER from yeast, *Homo sapiens*	On and off	Estrogen, 4-hydrox-ytamoxifen	It can be used for identification of Fos: an oncogene and alsoas a regulator for heterologous systems.
4.	Cry2-CIB1 from *Arabidopsis thaliana,* yeast as well as Light on from *Neurospora crassa*	On	Blue light	Cre-mediated DNA recombination along with transcriptional control by light
5.	CTA, CTS from *Comamonas testosteroni,* *Homo sapiens*	On and off	Food additives	Gene therapy
6.	ZF switches	On	*Mus musculus, Homo sapiens,* *Drosophila melanogaster*	Manipulation of phenotypes
7.	TIGR	Off	Temperature	Regulation of genes

2.2.3. Post-Transcriptional Switches

The activity or processing of mRNA is controlled. They have been developed using either RNAi or aptamers. The latter refers to folded, single-stranded nucleic acids that bind a target (Kim et al., 2008). Allosteric ribozymes combine aptamers with a ribozyme to work as a switch (Chen et al., 2010).

2.2.4. Ribozymes or Riboswitches for Compounds Against Pathogens

Ribozymes and riboswitches are the RNA-based devices that regulate genes and are important in controlling central metabolism, making them significant targets for antiparasitics and antibiotics (Berens, et al., 2015; Chappell, et al., 2015). Thiamine pyrophosphate riboswitches offer hope as antibacterial targets.

They regulate thiamine metabolism in several bacteria (Lünse, et al., 2014). A drug target in *Plasmodium falciparum,* dihydrofolate reductase-thymidylate, is modulated by activating glucosamine-induced ribozyme (Prommana, et al., 2013). A fluoride riboswitch reporter fusion gene was used in high-throughput screening (HTS), leading to the identification of compounds that enhance innate fluoride toxicity in *E. coli* and *Streptococcus mutants*. The screening approach for identifying chemical scaffolds that trigger riboswitch-mediated gene regulation involved fusing a reporter gene with a ligand-sensing riboswitch (Nelson, et al., 2015). It is observed that these synthetic riboregulators respond to various orthogonal input signals that regulate genes within the target cell. They also serve as programmable kill switches that can be used against various pathological microorganisms (Callura, et al., 2010).

2.2.5. Post-Translational Switches

The activity of a protein can be regulated by applying a destabilizing domain such as the HaloTag system. A dehalogenase tag is added to the protein that aids in its degradation in the presence of hydrophobic groups (Los GV et al., 2008). Another approach is intein-based post-translational switches that use the inherent splicing property of introns (Shi, 2005) Figure 2.2.

A summary of these approaches with their application in mammals is provided in Table 2.3.

Table 2.3. *Aptamers, Ribozymes, and Translational Switches in Mammals*

Switch Type	Name of the Switch	Type of Switch	Effector	Potential
Post-transcriptional switches	Aptamer RNAi that uses synthetic SELEX-derived aptamer	Off	Theophylline	To achieve regulation of gene expression.
	Aptazyme	Off	Theophylline	To achieve regulation of gene expression.
Post-translational switches	FM from humans	On	AP21998	For delivery of proteins in a rapid manner
	PROTAC	Off	proteolysis targeting chimeric molecules (PROTACS)	For knock out studies
	shield DD and shield LID	On and off respectively	shields	To exert controls over proteins

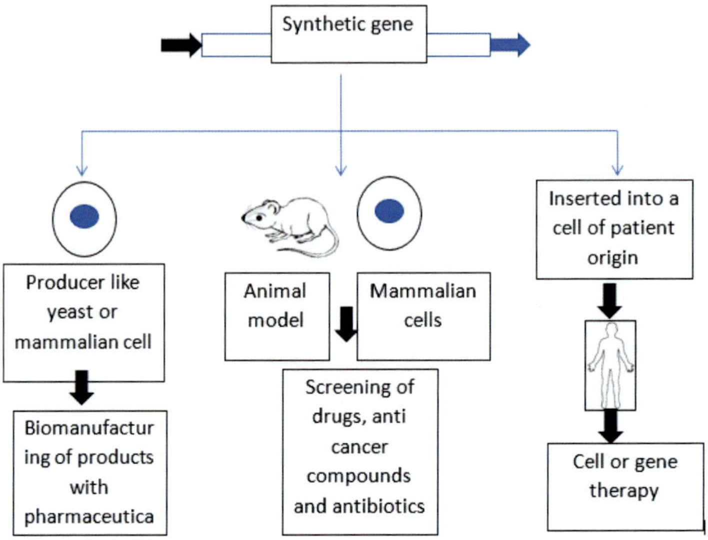

Figure 2.2. *Scheme of synthetic biology and medicine.*

Source: Made by the author.

2.3. SYNTHETIC BIOLOGY AND APPLICATION IN VACCINES

The use of conventional vaccines depends on attenuated live pathogenic organisms or subunits or inactivated microbes. Apart from issues of altering the specificity of a vaccine, the application of microbes in the vaccine is of concern (Jain, 2013). This is where a system of synthetic biology offers promise as a vaccine based on synthetic biology involves the synthesis of a nucleic acid of an immunogenic substance efficiently and within time along with the systems of their delivery (Kindsmüller & Wagner, 2011).

- A group in 2011 used liposomes in the form of engineered microbial cells to be used as liposome-based vaccines. The liposomes were used to create vaccines: antigen-expressing immunostimulatory liposomes (AnExILs). These molecules were associated with the induction of an immune response of the humoral system to a much greater extent as compared to controls. This system uses an amalgamation of DNA as vaccines to elicit an immune response using this synthetic biology tool. The same liposome can be applied for several pathogens to be used as vaccines and were reported to be of low risk as microbes are not involved (Amidi et al., 2011).

– In another study, peptide-based particles were designed of nano proportions using various software tools. These nanoparticles were found to undergo self-assembly in order to form structures that are isohedral in shape apart from showing biocompatibility (Raman et al., 2006).
– Subsequently, this system was applied in the design and construction of B cell immunodominant repeat epitope of the circumsporozoite protein of rodent malaria parasite *Plasmodium berghei*. This polypeptide-based vaccine against malaria was found to confer protection in mice models for half a year without the necessity of adjuvants (Kaba et al., 2009).

2.4. SYNTHETIC BIOLOGY TO TARGET PATHOGENS

- An arm of synthetic biology has been described as Phage-based development, such as phage display. This system involves the fusion of libraries of peptides to coat proteins of phages, yielding many results in the development of vaccines and drugs. Synthetic biology has used the T7 promoter of phages to aid in exercising control in both prokaryotic and eukaryotic cells. Other parts of phages, such as Cre, Bxb1, or PhiC31 recombinases, have been applied in the formulation of genetic circuits. In order to address infectious pathogens, modifications in the genome of several pathogens such as *Pseudomonas aeruginosa, Salmonella enterica, Shigella flexneri, Vibrio cholera*, and *E. coli* have been made (Kis et al., 2015).
- A synthetic network based on communication between cells can be used to evaluate the role of Quorum Sensing, which is used by bacteria for gene regulation, to use this system as a drug target. The synthesis of antibiotics in cells is subject to regulation by molecules such as acyl homoserine lactones or γ-butyrolactones. These molecules are associated with the regulation of efflux pumps that aid in the development of antibiotic resistance, which can be used as novel drug targets in pathogens (Trosset & Carbonell, 2015).

2.5. SYNTHETIC BIOLOGY AND DRUG DISCOVERY

The field of drug discovery and development has seen contributions from synthetic biology along with the use of next-generation sequencing approaches that supply information regarding minute changes in genomes reflective of polymorphisms. Synthetic biology offers the following advantages in the realm of drug discovery:

- The tools used for the engineering of biological pathways can be extended to an accurate regulation in such synthetic cells or tools.
- Cell lines can be engineered synthetically in order to serve as testing tools for drug candidates, causing a decrease in the time of screening.
- A combination of next-generation sequencing with synthetic biology can offer insights into the realm of personalized medicine.
- Therapeutic proteins can be produced synthetically, hastening the process of biotechnology apart from the use of regular host-based protein factories in recombinant DNA technology.
- The field of synthetic biology aids in being more economical (Reviewed by Jain, 2013).

The use of gene networks for the screening of drugs has an added advantage of circumventing the debates regarding ethical issues as the use of mammalian engineered cells has the same genetic material (Schmidt et al., 2009). For instance, the discovery of antibiotics was facilitated by the use of a circuit. A prodrug ethionamide is converted by a pathway in *Mycobacterium tuberculosis* into an active drug that is effective against the pathogen. This drug is applied in the final stages of defense against the Mycobacterium complex. The pathogen regulates the activity of the drug by the use of an ethionamide repressor (EthR) (Baulard et al., 2000). The use of such a circuit facilitated the testing of 2-phenyl ethyl butyrate that is used as a food additive as a drug to be used against multidrug-resistant *M. tuberculosis* (Weber et al., 2008).

The field of synthetic biology also finds application in the screening of cancer drugs. A synthetic genetic network based on cyclin-dependent kinase inhibitor p27Kip1 was reported in 2005 for arresting mammalian cells using the tTA system by tetracycline. The model was tested on Chinese hamster ovary cells by the removal of tetracycline, which arrested the cells in the G1 phase. Of these arrested cells, a few cells resumed the G1 phase to yield a population with both proliferating and proliferation-arrested cells. This system was used for testing chemicals targeting cancer cells (Gonzalez-Nicolini et al., 2004 and 2005). A further application of synthetic biology in this realm is the study of compounds that modulate cell membrane receptors such as G protein-coupled receptors (GPCRs). This class of molecules is used as drug targets, and augmenting the use of molecules that target these proteins shows promising results. A synthetic GPCR system was engineered to possess sensitivity to mechanical stress. The system was activated by the application of shear that can be used to test drug candidates that alter this property of shear (Kis et al., 2015).

2.6. IDENTIFICATION OF TARGETS

The process of drug discovery entails the use of techniques to identify the targets of a drug. The role of synthetic biology in the identification of drug targets has been reported by several studies. The development of the CRISPR/Cas9 system for introducing mutations in a genome or editing genomes has aided in this approach. Such mutations introduced can serve to identify targets of a drug. A research team in 2015 showed that a mutation in the XPO1 gene by CRISPR/Cas9 genome editing endowed such cells to be resistant to an anticancer drug called Selinexor. Thus, this tool of synthetically creating mutations in cells aided in the identification of the target of this drug as the XPO1 gene (Neggers et al., 2015).

A 2012 study used the system of bioengineering to construct a sensor to be used in the evaluation of drug targets. Using a database of biological parts and linkers, such a sensor was constructed as a cassette. Synthetic biology was also used to make a Magnetic Tweezer device for the manipulation of targets along with DNA substrates for testing drug targets for helicases from *Plasmodium* for the engineering of anti-malarial drugs (Firman et al., 2012).

2.7. SYNTHETIC BIOLOGY AND DISEASE MODELS

Models of disease refer to circuits that have been engineered to study the effects of a stimulus and the responses generated. Optogenetics involves switching a protein on and off using either light or synthetic photoswitches ligands (Trosset & Carbonell, 2015).

The field of optogenetics has been used to study patterns of circuits in animals as it circumvents the issues associated with knockout, such as hampered development. The technique involved activating orexin neurons with blue light. In sleep studies, mouse orexin neurons were studied with such an approach to reveal that these orexin neurons are functional during periods of waking and at a low-level during sleep. The use of inhibiting and activating sleep circuits during drug exposure can serve as a tool to develop drug targets for neurons (Rihel & Schier, 2013).

In the field of oncology, there are reports of the use of synthetic biology. A 2013 study engineered a system of BCR: B cell antigen receptor signaling by the use of J558L myeloma cells. The monitoring of tyrosine-related phosphorylation revealed similar patterns following stimulation using an antigen. Apart from serving as a model system for studying the BCR signaling pathway, the approach also revealed a new kinase involved called protein tyrosine kinase Fer (Harumiya et al., 2013).

2.8. PRODUCTION OF DRUGS BY METABOLIC ENGINEERING

The use of synthetic biology can aid in the production of molecules that can be used for pharmaceutical applications. The combination of synthetic biology as well as metabolic engineering can present advantages such as:

- Enzymes of pathways can be modified or other pathways can be introduced into producer cells.
- Chemical synthesis of certain drugs is challenging due to their chiral property.
- The use of microbes as factories has the ease of control, rapidity, and the use of more economic substrates (Reviewed by Trosset & Carbonell, 2015).

Drugs based on natural products include molecules such as polyphenols, polyketides, and isoprenoids among others. The group of isoprenoids is comprised of molecules active against several conditions such as malaria or cancer. The pathway for isoprenoid has been manipulated by synthetic biology in order to enhance its production. Two examples are discussed below:

- The synthesis of paclitaxel or Taxol (a drug used in chemotherapy) is extracted from its natural source at low levels while its synthesis is challenging. The use of synthetic biology for engineering the Taxol synthesis pathway in yeast yielded low amounts of product. Subsequently, the use of an alternative pathway from *S. acidocaldarius* as well as an enzyme from *Taxus chinensis* yielded a product with 40 times the yield. The development of tools led to the division of the entire pathway into units that were subjected to simultaneous variation to ascertain the optimum output. The yield of this approach reached 1 g/l. Despite the challenges, this field of synthetic biology offers the production of such anti-cancer drugs at a more economical cost from biological sources.
- The second example is that of artemisinin which has anti-malarial effects. Several approaches to synthetic biology have aided in increasing the concentration of this drug. From a yield of 100 mg/l of artemisinic acid reported from a redesigning of the mevalonate pathway of the host yeast cell and the addition of an amorphadiene synthase and cytochrome P450 from *Artemisia annua, subsequent work has increased the output. An overexpression of all the enzymes coupled with changing the bioreactor conditions brought up the yield to* >40 g/l amorphadiene (Immethun et al., 2013).

2.9. TOXICOLOGY

Drugs cause various side effects and often transform healthy cells, increasing the toxicity in cells (Zhang, et al., 2014). Such toxicity should be given consideration that can both limit the rate of production as well as aid in the design of new anti-pathogen drugs. A team in 2012 reported the development of an *E. coli* toxicity prediction web server based on a library of 166 compounds used for testing in the bacterium (Planson, et al., 2012).

2.10. USING CELLS FOR SCREENING

Structure-based designing of drugs and sequencing of the human genome have given insight into the discovery of important drugs. With the advent of these findings, it is easy to commence a therapeutic project with a pre-planned drug pathway. This is a more efficient and quick way compared to the lengthy process of drug target (Zheng, et al., 2013). In screening drugs, the disease phenotype is inserted in the desired cell-based assay. Though no mechanism of action is derived at this stage, several targets and pathways can be altered by the drug (Eder, et al., 2015).

Three commonly performed cell assays in drug discovery are:

- Cell viability assay that measures the capacity of a drug to eradicate a disease-causing cell,
- Cell signaling pathway assay where the component either inhibits or activates a metabolic pathway,
- Disease-related phenotypic assay, where the component speeds up the responses which lead to neurite outgrowth. These are used in further studies of Parkinson's disease, Alzheimer's disease, etc. (Zheng, et al., 2013).

2.11. APTAMERS IN HEALTH

Aptamers bind to targets ranging from small molecules to large proteins with high specificity and affinity. Studies show that when used as sensors of molecular chaperones, they modulate the lifetime and folding state while decreasing the tumorigenic and antiapoptotic activities observed in cancer cells (Gilbert, et al., 2013). A process for purification and selection combined in a single step helped in selecting DNA aptamers against eIF4e, which is an oncology target; characterizing aptamer candidates (Guo, et al., 2014).

Peptide aptamers are also called chemical antibodies as they bind with specificity and high affinity to protein targets. Chemica antibodies are also used to mark cell-specific receptors in immunotherapies (Sun, et al., 2014). They are fused to form a scaffold of stable protein for optimizing the biochemical function of multi-domain proteins and also for searching for partners in protein-protein interactions (Reverdatto, et al., 2015). A scaffold protein was developed using GST, FRB, and FKBP12 protein domains that bind when rapamycin, a small molecule, is absent (Miller, et al., 2007). Enhanced peptide aptamers from the combinatorial library were used to detect end products of glycation. This involved a yeast two-hybrid screen for the identification of peptide aptamers that bound to several receptor domains (Reverdatto, et al., 2013). Protein-protein interaction inhibitors can be approached by the selection of peptide aptamers through various combinations. AptaScreen uses peptide sequences that mark the surface loops of a given scaffold protein, using combinatorial libraries. This assay, also known as the HTS assay, recognizes the molecules which dislodge interactions involving cognate peptide aptamers and proteins (Bardou, et al., 2009). Experiments showing interference of peptide-aptamers are used for the identification of protein-protein interactions (Yeh, et al., 2013).

2.12. CASE STUDIES: SYNTHETIC BIOLOGY IN HUMAN HEALTH

The application of engineered cells or gene networks can be used for the benefit of human health. A few case studies are discussed below.

2.12.1. Synthetic Gene Circuits

- A team in 2011 reported the use of an RNAi-based logic circuit for the visualization of cancer cells. A set of endogenous miRNAs of HeLa cells were identified by a Boolean logic circuit that used miRNA. Once a cell meets the required numbers of miRNAs, a signal is engineered to initiate apoptosis in the cell. Such a system may be applied for the specific killing of damaged cells (Xie et al., 2011).

A 2012 study reported the construction of a synthetic memory circuit for the tracking of cell fates. The system uses upregulation of synthesis of an engineered transcription factor using a stimulus such as hypoxia, UV, or doxycycline. The factor uses a system of positive feedback to remain active in order to find application in the alterations of growth patterns along with gene expression. The system, along with synthetic switches, can aid in the visualization or tracking of cells in tumor regions or damaged tissues (Burrill et al., 2012).

2.12.2. Therapy

The understanding of aspects of metabolism along with the mechanisms that lead to disease can serve as tools in the correction of such disorders. SB in therapeutic studies aims at building a simple synthetic cell that has the genes essential for the cell's growth along with a genetic circuit that marks the phenotypic-based technique (Chiba, et al., 2012).

- Thymus-derived (T) lymphocytes are associated with cell-mediated immune pathways to target invading organisms. An advantage conferred by these cells is their extraction from patients to be subjected to engineering and subsequent reintroduction back into the patient. An RNA-based switch was engineered in 2010 that could be induced by a drug in order to alter the cascade of cytokines that in turn played a role in the cell division of T cells (Chen et al., 2010).
- In another study, a pharmaceutical product was used to treat a metabolic disorder: cardiovascular disease risk factors. The approach involved. The gene network was designed in order to augment the effects of Wytensin® in order to target such diseases. A drug used for high blood pressure drug guanabenz (Wytensin®) was used to activate chimeric trace-amine associated receptor 1. The drug along with cAMP activates the cAMP-response element binding protein (CREB1) which in turn activates a hormone called GLP-1-Fc-leptin. This hormone, in turn, regulates the levels of hyperglycemia and dyslipidemia. Thus, this system combines the therapeutic effects of the drug to achieve a three-pronged approach (Ye et al., 2013).
- The synthesis or metabolic engineering of key pathways can serve as a channel for the production of therapeutic agents that are derived from nature and also possess ideal pharmaceutical properties. For instance, a study in 2012 summarized an engineering of the isoprenoid pathway using synthetic biology that resulted in an *E. coli* that synthesized elevated levels of taxadiene that is subsequently converted to a key compound for cancer treatment: taxol. In another example, a plant *Artemisia annua* produces artemisinin used in Chinese medicine for the treatment of malaria. Synthetic biology was applied to cause flux from the isoprenoid pathway into that of the artemisinin synthesis pathway. The success of the approach is reflected in the drug now available at a commercial scale at an economical cost (Ye et al., 2012).

2.13. CONCLUSION

The Department of Defense of the United States issued a grant of 15.6 million US dollars in 2012 for research in the field of synthetic biology. In connection with this, the Defense Advanced Research Projects Agency is associated with 8 projects in this field. The same year saw an investment of USD 7.9 million for a tenure of 5 years by the UK Engineering and Physical Sciences Research Council. This project involves the development of a Syn-BIS synthetic biology project that is concerned with the designing of bioparts (Jain, 2013).

A 2012 study estimated that 300 universities in both Europe and the US include research institutes as well as companies and universities (Synthetic Biology, 2012) with the figures expected to increase. The research in synthetic biology was put at USD 800 million for the year 2011 and is estimated to reach USD 4 billion (Jain, 2013). The development of the field of synthetic biology has been applied in the field of drug design, screening of drugs, as well as the alterations in metabolic pathways. Studies involving the synthetic tools have aided in the models of disease and altered metabolites. The triggering of signals or pathways to activate or suppress certain molecules can aid in the production of pharmaceutical products such as the case of artemisinin. Additionally, several case studies have highlighted the roles of synthetic gene circuits in the targeting of several disease models as well as in the treatment of disease models such as hypertension or T-cell signaling. Thus, this field of engineering of biological materials using concepts of engineering has a long road to traverse in the field of medicine.

A few examples of successful metabolic engineering reports as summarized by Lee, Vo & Na (2018) are given in Table 2.4.

Table 2.4.

Methods		Description	Applications
Pathway construction	Synthetic engineered pathway	Rapid construction and optimization of multi-gene pathways for higher-titer production	n-Butanol FAEEs, FAMEs
	In silico pathway design		Succinic acid, 2, 4-dihydroxybutyric acid, and maleate
	Randomized BioBrick AssemblX		Lycopene production increased by 30%
	Chromosomal integration CRISPR-Cas-based method	Efficient and stable enzyme expression and higher production yield	(*R, R*)-2,3-butanediol Mevalonate and bisabolene titers increased by 41-fold and 5-fold, respectively

Enzyme expression optimization	RBS Calculator		Control over protein expression levels through ribosome-binding sites	Optimization of isoprenoid production
	RBS Designer			L-tyrosine production
	Red Libs			N-acetylneuraminate biosynthesis Lycopene synthesis
Pathway regulation	Dynamic regulation system		Dynamic regulation and balancing of metabolic systems as well as redirection of fluxes to achieve high production of desired protein	A fatty acyl-CoA biosensor FadR Fatty acids, aldehydes, butanol, and alkanes Isoprenoid biosynthetic
Spatial organization	Protein scaffolding		Concentrating intermediates and rapidly directing them through the metabolic pathway	77-fold increase in mevalonate production 5-fold increase in glucaric acid synthesis 3-fold increase in butyrate production 5-fold increase in resveratrol synthesis
	DNA and RNA scaffolds			50% reduction in production time of I-threonine Production of IAA Succinate Pentadecane
Synthetic components for expression regulation	Promoter tuning		Optimal regulation of the gene expression strength in a dynamic range up to hundreds-of-fold activation	-
	Promoter regulation			Production of Pep N protein at about 10–15% of the total cellular protein
	Synthetic transcr-iption factors	CRISPR-based regulation	Regulation of gene expression using a nuclease-deficient Cas9 (dCas9)	Production of 2.0 g/L β-carotene (fed-batch fermentation) Production control of P (3HB-co-4HB) in *E. coli*
		Engineering transcription factors	Engineering Ara C protein through altering effector specificity using a method of saturation mutagenesis	20-fold increased production of triacetic acid lactone in *E. coli*

		Chimeric transcr-iption factor	Fusion of the target metabolite recognition domain and the promoter regulatory domain	Tyrosine Isoprenoid
	Synthetic RNAs		Regulation of gene expression based on physicochemical models of RNA	Development of high-yield VB12 production strains Tight transcriptional regulation of the toxic protein SacB

Source: Lee H-M, Vo PNL, Na D. Advancement of metabolic engineering assisted by synthetic biology. Catalysts. 2018; 8(12):619. https://doi.org/10.3390/catal8120619:

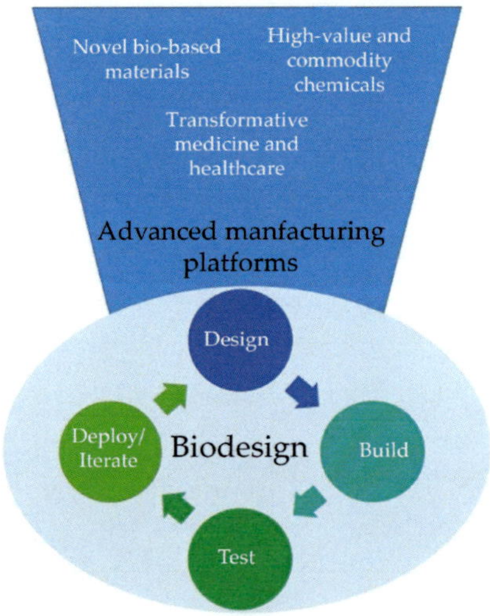

Figure 2.3.

Source: El Karoui M, Hoyos-Flight M & Fletcher L (2019) Future Trends in Synthetic Biology—A Report. Front. Bioeng. Biotechnol. 7:175. doi: 10.3389/fbioe.2019.00175:

"The technology of synthetic biology is currently accelerating at four times the rate of Moore's Law. It's been doing that since 2005, and it's likely to continue."

— Stewart Brand

This field is a double-edged path – the promise looks immense, but we cannot ignore the potential harm. For instance, the use of viruses as vectors can treat genetic maladies, but the potential of generating more deadly pathogens is uncertain! It becomes vital to introduce standardization to some degree in the field of DNA biology for reproducibility and accuracy. Another aspect is efficient data handling where the optimal and guided use of artificial intelligence (AI) and machine learning (ML) can help. The use of ethical approaches with responsible research becomes mandatory to avoid any harm. Figure 2.3 illustrates the use of metabolic engineering and synthetic biology as a "biodesign platform."

REFERENCES

1. Amidi, M., de Raad, M., Crommelin, D. J., Heinnik, W. E., & Mastrobattista, E. (2011). Antigen-expressing immunostimulatory liposomes as a genetically programmable synthetic vaccine. *Systems and Synthetic Biology, 5*, 21–31.
2. Bardou, C., Borie, C., Bickle, M., Rudkin, B. B., & Colas, P. (2009). Peptide aptamers for small molecule drug discovery. In P. Colas (Ed.), *Methods in Molecular Biology* (Vol. 535, pp. 373–388). Humana Press.
3. Baulard, A. R., Betts, J. C., Engohang-Ndong, J., Quan, S., McAdam, R. A., Brennan, P. J., & Besra, G. S. (2000). Activation of the pro-drug ethionamide is regulated in mycobacteria. *Journal of Biological Chemistry, 275*, 326–331.
4. Berens, C., Groher, F., & Suess, B. (2015). RNA aptamers as genetic control devices: The potential of riboswitches as synthetic elements for regulating gene expression. *Biotechnology Journal, 10*(2), 246–257.
5. Burrill, D. R., Inniss, M. C., Boyle, P. M., & Silver, P. A. (2012). Synthetic memory circuits for tracking human cell fate. *Genes & Development, 26*, 1486–1497.
6. Callura, J. M., Dwyer, D. J., Isaacs, F. J., Cantor, C. R., & Collins, J. J. (2010). Tracking, tuning, and terminating microbial physiology using synthetic riboregulators. *Proceedings of the National Academy of Sciences of the United States of America, 107*(36), 15898–15903.
7. Chappell, J., Takahashi, M. K., & Lucks, J. B. (2015). Creating small transcription activating RNAs. *Nature Chemical Biology, 11*(3), 214–220.
8. Chen, Y. Y., Jensen, M. C., & Smolke, C. D. (2010). Genetic control of mammalian T-cell proliferation with synthetic RNA regulatory systems. *Proceedings of the National Academy of Sciences of the United States of America, 107*, 8531–8536.
9. Chiba, T., Tsuchiya, T., Mori, R., & Shimokawa, I. (2012). Protein reporter bioassay systems for the phenotypic screening of candidate drugs: A mouse platform for anti-aging drug screening. *Sensors, 12*(2), 1648–1656.
10. Chopra, P., & Kamma, A. (2006). Engineering life through synthetic biology. *In Silico Biology, 6*, 401–410.
11. Curran, K. A., & Alper, H. S. (2012). Expanding the chemical palate of cells by combining systems biology and metabolic engineering. *Metabolic Engineering, 14*(4), 289–297.
12. Du, J., Shao, Z., & Zhao, H. (2011). Engineering microbial factories for

synthesis of value-added products. *Journal of Industrial Microbiology & Biotechnology, 38*(8), 873–890.

13. Eder, J., & Herrling, P. L. (2015). Trends in modern drug discovery. *Handbook of Experimental Pharmacology*, 1–20.
14. El Karoui, M., Hoyos-Flight, M., & Fletcher, L. (2019). Future trends in synthetic biology—A report. *Frontiers in Bioengineering and Biotechnology, 7*, Article 175. doi: 10.3389/fbioe.2019.00175.
15. Firman, K., Evans, L., & Youell, J. A. (2012). Synthetic biology project–Developing a single-molecule device for screening drug-target interactions. *FEBS Letters, 586*(15), 2157–2163.
16. Fussenegger, M., Morris, R. P., Fux, C., et al. (2000). Streptogramin-based gene regulation systems for mammalian cells. *Nature Biotechnology, 18*(11), 1203–1208.
17. Gibert, B., Simon, S., Dimitrova, V., Diaz-Latoud, C., & Arrigo, A.-P. (2013). Peptide aptamers: Tools to negatively or positively modulate HSPB1(27) function. *Philosophical Transactions of the Royal Society B: Biological Sciences, 368*(1617), 20120075–20120083.
18. Gonzalez-Nicolini, V., & Fussenegger, M. (2004). A novel mammalian cell-based approach for the discovery of anticancer drugs with reduced cytotoxicity on non-dividing cells. *Investigational New Drugs, 22*, 253–262.
19. Gonzalez-Nicolini, V., & Fussenegger, M. (2005). In vitro assays for anticancer drug discovery: A novel approach based on engineered mammalian cell lines. *Anticancer Drugs, 16*, 223–228.
20. Gossen, M., Freundlieb, S., Bender, G., Muller, G., Hillen, W., & Bujard, H. (1995). Transcriptional activation by tetracyclines in mammalian cells. *Science, 268*, 1766–1769.
21. Guo, W. M., Kong, K. W., Brown, C. J., et a□ (2014). Identification and characterization of an eIF4e DNA aptamer that inhibits proliferation with high throughput sequencing. *Molecular Therapy Nucleic Acids, 3*(12), e217–e227.
22. Hara, K. Y., Araki, M., Okai, N., Wakai, S., Hasunuma, T., & Kondo, A. (2014). Deve□opment of bio-based fine chemica□production through synthetic bioengineering. *Microbial Cell Factories, 13*(1), 173–191.
23. Harris, C. J., Hill, R. D., Sheppard, D. W., Slater, M. J., & Stouten, P. F. W. (2011). The design and application of target-focused compound libraries. *Combinatorial Chemistry & High Throughput Screening, 14*(6), 521–531.
24. Harumiya, S., Yoshino, A., Hayashizaki, K., Mizuno, K., Yakura, H., &

Adachi, T. (2013). A system for reconstructing B cell antigen receptor signaling in the mouse myeloma J558L cell line. *Archives of Biochemistry and Biophysics, 533*(1–2), 18–24.

25. Hu, Y., Gupta-Ostermann, D., & Bajorath, J. (2014). Exploring compound promiscuity patterns and multi-target activity spaces. *Computational and Structural Biotechnology Journal, 9*, e201401003.

26. Immethun, C. M., Hoynes-O'Connor, A. G., Balassy, A., & Moon, T. S. (2013). Microbial production of isoprenoids enabled by synthetic biology. *Frontiers in Microbiology, 4*, 75. doi:10.3389/fmicb.2013.00075.

27. Jain, K. K. (2013). Synthetic biology and personalized medicine. *Medical Principles and Practice, 22*, 209–219.

28. Kaba, S. A., Brando, C., Guo, Q., Mittelholzer, G., Raman, S., Tropel, D., & Lanar, D. E. (2009). A nonadjuvanted polypeptide nanoparticle vaccine confers long-lasting protection against rodent malaria. *Journal of Immunology, 183*, 7268–7277.

29. Keasling, J. D. (2012). Synthetic biology and the development of tools for metabolic engineering. *Metabolic Engineering, 14*(3), 189–195.

30. Keung, A. J., Joung, J. K., Khalil, A. S., & Collins, J. J. (2015). Chromatin regulation at the frontier of synthetic biology. *Nature Reviews Genetics, 16*(3), 159–171.

31. Kim, D. S., Gusti, V., Dery, K. J., & Gaur, R. K. (2008). Ligand induced sequestering of branchpoint sequence allows conditional control of splicing. *BMC Molecular Biology, 9*, 23.

32. Kindsmüller, K., & Wagner, R. (2011). Synthetic biology: Impact on the design of innovative vaccines. *Human Vaccines, 7*, 658–662.

33. Kis, Z., Pereira, H. S. A., Homma, T., Pedrigi, R. M., & Krams, R. (2015). Mammalian synthetic biology: Emerging medical applications. *Journal of the Royal Society Interface, 12*, 20141000. https://doi.org/10.1098/rsif.2014.1000.

34. Kistner, A., Gossen, M., Zimmermann, F., Jerecic, J., Ullmer, C., Lubbert, H., & Bujard, H. (1996). Doxycycline-mediated quantitative and tissue-specific contro□of gene expression in transgenic mice. *Proceedings of the National Academy of Sciences of the United States of America, 93*(10), 933–938.

35. Lee, H.-M., Vo, P. N. L., & Na, D. (2018). Advancement of metabolic engineering assisted by synthetic biology. *Catalysts, 8*(12), 619. https://doi.org/10.3390/catal8120619.

36. Lee, S. Y. Y., Kim, H. U. U., Park, J. H. H., Park, J. M. M., & Kim, T. Y. Y. (2009). Metabolic engineering of microorganisms: General strategies and drug production. *Drug Discovery Today, 14*(1–2), 78–88.

37. Los, G. V., et al. (2008). Halotag: A novel protein labelling technology for cell imaging and protein analysis. *ACS Chemical Biology, 3*, 373–382.

38. Lünse, C. E., Scott, F. J., Suckling, C. J., & Mayer, G. (2014). Novel TPP-riboswitch activators bypass metabolic enzyme dependency. *Frontiers in Chemistry, 2*, Article 53.

39. Maruta, H. (2009). From chemotherapy to signal therapy (1909–2009): A century pioneered by Paul Ehrlich. *Drug Discovery and Therapy, 3*(2), 37–40.

40. Miller, R. A., Binkowski, B. F., & Belshaw, P. J. (2007). Ligand-regulated peptide aptamers that inhibit the 5′-AMP-activated protein kinase. *Journal of Molecular Biology, 365*(4), 945–957.

41. Mulinari, S. (2014). The specificity triad: Notions of disease and therapeutic specificity in biomedical reasoning. *Philosophy, Ethics, and Humanities in Medicine, 9*, 14.

42. Neggers, J. E., Vercruysse, T., Jacquemyn, M., et al. (2015). Identifying drug-target selectivity of small-molecule CRM1/XPO1 inhibitors by CRISPR/Cas9 genome editing. *Chemical Biology, 22*(1), 107–116.

43. Nelson, J. W., Plummer, M. S., Blount, K. F., Ames, T. D., & Breaker, R. R. (2015). Small-molecule fluoride toxicity agonists. *Chemistry & Biology, 22*(4), 527–534.

44. Nielsen, J., & Keasling, J. D. (2011). Synergies between synthetic biology and metabolic engineering. *Nature Biotechnology, 29*(8), 693–695.

45. Planson, A.-G., Carbonell, P., Paillard, E., Pollet, N., & Faulon, J.-L. (2012). Compound toxicity screening and structure–activity relationship modeling in Escherichia coli. *Biotechnology and Bioengineering, 109*(3), 846–850.

46. Prommana, P., Uthaipibull, C., Wongsombat, C., et al. (2013). Inducible knockdown of Plasmodium gene expression using the glmS ribozyme. *PLoS One, 8*(8), e73783–e73793.

47. Rajagopal, R. (2012). Bio-based chemicals: In need of innovative strategies. *Chemical Weekly, 2*(4), 195–200.

48. Raman, S., Machaidze, G., Lustig, A., Aebi, U., & Burkhard, P. (2006). Structure-based design of peptides that self-assemble into regular polyhedral nanoparticles. *Nanomedicine, 2*, 95–102.

49. Reverdatto, S., Burz, D. S., & Shekhtman, A. (2015). Peptide aptamers:

Development and applications. *Current Topics in Medicinal Chemistry, 15*(12), 1082–1101.

50. Reverdatto, S., Rai, V., Xue, J., Burz, D. S., Schmidt, A. M., & Shekhtman, A. (2013). Combinatorial library of improved peptide aptamers, CLIPs to inhibit RAGE signal transduction in mammalian cells. *PLoS One, 8*(6), e65180–e65195.
51. Rihel, J., & Schier, A. F. (2013). Sites of action of sleep and wake drugs: Insights from model organisms. *Current Opinion in Neurobiology, 23*(5), 831–840.
52. Schmidt, M., Ganguli-Mitra, A., Torgersen, H., Kelle, A., Deplazes, A., & Biller-Andorno, N. (2009). A priority paper for the societal and ethical aspects of synthetic biology. *Systems and Synthetic Biology, 3*, 3–7.
53. Shi, J., & Muir, T. W. (2005). Development of a tandem protein trans-splicing system based on native and engineered split inteins. *Journal of the American Chemical Society, 127*, 6198–6206.
54. Sun, H., Zhu, X., Lu, P. Y., Rosato, R. R., Tan, W., & Zu, Y. (2014). Oligonucleotide aptamers: New tools for targeted cancer therapy. *Molecular Therapy Nucleic Acids, 3*(8), e182–e196.
55. Synthetic Biology Project. (2012). *Mapping the Emerging Synthetic Biology Landscape*. Woodrow Wilson International Center for Scholars, Washington, DC.
56. Trosset, J.-Y., & Carbonell, P. (2015). Synthetic biology for pharmaceutical drug discovery. *Drug Design, Development and Therapy, 9*, 6285–6302. https://doi.org/10.2147/DDDT.S58049.
57. Weber, W., Schoenmakers, R., Keller, B., Gitzinger, M., Grau, T., Daoud-El Baba, M., & Fussenegger, M. (2008). A synthetic mammalian gene circuit reveals antituberculosis compounds. *Proceedings of the National Academy of Sciences of the United States of America, 105*, 9994–9998.
58. Wise de Valdez, M. R., Nimmo, D., Betz, J., Gong, H. F., James, A. A., Alphey, L., & Black, W. C. 4th. (2011). Genetic elimination of dengue vector mosquitoes. *Proceedings of the National Academy of Sciences of the United States of America, 108*, 4772–4775.
59. Xie, Z., Wroblewska, L., Prochazka, L., Weiss, R., & Benenson, Y. (2011). Mu ti-input RNAi-based ogic circuit for identification of specific cancer cells. *Science, 333*, 1307–1311.
60. Ye, H. F., Charpin-El Hamri, G., Zwicky, K., Christen, M., Folcher, M., & Fussenegger, M. (2013). Pharmaceutically controlled designer circuit for the treatment of the metabolic syndrome. *Proceedings of the National Academy of Sciences of the United States of America, 110*, 141–146.

61. Ye, V. M., & Bhatia, S. K. (2012). Metabolic engineering for the production of clinically important molecules: Omega-3 fatty acids, artemisinin, and taxol. *Biotechnology Journal, 7*, 20–33.
62. Yeh, J. T.-H., Binari, R., Gocha, T., Dasgupta, R., & Perrimon, N. (2013). PAPTi: A peptide aptamer interference toolkit for perturbation of protein–Protein interaction networks. *Scientific Reports, 3*(1), 1156–1164.
63. Zhang, Q., Bhattacharya, S., Conolly, R. B., Clewell, H. J., Kaminski, N. E., & Andersen, M. E. (2014). Molecular signaling network motifs provide a mechanistic basis for cellular threshold responses. *Environmental Health Perspectives, 122*(12), 1261–1270.
64. Zheng, W., Thorne, N., & McKew, J. C. (2013). Phenotypic screens as a renewed approach for drug discovery. *Drug Discovery Today, 18*(21–22), 1067–1073.

Chapter 3

FEW PLATFORMS FOR ENGINEERING AND PRODUCTION: THE QUEST FOR FUELS AND PRECURSORS

Shiv Sanjeevi Sripathi

"With moth cytochrome C there are 30 differences and 74 identities. With bread yeast and humans, there are about 45 amino acids that are different and about 59 that are identical. Think how close together man and this other organism, bread yeast, are. What is the probability that in 59 positions the same choice out of 20 possibilities would have been made by accident? It is impossibly small. There is, there must be, a developmental explanation of this. The developmental explanation is that bread yeast and man have a common ancestor, perhaps two billion years ago. And so we see that not only are all men brothers, but men and yeast cells, too, are at least close cousins, to say nothing about men and gorillas or rhesus monkeys. It is the duty of scientists to dispel ignorance of such relationships."

—— Linus Pauling

'The Social Responsibilities of Scientists and Science,' The Science Teacher (1933), 33, 15.

3.1. PRODUCING FUELS AND CHEMICALS USING "CELL FACTORIES"

The quest for alternatives to crude oils has led to the use of several host organisms to produce lipids. Several host systems have been employed like algae, fungi, plants, etc. If we consider yeasts, fatty acid production entails a carbon source and these synthesized fatty acids (often called FA) yield significant ATP upon hydrolysis. The system of lipid production in yeast is shown in Figure 3.1 (Table 3.1).

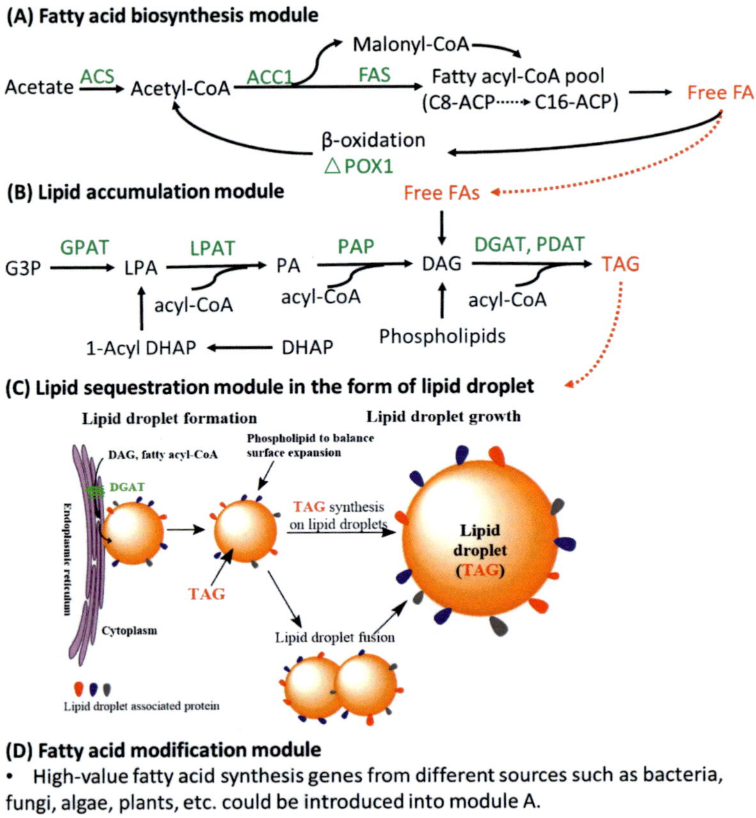

Figure 3.1

Source: Jiang W, Li C, Li Y, Peng H. Metabolic engineering strategies for improved lipid production and cellular physiological responses in yeast saccharomyces cerevisiae. Journal of Fungi. 2022; 8(5):427. https://doi.org/10.3390/jof8050427:

The use of metabolic engineering approaches in *Saccharomyces cerevisiae* for producing bioethanol becomes pertinent given that the organism is "Generally Recognized as Safe (GRAS)," grows rapidly, and shows resistance to various forms of stress. Topaloğlu et al. (2023) reviewed the scheme of approaches in *Saccharomyces cerevisiae* for producing bioethanol and the tools employed (Figure 3.2).

Top: Metabolic engineering of baker's yeast.

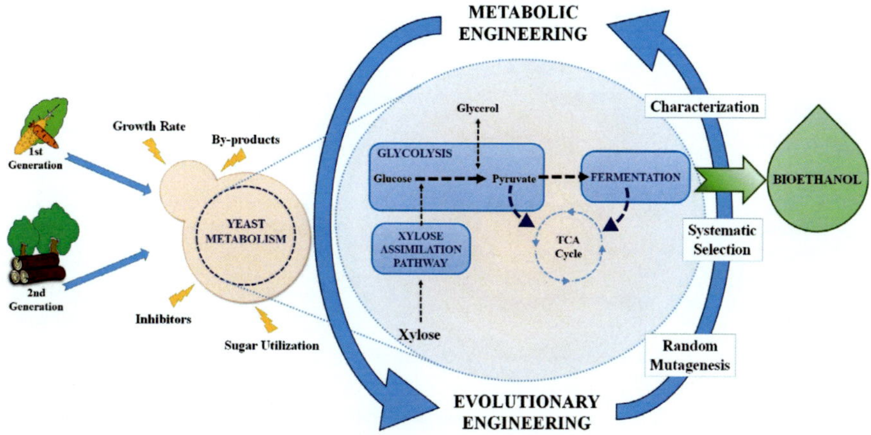

Bottom: Approaches employed for engineering

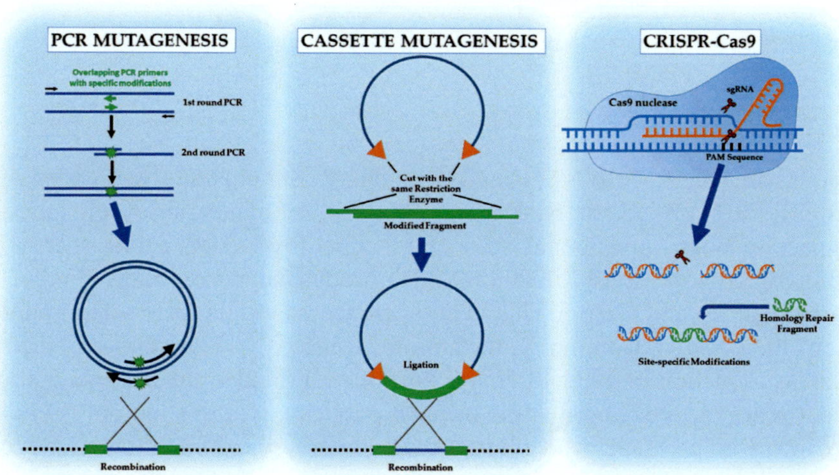

Figure 3.2

Source: Topaloğlu A, Esen Ö, Turanlı-Yıldız B, Arslan M, Çakar ZP. From Saccharomyces cerevisiae to Ethanol: Unlocking the Power of Evolutionary Engineering in Metabolic Engineering Applications. Journal of Fungi. 2023; 9(10):984. https://

A slew of metabolic engineering approaches has been applied to enable yeast like *S. cerevisiae* to use several raw materials as an alternative to the traditional feedstocks. Madhavan et al. (2017) presented a scheme of transporter and enzyme genes that have been engineered for substrate use Figure 3.3.

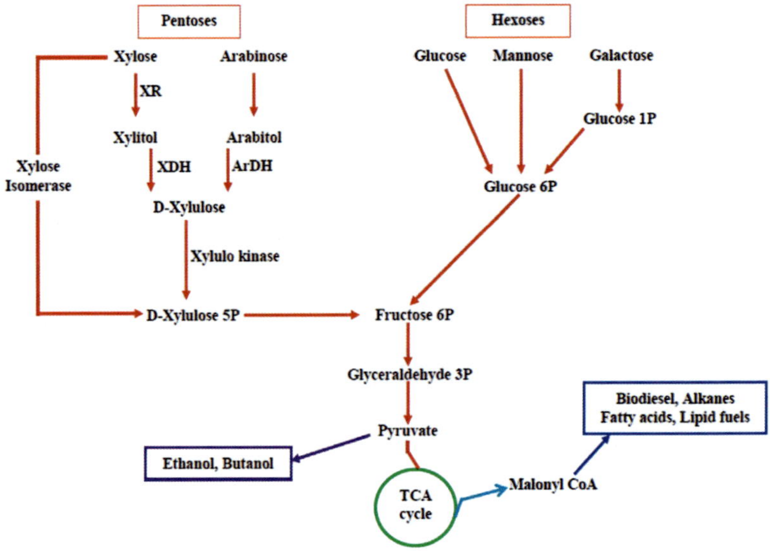

Figure 3.3

Source: Madhavan A, Jose AA, Binod P, Sindhu R, Sukumaran RK, Pandey A and Castro GE (2017) Synthetic Biology and Metabolic Engineering Approaches and Its Impact on Non-Conventional Yeast and Biofuel Production. Front. energy Res. 5:8. doi: 10.3389/fenrg.2017.00008:

Table 3.1. Many Lipid Production Systems Warrant Extensive Work as Documented in the Table

Sources	Host Model	Pros	Cons
Fungi	*S. cerevisiae*	robust genetically tractable easy to culture industrial usage	low lipid yield not yet commercialized
Bacteria	*E. coli*	short generation time genetically tractable industrial usage	low lipid yield no stable lipid body
Microalgae	*Chlorella*	high oil productivity CO_2 fixation	low biomass concentrations high costs for oil recovery open pond culture issue
Plant	*Arabidopsis thaliana*	high oil yield CO_2 fixation	difficulties to industrialize deforestation greenhouse gas emission

Source: Jiang W, Li C, Li Y, Peng H. Metabolic engineering strategies for improved lipid production and cellular physiological responses in yeast saccharomyces cerevisiae. Journal of fungi. 2022; 8(5):427. https://doi.org/10.3390/jof8050427:

To give an example of chemical production, 3-hydroxypropionic acid (3-HP) is employed for producing acrylic acid and poly(3-hydroxypropionic acid) (P[3-HP]). The former is used in paper, paints, diapers, etc., and the latter finds use in surgical equipment. Given the challenges of the traditional system of 3-HP synthesis, the cost and resource limitations have led to research on producing it by other approaches. The malonyl-CoA reductase (MCR) pathway and the β-alanine pathway have been metabolically engineered in yeast to produce 3-HP, where the latter approach has been reported as economically viable. Ji et al. (2018) summarized these approaches as shown in Figure 3.4.

Key: The MCR pathway is in green, and the β-alanine pathway is in purple: G3P, glyceraldehyde 3-phosphate; 3PG, 3-phosphoglycerate; GYC, glyoxylate cycle; TCA, tricarboxylic acid cycle; AAT2, aspartate aminotransferase 2; ACC1**, Snf1 deregulated acetyl-CoA carboxylase (two site mutations, one is Ser659→Ala, the other is Ser1157→Ala); ACS, acetyl-CoA synthetase; ACSse, acetylation-insensitive acetyl-CoA synthetase from *Salmonella enterica*; ADH2, alcohol dehydrogenase 2; ALD6, NADP-dependent aldehyde dehydrogenase; ALT, alanine aminotransferase; BcBAPAT, β-alanine-pyruvate aminotransferase from *Bacillus cereus*; CIT, citrate synthase; EcHPDH, 3-hydroxypropionate dehydrogenase from *E. coli;* GAPN, NADP-dependent G3P dehydrogenase; CaMCR, malonyl-CoA reductase from *Chloroflexus aurantiacus*; MLS1, cytosolic malate synthase; PDC1, pyruvate decarboxylase; PDH, pyruvate dehydrogenase complex; PYC1/2, pyruvate carboxylase 1/2; TcPAND, aspartate decarboxylase from *Tribolium castaneum*.

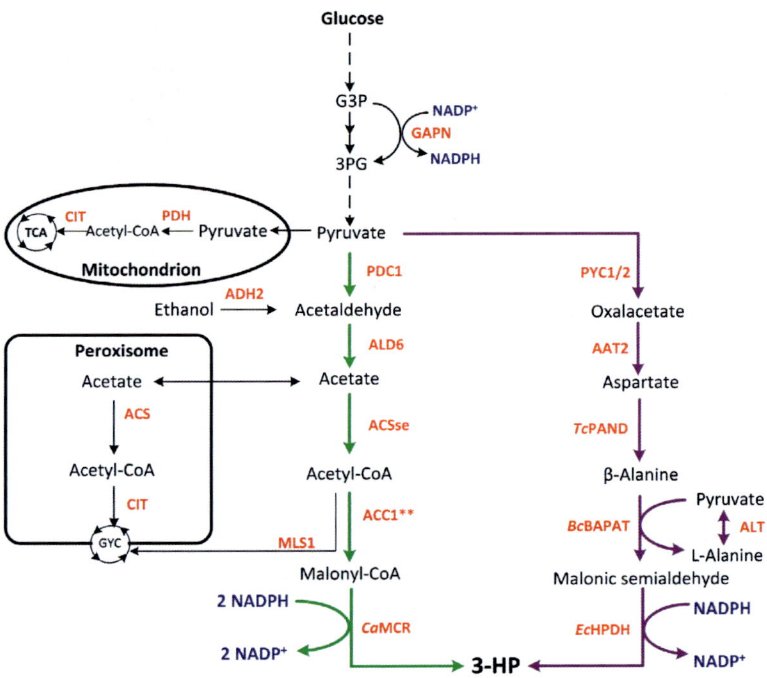

Figure 3.4.

Source: Ji R-Y, Ding Y, Shi T-Q, Lin L, Huang H, Gao Z & Ji X-J (2018) Metabolic Engineering of Yeast for the Production of 3-Hydroxypropionic Acid. Front. Microbiol. 9:2185. doi: 10.3389/fmicb.2018.02185:

accordance with accepted academic practice. No use, distribution or reproduction is permitted which does not comply with these terms.

In a 2023-published study, Liu et al., used the malonyl-CoA pathway in *Yarrowia lipolytica* to produce 3-HP by overexpressing its endogenous ACC and adding the genes MCR from *C. aurantiacus*, and ACS*L*641*P*-641 from *Salmonella enterica*. The study was reported as the first to produce 3-HP in *Y. lipolytica.*

The 3-HP production pathway in *Yarrowia lipolytica* as made by the researchers is shown in Figure 3.5.

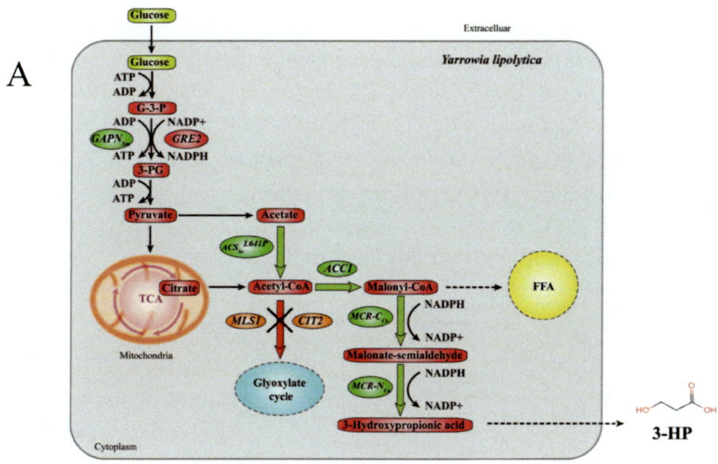

Figure 3.5

Source: Liu S, Sun Y, Wei T, Gong D, Wang Q, Zhan Z, Song J. Engineering 3-hydroxypropionic acid production from glucose in yarrowia lipolytica through malonyl-coa pathway. Journal of fungi. 2023; 9(5):573. https://doi.org/10.3390/jof9050573: © 2023 by the authors. Licensee MDPI, Basel, Switzerland. This article is an open access article distributed under the terms and conditions of the Creative Commons Attribution (CC BY) license (https://creativecommons.org/licenses/by/4.0/).

Further, the expression of several genes was knocked out to augment 3-HP. The approach as illustrated by the authors is given in Figure 3.6 (Tables 3.2 and 3.3).

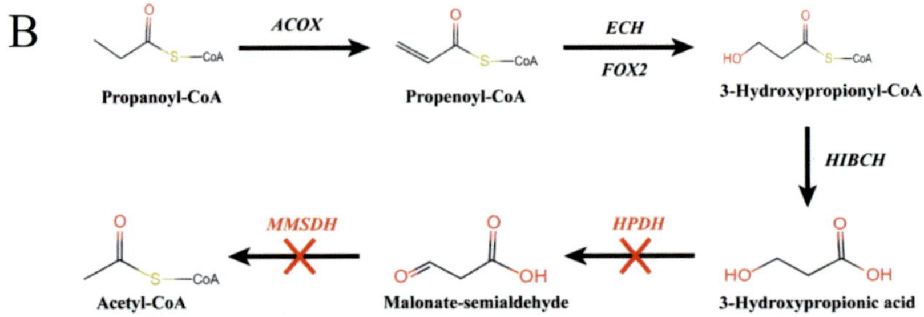

Figure 3.6.

Source: Liu S, Sun Y, Wei T, Gong D, Wang Q, Zhan Z, Song J. engineering 3-hydroxypropionic acid production from glucose in yarrowia lipolytica through malonyl-coa pathway. Journal of fungi. 2023; 9(5):573. https://doi.org/10.3390/jof9050573:

Table 3.2. The Strains and Plasmids Used in This Work

Strain	Description
E. coli JM109	For construction of recombinant plasmids
Y. lipolytica Po1f	MATa, Leu2−, Ura3−
Po1f-wMCR	Po1f cells harboring p1312-wMCR
Po1f-tMCR	Po1f cells harboring p1312-tMCR
Po1f-NC-00	Po1f cells harboring p1312-tMCRN-tMCRC
Po1f-NC-01	Po1f cells harboring p1312-NC-GAPN*Bc*
Po1f-NC-02	Po1f cells harboring p1312-NC-GAPN*Sm*
Po1f-NC-03	Po1f cells harboring p1312-NC-GAPN*Sm* and p1269-ACC1
Po1f-NC-04	Po1f cells harboring p1312-NC-GAPN*Sm* and p1269-ACC1-ACC1
Po1f-NC-05	Po1f cells harboring p1312-NC-GAPN*Sm* and p1269-ALD6
Po1f-NC-06	Po1f cells harboring p1312-NC-GAPN*Sm* and p1269--*SL*641*PSe* 641
Po1f-NC-07	Po1f cells harboring p1312-NC-GAPN*Sm* and p1269--*SL*641*PSe* 641-ACS*L*641*PSe* 641
Po1f-NC-08	Po1f cells harboring p1312-NC-GAPN*Sm* and p1269--C1-ACS*L*641*PSe* 641
Po1f-NC-09	Po1f-NC-08 transformed with JME4580-MLS1-sgRNA and fragment of donor-MLS1
Po1f-NC-10	Po1f-NC-08 transformed with JME4580-CIT2-sgRNA and fragment of donor-CIT2

Po1f-NC-11	Po1f-NC-09 transformed with JME4580-CIT2-sgRNA and fragment of donor-CIT2
Po1f-NC-12	Po1f-NC-11 transformed with JME4580-HPDH-sgRNA and fragment of donor-HPDH
Po1f-NC-13	Po1f-NC-11 transformed with JME4580-MMSDH-sgRNA and fragment of donor-MMSDH
Po1f-NC-14	Po1f-NC-12 transformed with JME4580-MMSDH-sgRNA and fragment of donor-MMSDH

Table 3.3.

Plasmid	Description
pINA1312	*Y. lipolytica*-integrative plasmid, hp4d promoter, XPR2 terminator, ura3d1 selection marker, Km*R*
pINA1269	*Y. lipolytica*-integrative plasmid, hp4d promoter, XPR2 terminator, LEU2 selection marker, Amp*R*
pTrc99a-wMCR	pTrc99a plasmid contains wild MCR gene from *C. aurantiacus*
p1312-wMCR	pINA1312 plasmid contains wild MCR gene from *C. aurantiacus*
p1312-tMCR	pINA1312 plasmid contains codon-optimized MCR gene from *C. aurantiacus*
p1312-tMCRN-tMCRC	pINA1312 plasmid contains codon-optimized MCR-N gene cassette and MCR-C gene cassette
p1312-NC-GAPN*Bc*	pINA1312 plasmid contains codon-optimized MCR-N/C gene cassettes and codon optimized GAPN gene cassette from *Bacillus cereus*
p1312-NC-GAPN*Sm*	pINA1312 plasmid contains codon-optimized MCR-N/C gene cassettes and codon optimized GAPN gene cassette from *Streptococcus mutans*
p1269-ACC1	pINA1269 plasmid contains ACC1 gene from *Y. lipolytica*
p1269-ACC1-ACC1	pINA1269 plasmid contains two ACC1 gene cassettes
p1269-ACS*L641PSe* 641	pINA1269 plasmid contains codon-optimized mutant ACS*L641P* 641 gene from *Salmonella enteric*

p1269-ACS*L*641*PSe* 641-ACS*L*641*PSe* 641	pINA1269 plasmid contains two ACS*L*641*PSe* 641 gene cassettes
p1269-ALD6	pINA1269 plasmid contains ALD6 gene from *Y. lipolytica*
p1269-ACC1-ACS*L*641*PSe* 641	pINA1269 plasmid contains ACC1 gene cassette and mutant ACS*L*641*PSe* 641 gene cassette
JME4580	Y. lipolytica CRISPR/Cas9 plasmid, HPHex marker for sgRNA cloning, RFP, Amp*R*
JME4580-MLS1-sgRNA	JME4580 plasmid contains sgRNA of MLS1 gene instead of RFP expression gene
JME4580-CIT2-sgRNA	JME4580 plasmid contains sgRNA of CIT2 gene instead of RFP expression gene
JME4580-HPDH-sgRNA	ME4580 plasmid contains sgRNA of HPDH gene instead of RFP expression gene
JME4580-MMSDH-sgRNA	JME4580 plasmid contains sgRNA of MMSDH gene instead of RFP expression gene
pCE-Zero	For constructing CRISPR/Cas9 donor fragment, Amp*R*, Km*R*, linearized
donor-MLS1	pCE-Zero plasmid contains donor fragment of 1kb upstream and 1kb downstream of MLS1 gene (YALI0D19140g)
donor-CIT2	pCE-Zero plasmid contains donor fragment of 1kb upstream and 1kb downstream of CIT2 gene (YALI0E00638g)
donor-HPDH	pCE-Zero plasmid contains donor fragment of 1kb upstream and 1kb downstream of HPDH gene (YALI0F02607g)
donor-MMSDH	pCE-Zero plasmid contains donor fragment of 1kb upstream and 1kb downstream of MMSDH gene (YALI0C01859g)

Source: Liu S, Sun Y, Wei T, Gong D, Wang Q, Zhan Z, Song J. Engineering 3-hydroxypropionic acid production from glucose in yarrowia lipolytica through malonyl-coa pathway. Journal of fungi. 2023; 9(5):573. https://doi.org/10.3390/jof9050573:

The 3-HP production in recombinant strain Po1f-NC-14 was 1.128 $g{\cdot}L^{-1}$ in shake flask fermentation, whi◻e the yie◻d in fed-batch fermentation was 16.23 $g{\cdot}L^{-1}$.

The results obtained by the authors have been presented in Figures 3.7 and 3.8.

Top: the quantitative yield of 3-HP by the engineered yeast strains.

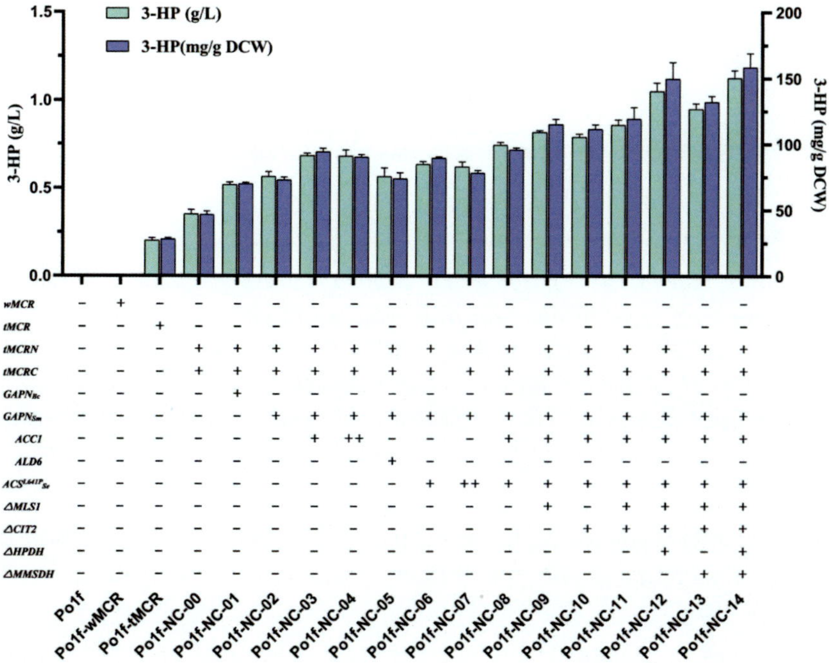

Figure 3.7

Source: Liu S, Sun Y, Wei T, Gong D, Wang Q, Zhan Z, Song J. Engineering 3-hydroxypropionic acid production from glucose in yarrowia lipolytica through malonyl-coa pathway. Journal of fungi. 2023; 9(5):573. https://doi.org/10.3390/jof9050573:

Bottom: fed-batch fermentation of Po1f-NC-14 strain for 3-HP production

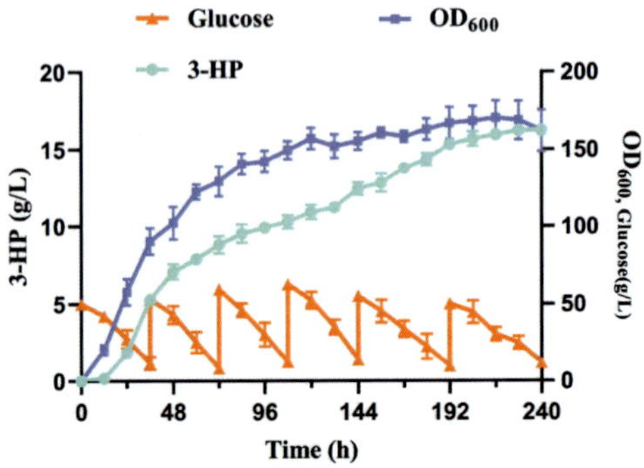

Figure 3.8

Source: Liu S, Sun Y, Wei T, Gong D, Wang Q, Zhan Z, Song J. Engineering 3-hydroxypropionic acid production from glucose in yarrowia lipolytica through malonyl-coa pathway. Journal of fungi. 2023; 9(5):573. https://doi.org/10.3390/jof9050573: © 2023 by the authors. Licensee MDPI, Basel, Switzerland. This article is an open access article distributed under the terms and conditions of the Creative Commons Attribution (CC BY) license (https://creativecommons.org/licenses/by/4.0/).

These results were found to be on par with other yeast production cells, serving as a stepping stone for further research on this system for 3-HP production.

A summary of metabolic engineering approaches for 3-HP production, as compiled by Ji et al. (2018), has been presented in Table 3.4.

Table 3.4

Metabolic engineering strategies	Beneficial effect for 3-HP production	Reference
Increasing the supply of malonyl-CoA: Overexpression the enzymes (ALD6, ACSse, ADH2) for acetyl-CoA accumulation; Deleting the MLS1 to block the consumption of acetyl-CoA; Overexpression the ACC1 catalyzing acetyl-CoA to malonyl-CoA. Increasing the supply of NADPH: Overexpression the GAPN catalyzing the formation of extra NADPH. Establishing the MCR pathway: Overexpression the *Ca*MCR.	3-HP production was increased to 463 mg/L	Chen et al., 2014

Generating the ACC1 mutant (ACC1Ser659Ala Ser1157Ala) through mutating the potential phosphorylation sites to abolish the post-translational regulation Increasing the ACC1 activity by overexpression of the ACC1 mutant	3-HP production was up to ~2.2-fold more than that of the wild-type ACC1	Shi et al., 2014
Constructing the β-alanine pathway by overexpressing AAT2, PYC1, PYC2, ALT, BcBAPAT, EcHPDH, and multiple copies of TcPAND	13.7 g/L 3-HP was generated through the constructed β-alanine pathway from glucose in fed-batch fermentation at low pH.	Borodina et al., 2015
Increasing the supply of acetyl-CoA by overexpressing *PDC1*, *ALD6*, and ACSse; Engineering the cofactor specificity of the GAPN to increase the production of NADPH at the expense of NADH and thus improve 3-HP production and reduce formation of glycerol as by-product; Integrating multiple copies of *Ca*MCR and ACC1 mutant genes into the genome.	3-HP was produced at a titer 7.37 g/L in a carbon-limited fed-batch fermentation	Kildegaard et al., 2016
Developing a malonyl-CoA biosensor based on the bacterial transcription factor FapR to monitor and precisely control the intracellular malonyl-CoA concentration	3-HP titer was enhanced i by 120%	Li et al., 2015
A hierarchical dynamic control strategy to control the expression level of *Ca*MCR depending on the intracellular malonyl-CoA concentration: The upper level of control was to dynamically downregulate fatty acid biosynthesis using HXT1 promoter. The lower level was based on the malonyl-CoA biosensor.	3-HP production was increased by 10-fold	David et al., 2016
Improving the availability of malonyl-CoA through down-regulating lipid synthesis. Manipulating the phospholipid synthesis transcriptional regulators including Ino2p, Ino4p, Opi1p, and a series of synthetic Ino2p variants, combining with studying the inositol and choline effect.	3-HP production was increased by 9-fold	Chen et al., 2017

Identifying and characterizing promoters that depend on glucose concentration for use as dynamic control elements. Identifying 34 candidate promoters that strongly responded to glucose presence or absence. A subset of promoters, pADH2, pICL1, and pHXT7, were demonstrated as suitable for dynamic control of 3-HP production.	Regulating the 3-HP pathway by the ICL1 promoter resulted in 70% improvement of 3-HP titer in comparison to PGK1 promoter.	Maury et al., 2018

AAT2, aspartate aminotransferase 2; ACC1, acetyl-CoA carboxylase; ACSse, acetylation-insensitive acetyl-CoA synthetase from Salmonella enterica; ADH2, alcohol dehydrogenase; ALD6, NADP-dependent aldehyde dehydrogenase; ALT, alanine aminotransferase; BcBAPAT, β-alanine-pyruvate aminotransferase from Bacillus cereus; EcHPDH, 3-hydroxypropionate dehydrogenase from E. coli; GAPN, NADP-dependent G3P dehydrogenase; CaMCR, malonyl-CoA reductase from Chloroflexus aurantiacus; MLS1, cytosolic malate synthase; PDC1, pyruvate decarboxylase; PYC1/2, pyruvate carboxylase 1/2; TcPAND, aspartate decarboxylase from Tribolium castaneum.

Source: Ji R-Y, Ding Y, Shi T-Q, Lin L, Huang H, Gao Z & Ji X-J (2018) Metabolic engineering of yeast for the production of 3-hydroxypropionic acid. Front. microbiol. 9:2185. doi: 10.3389/fmicb.2018.02185:

In addition to the gold standard, *Saccharomyces cerevisiae*, other yeast organisms can also be subjected to engineering given the availability of genetic tools and OMICs approaches. Madhavan et al., (2017) presented an excellent scheme of amalgamating data for yeast microbial factories, which is shown in Figure 3.9.

To give an example, Chen et al. (2014) engineered the beta-oxidation pathway of *Saccharomyces cerevisiae* grown on fatty acid-rich feedstock to augment the ratio of medium-chain fatty acids (MCFAs). The approach entailed the deletion of *POX1* that coded for the acyl-CoA oxidase of the haploid deletion strain *Δpox1*. The scheme of metabolic engineering as illustrated by the authors is given in Figure 3.10.

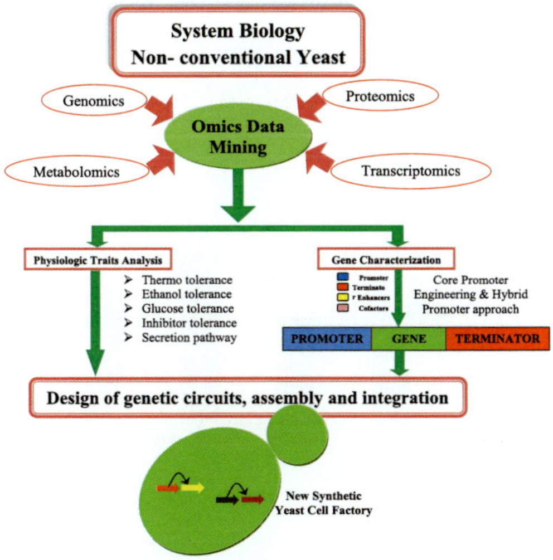

Figure 3.9

Source: Madhavan A, Jose AA, Binod P, Sindhu R, Sukumaran RK, Pandey A and Castro GE (2017) Synthetic biology and metabolic engineering approaches and its impact on non-conventional yeast and biofuel production. Front. energy Res. 5:8. doi: 10.3389/fenrg.2017.00008: © 2017 Madhavan, Jose, Binod, Sindhu, Sukumaran, Pandey and Castro. This is an open-access article distributed under the terms of the Creative Commons Attribution License (CC BY). The use, distribution or reproduction in other forums is permitted, provided the original author(s) or licensor are credited and that the original publication in this journal is cited, in accordance with accepted academic practice. No use, distribution or reproduction is permitted which does not comply with these terms.

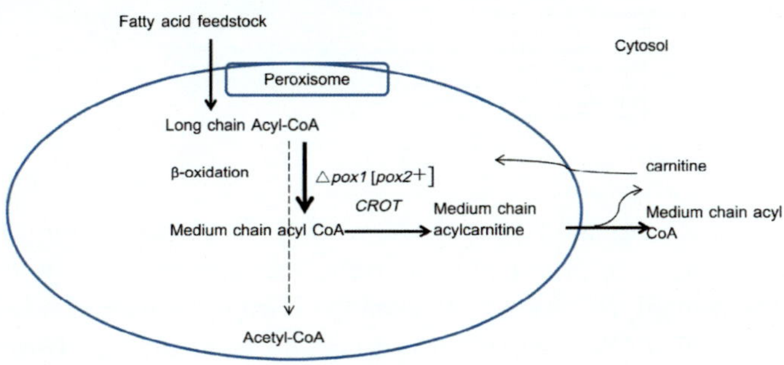

Figure 3.10

Source: Chen, L., Zhang, J., Chen, W. N. (2014). Engineering the Saccharomyces cerevisiae β-Oxidation Pathway to Increase Medium Chain Fatty Acid Production as Potential Biofuel. PLoS ONE 9(1): e84853. https://doi.org/10.1371/journal.

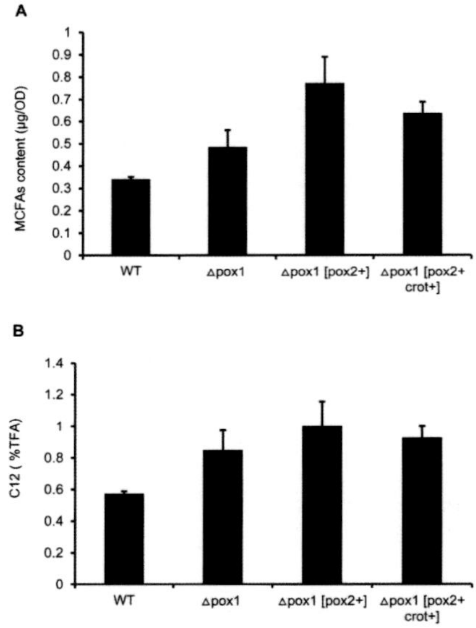

Fatty acid type	WT	Δ*pox1*	Δ*pox1* [*pox2+*]	Δ*pox1*[*pox2+ crot+*]
C12:0	0.341	0.486	0.772	0.637
C14:0	1.183	1.590	2.143	1.855
C16:1	0.881	1.384	1.878	1.788
C16:0	30.623	27.675	35.794	32.535
C18:1	1.449	2.977	3.334	3.170
C18:0	24.889	22.959	32.890	28.548
C20:0	0.120	0.143	0.202	0.227
Total fatty acids	59.486	57.214	77.014	68.760

[a]Data represent fatty acid composition in µg/OD cell when WT (wild-type) and engineered strains were cultured in YNBD$_{0.5}$O$_3$ medium. The values are the means from three experiments examining the cell extracts at 24 h. The standard deviations were <5% of the values.
doi:10.1371/journal.pone.0084853.t001

Figure 3.11.

Source: Chen, L., Zhang, J., Chen, W. N. (2014). Engineering the saccharomyces cerevisiae β-oxidation pathway to increase medium chain fatty acid production as potential biofuel. PLoS ONE 9(1): e84853. https://doi.org/10.1371/journal.pone.0084853:

Given this defect, the *Δpox1* [*pox2+*] strain organism was challenged in terms of growth. In order to facilitate the transport of MCFAs blocked in the peroxisomes, the organism encoded *CROT*, coding for carnitine O-octanoyl transferase, which facilitated its (Δ*pox1* [*pox2+, crot+*] strain) growth. Both Δ*pox1 [pox2+]* and *Δpox1* [*pox2+, crot+*] strains showed an increased fatty acid accumulation by 29.5% and 15.6%, respectively compared to the wild-type when grown on oleic acid-based media.

The MFCA production by the metabolically engineered strains vs. in the wild-type has been shown in Figure 3.11 as obtained by the researchers.

The accumulation of MFCA in in Δ*pox1 [pox2+]* and Δ*pox1* [*pox2+, crot+*] rose by factors of 2.26- and 1.87-fold, while the content in the culture media were boosted by 3.29-fold and 3.34-fold, respectively (Figure 3.12).

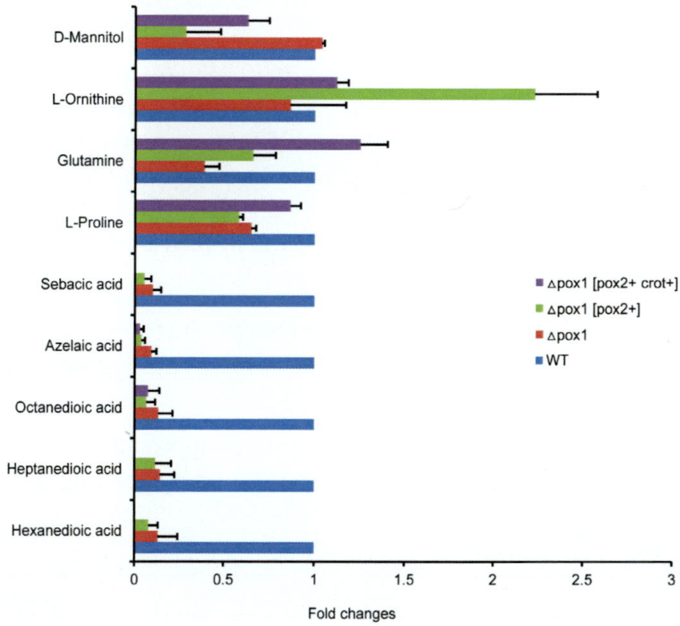

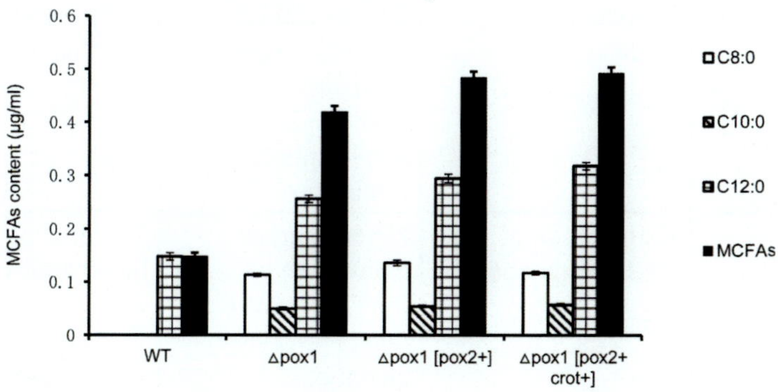

Fold changes of fatty acids	Δ*pox1*	Δ*pox1* [*pox2+*]	Δ*pox1*[*pox2+ crot+*]
C12:0	1.422	2.262	1.866
C14:0	1.344	1.812	1.568
C16:1	1.570	2.132	2.029
C16:0	0.904	1.169	1.062
C18:1	2.054	2.300	2.187
C18:0	0.923	1.322	1.147
C20:0	1.194	1.681	1.889
Total fatty acids	0.962	1.29	1.156

[b]The fold change data for the engineered strains were analyzed compared to the WT (wild-type) strain. The amount of fatty acids in the WT strain was set as 1. The values are the means from three experiments examining cell extracts at 24 h. The standard deviations were <5% of the values.
doi:10.1371/journal.pone.0084853.t002

Figure 3.12

Source: Chen, L., Zhang, J., Chen, W. N. (2014). Engineering the saccharomyces cerevisiae β-oxidation pathway to increase medium chain fatty acid production as potential biofuel. PLoS ONE 9(1): e84853. https://doi.org/10.1371/journal.pone.0084853: © 2014 Chen et al. This is an open-access article distributed under the terms of the Creative Commons Attribution License, which permits unrestricted use, distribution, and reproduction in any medium, provided the original author and source are credited.

The researchers highlighted the application of β-oxidation pathway engineering to increase the MFCA content, which can augment the value of such fatty acid feedstocks.

Table 3.5 reviews metabolic engineering approaches for bioethanol production in *Saccharomyces cerevisiae* as compiled by Topaloğlu et al. (2023).

Table 3.5.

Purpose		Modification	Improvement
Lowering ATP Yield	Introduction of futile cycle	Expression of *E. coli* FBPase insensitive to fructose-2,6-bisphosphate inhibition	8.8% higher bioethanol yield
	Introduction of ATPase enzyme	Heterologous expression of the F1 part of *E. coli* H+-ATPase enzyme	10% higher bioethanol yield

	Increasing the unspecific alkaline phosphatase activity	Overexpression of PHO8 gene	13% higher bioethanol yield
	Introduction of Entner-Doudoroff (ED) pathway	Expression of KDPG (2-keto-3-deoxy-6-phosphogluconate) aldolase from *E. coli*	-
	Relocation of sucrose hydrolysis from the extracellular space to the cytosol	Engineering the promoter and 5′ coding sequences of SUC2 gene	4% higher bioethanol yield
Sustainable Reduction of Glycerol Formation	Fine-tuning of glycerol 3-phosphate dehydrogenases (GPDH)	Introducing lower-strength TEF1 promoters to GPD1 and GPD2 genes	5% higher bioethanol yield
	Fine-tuning of glycerol 3-phosphate dehydrogenase	Engineering the promoter of GPD1 in a gpd2Δ background	7% higher bioethanol yield
	Introduction of direct cofactor regulation strategies	Expressing B. cereus gapN gene, *E. coli* frdA gene and mhpF gene independently	Increased bioethanol yield
	Engineering of ammonium assimilation	Overexpression of GLN1 and GLT1 genes, deletion of GDH1 gene	10% higher bioethanol yield
	Introduction of Calvin-cycle enzymes	Deletion of GPD2 and heterologous expression of PRK and RuBisCO	15% higher bioethanol yield

Prevention of Bacterial Contamination	Introduction of phage endolysin on the cell surface	Anchoring recombinant peptidoglycan hydrolase, endolysin LysKB317, by using cell surface display	83.8% decrease in bacterial cell counts
Introduction of Xylose Catabolism	Construction of xylose reductase-xylitol dehydrogenase (XR-XDH) pathway	Genomic integration of XYL1 and XYL2 genes from *S. stipitis*, XKS1 from *S. cerevisiae*, BGL1 from *A. aculeatus*, and GXS1 from C. intermedia	Xylose consumption rate of 6.62 g/L/h and an ethanol yield of 0.394
	Construction of xylose reductase-xylitol dehydrogenase (XR-XDH) pathway	Expression of XYL1.2 from *S. passalidarum* and XYL2 from *S. stipitis*	0.40 g g^{-1} CDW ethanol yield and 0.33 g g^{-1} CDW h^{-1} productivity
	Construction of xylose isomerase (XI) pathway	Expression of XI gene (xylA) from the bacterium *B. cenocepacia*	5-fold increase in xylose consumption and over 1.5-fold increase in ethanol production
	Construction of xylose isomerase (XI) pathway	Expression of XI from the bacterium *P. acidipropionici* with the co-expression of GroEL-GroES chaperonin complex from *E. coli*	Yield of 0.44 g ethanol/g xylose
Improving Xylose Catabolism	Lowering xylitol production	Deletion of GRE3 gene in *S. cerevisiae*	Increased ethanol yield of 0.47 g/g of total sugars during fermentation of corn-cob hydrolysate

	Introduction of a synthetic reductive PPP for carbon dioxide recycling	Expression of RuBisCO and PRK enzymes in a *S. cerevisiae* strain harboring the XR/ XDH pathway	Increased ethanol yield and reduced release of carbon dioxide
Increasing Accessibility of Lignocellulosic Biomass	Conversion of cellulose into glucose	Expression of cassette carrying a cellulase gene from *A. gigas* Spix	37.7-fold higher ethanol yield
Increasing Stress Tolerance	Increasing osmotolerance	Overexpression of TPS1 and TPS2 genes in *S. cerevisiae*	Increased ethanol yield and Osmo tolerance, decreased glycerol production
	Increasing thermotolerance	Overexpression of RSP5 gene	Thermotolerance at 41°C and ability to tolerate higher temperatures.
	Increasing ethanol tolerance	Overexpression of a truncated version of the MSN2 gene in an industrial fuel ethanol strain	Increased ethanol tolerance (up to 14%)
	Increasing acetate tolerance	Overexpression of HAA1 gene	The addition of acetate at 0.5% (w/v, pH 4.5) does not inhibit ethanol production
	Increasing HMF tolerance	Overexpression of ADH1 and ADH6 genes	Higher specific ethanol productivity in the presence of HMF
	Increasing acetic acid tolerance	Deletion of ADY2 gene	14.7% increase in ethanol yield, in the presence of 3.6 g/L acetic acid
	Increasing coniferyl aldehyde tolerance	Overexpression of ATR1 and FLR1 genes	Increased coniferyl aldehyde tolerance

Source: Topaloğlu A, Esen Ö, Turanlı-Yıldız B, Arslan M, Çakar ZP. From Saccharomyces cerevisiae to ethanol: unlocking the power of evolutionary engineering in metabolic engineering applications. Journal of fungi. 2023; 9(10):984. https://doi.

Jiang, Li & Peng (2022) reviewed several metabolic engineering approaches in yeast that are shown in Table 3.6 as compiled by them.

Table 3.6.

Gene/Enzyme Modification	Remarks/Achievements
↑ ACC1	↑ 58%, 6.8% lipids
(1)↑ ACC1 (2)↑ ACC1ser659ala,ser1157ala	(1) Not significant (2) ↑ 65% FAs
ACC1ser1157ala	↑ 3-fold FAs
ACC1ser659ala, ser1157ala	Not significant
↑ FAS1, FAS2	↑ 30% lipid content, 70.6 mg/L (5.6% CDW)
↑ ACS1, ACS2	↑ 2–5 × acetyl-CoA level
↑ SeACSL641P	↑ α-santalene
↑ ACS1	↑ acyl-CoA level ↑ 8–23% amorphadiene
↑ SeACSL641P, ADH2, ALD6, WS2	↑ 3 × FAEE, 408 ± 270 ug/g CDW
(1) ∆ POX1, ↑ POX2; (2) ∆ POX1, ↑ POX2, ↑ crot	(1) ↑ 29.5% total FAs, 2.26 × intracellular MCFAs, 3.29 × extracellular MCFAs; (2) ↑ 15.6% total FAs, 1.87 × intracellular MCFAs, 3.34 × extracellular MCFAs
∆ POX1	↑ 4 × FFAs
∆β-oxidation, ∆ACSs, ∆ADH1, ↑ thioesterases, ↑ ACC1, ↑ acetyl-CoA	↑ 2 ×, 140 mg/L FAs
(1) ∆β-oxidation, ∆FAA2, PXA1, POX1; (2) ∆ACSs, FAA1, FAA4, FAT1; (3) Combine (1) & (2)	(1) ↑ intracellular FAs up to 55%; (2) ↑ extracellular FFAs to 490 mg/L; (3) ↑ 1.3 g/L extracellular FFAs
∆FAA1∆FAA4, ↑ acyl-CoA thioesterase ACOT5 (Acot5s)	↑ 6.43 × FFAs, 500 µg/mL, ↑ UFA ratio (42% > 0 in WT), ACOT5 helps restore cell growth
(1) ∆ARE1∆ DGA1 ∆ARE2∆LRO1; (2) ∆POX1; (3) Combine (1) & (2)	(1) 3 × FFA; (2) 4 × FF; (3) 5 × FFA

↑ ACC1, ↑ FAS1, ↑ FAS2	>17% DCW lipids, ↑ 4× than WT
↑ At-Gh13LPAAT5	↑ 25–31% in palmitic acid and oleic acid; ↑ 16–29% TAG
↑ DGAT	↑ 3–9 × TAG, (25- 80 nmol TAG/ mg DCW)
↑ Dga1p (YOR245c)	↑ 70–90 × DGAT activity in LDs; ↑ 2–3 × in ER.
↑ DGAT1, ↑ N-terminal tag	↑ 53% × TAG, 28% × total FAs, 453 mg FAs/L
↑ PDAT	↑ TAG, 2 × (log phase), 40% × (stationary phase), identified PDAT gene YNR008w
↑ LuPDCT1, LuPDCT2, ↑ FAD2, FAD3	↑ PUFAs (linoleic acid (18:2 cisΔ9,12), α-linolenic acid (18:3 cisΔ9, 12, 15)) levels in phosphatidylcholine (PC), DAG, and TAG
ΔSLC1 (YDL052C); ΔTGL3 (YMR313C)	↑ LD content
ΔHMG2 (YLR450W); ΔDGA1(YOR245C); ΔERG4 (YGL012W); ΔERG5 (YMR015C); ΔARE (YNR019W); ΔSIT4 (YDL047W); ΔREG1 (YDR028C); ΔSAP190 (YKR028W)	↓ LD content
ΔPAH1	↓ 63% × LDs number, total lipids stable
↑ WS2, ACB1, GAPN	↑ 7.7×, 48 mg/L FAEE
↑ ACC1, FAS1, FAS2, terminal "converting enzymes"	400 mg/L FFA, 100 mg/L fatty alcohols, 5 mg/L FAEE
(1) ↑ ACL, ΔIDH1; (2) ↑ ACL, ΔIDH2; (3) ↑ ACL, ΔIDH1, 2	(1) ↑ 80% C16:1, ↑60% C18:1; (2) ↑ 60% C16:1, ↑45% C18:1; (3) ↑ 92% C16:1, ↑77% C18:1
(1) ↑ Reversed β–oxidation pathway, SeACSL641P, ΔADH1, 4, ΔGPD1,2; (2) ↑ Reversed β-oxidation pathway, EEB1 or ETH1	(1) ↑ medium-chain FAEEs (0.011 g/L FFA, C16, C18); (2) ↑ FAEE (C4–C10, 0.75 g/L)

(1)↑ WS2, ADH2, ALD6, SeACSL641P; (2)↑ WS2, xpkA, ack, pta	(1) ↑ 3 × FAEE (408 ± 270 μg/g, DCW) (2) ↑ 1.7-fold FAEE (5100 ± 509 μg/g, DCW).
↑ WS2, Δ FAA2, ΔACB1, ΔPXA2	↑ 17×, 25 mg/L FAEE
Δβ-oxidation, ΔFAA2, ΔPXA1 & ΔPOX1, ΔACSs, ΔFAA1, 4 & ΔFAT1, ↑DGA1 & ↑TGL3	2.2 g/L extracellular FFAs
(1) ΔSNF2; (2) ↑LEU2, ΔSNF2; (3) ↑DGA1, ΔSNF2; (4) ↑LRO1, ΔSNF2; (5) ↑FAA3, ↑DGA1, ΔSNF2	(1) ↑ lipid; (2) ↑ growth and lipid accumulation; (3) ↑ lipids; (4) ↓ lipids; (5) ↑ lipids, 30% lipids content, mainly TAG (add exogenous FAs).
↑ ACC1ser659ala, ser1157ala, ↑PAH1, ↑DGA1, Δ GUT2, ΔARE1, ΔPXA1, ΔPOX1, ΔFAA2, ΔTGL3, 4, 5	254 mg TAG/g DCW, 27.4% of the maxi theoretical yield
Engineering ScFAS, bacterial type I FAS, directed evolution of membrane transporter Tpo1, strain adaptive laboratory evolution	↑ 250-fold, >1 g/L Medium-chain fatty acids (C6–C12)
↑ RtME, ↑MDH3, ↑CTP1, ↑MmACL, ↑RtFAS, ↑ACC1, ↑tesA, ΔPOX1, ΔFAA1, 4	10.4 g/L extracellular FFA
↑ Cytosolic acetyl-CoA, ↑ NADPH supply, ↑ FA biosynthesis, Δethanol pathway, mutate pyruvate kinase and direction evolution: ↑MPC, RtCIT1, ME, PYC1, YHM2, MDH3, RtFAS, ACC1, AnACL, MmACL, IDP2, ↓PGI1, IDH2, ↑ ZWF1, GND1, TKL1, TAL1, etc.	33.4 g/L extracellular FFA, the highest titer reported to date in *S. cerevisiae*

Key: ACB1, acyl-CoA binding protein; ACS, acetyl-CoA synthase; ACC, acetyl-CoA carboxylase; ADH, alcohol dehydrogenase; ACP, acyl carrier protein; CDW, cell dry weight; DAG, diacylglycerol; DGAT, acyl-CoA: diacylglycerol acyltransferase; DHAP, dihydroxyacetone phosphate; DHA, docosahexaenoic acid (22:6 ω-3); EPA, eicosapentaenoic acid (20:5 ω-3); ER, endoplasmic reticulum; FAEE(s), fatty acid ethyl ester(s); FAME, fatty acid methyl ester; FADs, fatty acid desaturases; FAS, fatty acid synthase; FFA(s), free fatty acid(s); G3P, glycerol-3-phosphate; GPAT, G3P acyltransferase; GAPN, NADP+-dependent glyceraldehyde-3-phosphate dehydrogenase; GLA, γ-linolenic acid (18:3 ω-6); GUT2, glycerol-3-phosphate dehydrogenase gene; LDs, lipid droplets; LPA, lysophophatidate; LPAT, lysophophatidate acyl-transferase; ME, malic enzyme; MCFA, medium chain fatty acid; PDAT, phospholipid: diacylglycerol acyltransferase; PA, phosphatidate; PAP, phosphatidate phosphatase; PDAT, phospholipid: diacylglycerol acyltransferase; PDCT, phosphatidylcholine diacylglycerol cholinephosphotransferase; POX, peroxisoma□β-oxidation; PUFA, po□yunsaturated fatty acid; tesA, encode thioesterase I; UFA, unsaturated fatty acid; '↑,' overexpression or hetero□ogous expression, increase; '↓': downregu□ation or reduce; 'Δ': de□etion or knockout, '×,' times by folds.

Source: Jiang W, Li C, Li Y, Peng H. Metabolic engineering strategies for improved lipid production and cellular physiological responses in yeast saccharomyces cerevisiae. Journal of fungi. 2022; 8(5):427. https://doi.org/10.3390/jof8050427:

Among other alternatives of fuels, butanol has received much attention given its closeness with gasoline, lower corrosivity, and higher energy density (27 MJ/L) vs. ethanol (19.6 MJ/L). The promising aspects of this fuel in terms of its economic aspects and CO_2 emissions have been illustrated by Nawab et al. (2020) in Figure 3.13.

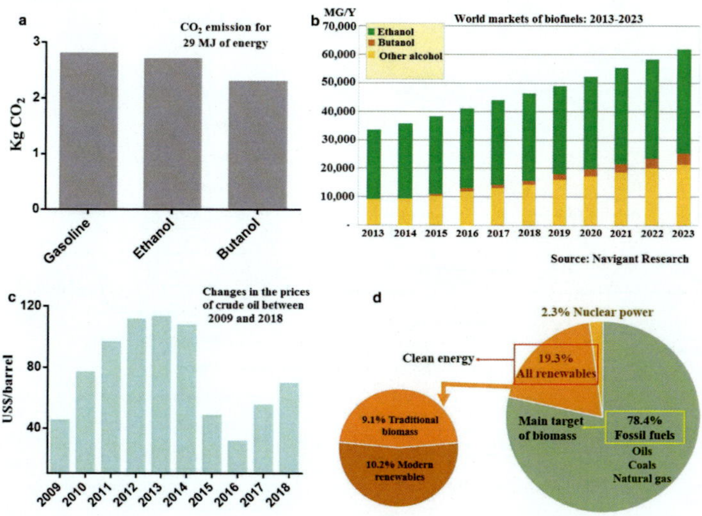

Figure 3.13

Source: Nawab, S., Wang, N., Ma, X. et al. Genetic engineering of non-native hosts for 1-butanol production and its challenges: a review. Microb Cell Fact 19, 79 (2020). https://doi.org/10.1186/s12934–020–01337-w:

The traditional production of 1-butanol is from Clostridial species. However, challenges like unwanted products, slow growth, and genetic manipulation difficulties have led researchers to employ non-native organisms to produce 1-butanol. For instance, a slew of heterologous genes has been expressed in yeast for producing the alcohol (Figures 3.14 and 3.15).

Top: Heterologous hosts and various feedstocks for 1-butanol production.

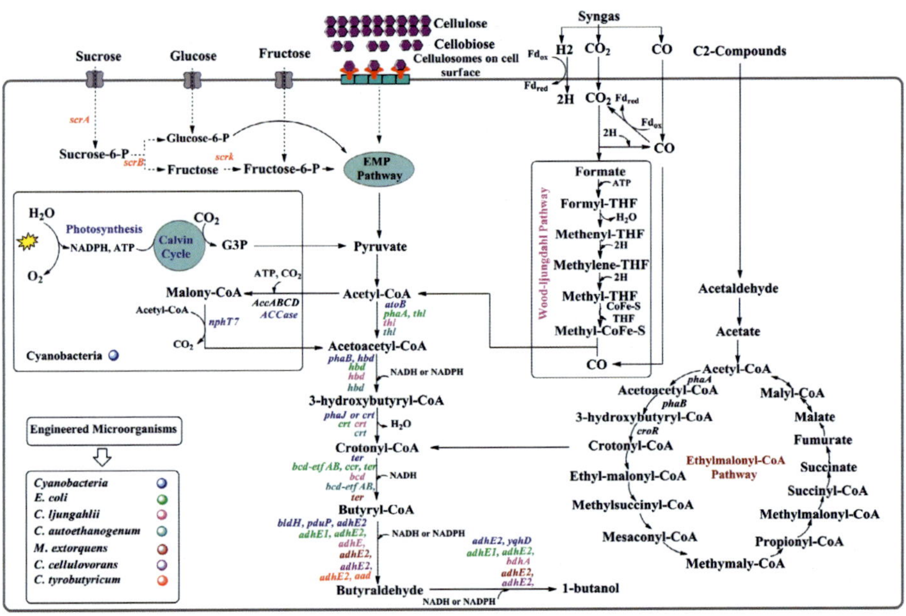

Figure 3.14

Source: Nawab, S., Wang, N., Ma, X. et al. Genetic engineering of non-native hosts for 1-butanol production and its challenges: a review. Microb cell Fact 19, 79 (2020). https://doi.org/10.1186/s12934–020–01337-w: Open Access This article is licensed under a Creative Commons Attribution 4.0 International License, which permits use, sharing, adaptation, distribution and reproduction in any medium or

Bottom: Metabolic engineering of yeast for 1-butanol production (black-host genes; red-heterologous genes).

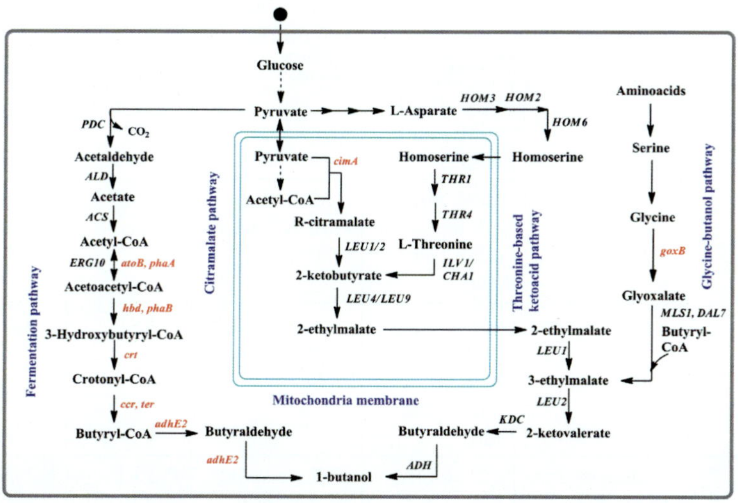

Figure 3.15

Source: Nawab, S., Wang, N., Ma, X. et al. Genetic engineering of non-native hosts for 1-butanol production and its challenges: a review. Microb Cell Fact 19, 79 (2020). https://doi.org/10.1186/s12934–020–01337-w:

Table 3.7 compiled by Nawab et al., (2020) summarizes how organisms other than the native Clostridial producers were engineered to produce 1-butanol.

Table 3.7.

Host	Substrate	**Genes overexp-ressed**	**Knockout genes**	**Promoter**	**1-butanol titer**
E. coli	Glucose and glycerol	atoB, thl, hbd, crt, ccr, bcd, etfAB and adhE2	Δ□dhA, ΔadhE, ΔfrdBC, Δfnr, Δpta, ΔpntA and Δpf□B	PLlacO1	552 mg/L
E. coli	Glucose	thl, hbd, crt, bcd, etfAB, adhE1 and adhE2	None	Ptac	1.2 g/L
E. coli	Glucose and glycerol	atoB, hbd, crt, ter, adhE2 and fdh	ΔadhE, Δ□dhA, ΔfrdBC, Δpta, and Δfnr	PLlacO1	15 g/L
E. coli	Glycerol and glucose or galactose	aceEF-lpdfbr, atoB, hbd, crt, ter, adhE2 and fdh1	Δ□dhA, ΔadhE, ΔfrdBC and Δpta	PatoB	6.8 g/L
E. coli	Glucose and butyrate	glf, atoDA, acs, and adhE2	Δ□dhA, ΔadhE, ΔfrdA, and Δpta	PλPRPL and PλPL	6.2 g/L
E. coli	Glucose	glf, atoDA, phaA, hbd, crt, ter and adhE2	Δ□dhA, ΔadhE, ΔfrdA, and Δpta	PλPRPL and PλPL	5.5 g/L
E. coli	Cellulose hydrolysate	glf, atoDA, fdh1, phaA, hbd, crt, ter and adhE2	Δ□dhA, ΔadhE, ΔfrdA, and Δpta	Ptrc and PλPL	2.8 g/L
E. coli	Cellulose hydrolysate	glf, phaA, hbd, crt, ter and adhE2	Δ□dhA, ΔadhE, ΔfrdA, and Δpta	PλPRPL and PλPL	5.8 g/L

E. coli	Glucose	thl, crt, bcd, etfA, etfB, hbd, adhE2 and ompC-tmt	None	Ppr and PT7	320 mg/L
E. coli	Glucose and glycerol	fdh, pduP, panK, atoB, hbd, crt, ter and adhE2	ΔldhA, ΔadhE, ΔfrdA, and Δpta	PLlacO1	18.3 g/L
E. coli	Glucose	atoB, hbd, crt, ter and adhE2	ΔadhE, ΔfrdABCD, ΔfdhF, Δmdh, ΔyqhD, Δhyc-hyp, ΔmaeB, ΔeutE, Δpta, ΔackA, ΔpykA, ΔldhA, ΔyciAI, ΔpoxB, ΔyagM	miniPtac, PfdhF, PT7, PydfZ and PadhE	20 g/L
C. tyrobutyricum	Glucose	aad and adhE2	Δack and Δptb	Pthl, Paad, and Pptb	10 g/L
C. tyrobutyricum		adhE1 and adhE2	Δcat1	Pcat1	26.2 g/L
C. tyrobutyricum	Mannitol	aad and adhE2	Δack and Δptb	Pthl, Paad, and Pptb	16 g/L
C. tyrobutyricum	Glucose and mannitol	adhE2	None	Pthl	6.8 g/L and 20.5 g/L
C. tyrobutyricum	Glucose	adhE2	Δack	Pthl	16.68 g/L
C. tyrobutyricum	Glucose	fdh and adhE2	None	Pthl	12.34 g/L
C. tyrobutyricum	Sucrose and sugarcane juice	scrA, scrB, scrK and adhE2	Δack	Pthl	18.8 g/L
C. cellulovorans	Cellulose	adhE2	None	Pthl	1.42 g/L
C. cellulovorans	Corn cob	adhE2	None	Pthl	3.37 g/L
C. cellulovorans	Desheled corn cobs	adhE1 and ctfAB and adc	None	Pthl	3.47 g/L

C. cellulolyticum		atoB, hbd, crt, bcd and adhE2	None	Pthl	0.12 g/L
C. ljungahlii	Syngas	thlA, hbd, crt, bcd, adhE, and bdhA	None	Pptb	0.15 g/L
C. autoethanogenum	Syngas	thlA, hbd, crt, bcd and etfAB	None	Ppta-ack	1.54 g/L
M. extorquens	Ethylamine	ter and adhE2	None	PmxaF, Plac, Pmetal_ 3616, Pmetal_ 002	15.2 mg/L
K. pneumonia	Glycerol	ter, bdhB, and bdhA	None	Ppk	15.03 mg/L
K. pneumonia	Glycerol	Kivd	None	Ppk	28.7 mg/L
K. pneumonia	Glycerol	leuABCD, kivd, adhE1, fdh, gdh and udh	None	Ptac	100 mg/L
T. saccha-rolyticum	Xylose	thl, hbd, crt, bcd, etfA, etfB and dhE2	Δldh, Δpta, and Δack	Ppta-ack, PgapDH, and Pcbp	1.05 g/L
P. furiosus	Maltose	thl, hbd, crt, ter, bad, bdh, adhA and adhE2	None	Pporγ	70 mg/L
L. brevis	Glucose	crt, bcd, etfB, etfA, and hbd	None	Pbcs-operon	300 mg/L
P. putida	Glucose and glycerol	thl, hbd, crt, etfAB, bcd, bdhb and adhE1	None	Plac and Ptac-lac	44 and 122 mg/L
B. subtilis	Glucose and glycerol	thl, crt, etfAB, hbd, and adhE2	None	Phyper-spank	23 and 24 mg/L
Synechococcus 7942	CO_2	atoB, hbd, crt, ter and adhE2	None	PLlacO1 and Ptrc	14.5 mg/L

Synechococcus 7942	CO_2	atoB, nphT7, phaB, hbd, crt, phaJ, ter, bldh, yqhD, and adhE2	None	PLlacO1 and Ptrc	30 mg/L
Synechococcus 7942	CO_2	nphT7, phaB, crt, phaJ, ter, pduP and yqhD	None	PLlacO1 and Ptrc	404 mg/L
Synechococcus 7942	CO_2	accase, nphT7, phaB, phaJ, ter, pduP and yqhD	ΔaldA	PLlacO1 and Ptrc	418.7 mg/L
Synechocystis PCC 6803	CO_2	nphT7, phaA, phaB, Slr0942, Slr1192, crt, phaJ, ccr, fadB, ter, pduP, yjgB, yqhD, mhpF,	ΔphaEC, ΔackA, Δslr0168 and Δach	Ptrc20, Pcpc560, PpsbA2	4.8 g/L
S. cerevisiae	Galactose	atoB, erg10, thl, phaA, phaB, hbd, crt, ccr and adhE2	None	Pgal1 and Pgal10	2.5 mg/L
S. cerevisiae	Glucose	ald6, adh2, erg10, acs, hbd, crt, ter adhE2	Δcit2 and Δmls1	Ptef1 and Ppgk1	16.3 mg/L
S. cerevisiae	Glucose	thl, hbd, crt, etfA, etfB, ad, aad, and ter	Δgpd1 and Δgpd2	Pgap	14.5 mg/L
S. cerevisiae	Glucose	pdh, acs, acl, thl, hbd, crt, ter, bad, bdhB, mhpF, eutE, adhE, and adhE2,	Δadh1, Δadh4, Δgpd1, Δgpd2, Δcit2 and Δmls1		120 mg/L
S. cerevisiae	Glycine and glucose	goxB, mls1, dal7 and leu2	Δmls1 and Δleu2	Ptpi1	92 mg/L

S. cerevisiae	Glucose	hom3, hom2, hom6, thr1, thr4, ilv1, cha1, leu4, leu1, adh1, adh2, adh3, adh4, adh5, adh6, bdhB, kivD and aro10m	Δadh1, Δi□v1, Δi□v2, Δi□v3, Δi□v6, Δ□eu1, Δ□eu4 and Δ□eu9,	Ptef1 and Ptpi1	242.8 mg/L
S. cerevisiae	Glucose	hom3, hom2, hom6, thr1, thr4, leu4, leu1, leu2, leu5, leu9, NFS1, ilv1, cimA, adh7, aro10m and kivD	Δadh1	Pgpd1	835 mg/L
Y. lipolytica		EutE, ETR1, ylact1, ylact2, ylhbd, ylcrt, YlGPD, YlMDH, hom6 and YlBDHs	ΔYALI0 E17787g	Php4d	123 mg/L
A. adeninivorans	Starch	thl, hbd, crt, bdh, and adhE2	Δaacp, Δaact, Δaaco, and Δagpd1	Paynr1, Payni1, Paynt1, and Ptef1	20 g/L

Key: atoB, acetyl-CoA acetyltransferase; thl, thiolase; hbd, 3-hydroxybutyryl-CoA dehydrogenase; crt, crotonase; bcd, butyryl-CoA dehydrogenase; etfAB, electron transfer flavoproteins; adhE1, aldehyde/alcohol dehydrogenase; adhE2, butyraldehyde-butanol dehydrogenase; ter, trans-enoyl-CoA reductase; fdh, formate dehydrogenase; aceF, dihydrolipoamide acetyltransferase; lpd, dihydrolipoamide dehydrogenase; glf, glucose facilitator; atoDA, acetoacetyl-CoA transferase; acs, acetyl-CoA synthetase; phaA, β-ketothiolase; ompC-TMT, outer-membrane targeted tilapia metallothionein; pduP, CoA-acylating aldehyde dehydrogenase; panK, pantothenate kinase; scrA, sucrose-specific PTS; scrB,sucrose-6-phosphate hydrolase or sucrose; scrK, fructokinase; bdh, butanol dehydrogenase; bdhA, butanol dehydrogenase; bdhB, Butanol dehydrogenase B; kivd, ketoisovalerate decarboxylase; leuABCD, 2-isopropylmalate synthase; adc, acetoacetate decarboxylase; gdh, glucose dehydrogenase; udh, pyridine nucleotide transhydrogenase, bad, butyraldehyde dehydrogenase; accase, acetyl-CoA carboxylase; nphT7, acetoacetyl CoA synthase; phaB, Acetoacetyl-CoA reductase; phaJ, acetoacetyl-CoA reductase;

ach, acetyl-CoA hydrolase; bldh, butyraldehyde dehydrogenase; aldA and yqhD, alcohol dehydrogenase; erg10, thiolase; ccr, crotonyl-CoA reductase; adh1, adh2, adh3, adh4, adh5, adh6 and adh7, alcohol dehydrogenase; ald6, acetaldehyde dehydrogenase; ad, aldehyde dehydrogenase; aad, aldehyde-alcohol dehydrogenase; pdh, pyruvate dehydrogenase; acs, acetyl-CoA synthetase; acl, citrate lyase; mhpF, acetaldehyde dehydrogenase; eutE, aldehyde dehydrogenase; goxB, glycine oxidase; leu1 and leu2, 3-isopropylmalate dehydrogenase; leu4 and leu9, 2-isopropylmalate synthase, mls1 and dal7, malate synthase; hom3, aspartokinase; hom2, aspartate-semialdehyde dehydrogenase; hom6, homoserine dehydrogenase; thr1, homoserine kinase; thr4, threonine synthase; ilv1/cha1 threonine deaminase; kivd, α-ketoisovalerate decarboxylase; nfs1, cysteine desulfurase; cimA, citramalate synthase; ldhA, lactate dehydrogenase; frdBC, fumarate reductase; pta, phosphate acetyltransferase; fnr, fumarate and nitrate reductase; pfl, pyruvate formate lyase; pntA, pyridine nucleotide transhydrogenase; frdA, fumarate reductase flavoprotein subunit; ack, acetate kinase; cit2, peroxisomal citrate synthase; gpd1 and gpd2, glycerol-3-phosphate dehydrogenase; aacp, peroxisomal acyl-carnitine permease; aact, acyl-carnitine transferase; aaco, acyl-CoA oxidase

Source: Nawab, S., Wang, N., Ma, X. et al. Genetic engineering of non-native hosts for 1-butanol production and its challenges: a review. Microb cell Fact 19, 79 (2020). https://doi.org/10.1186/s12934–020–01337-w:

The industria□y important chemica□ 1,2-Propanedio□ (1,2-PDO) finds use in resins, detergents, food, cosmetics, etc. Several microbes have been documented to produce this compound that have been summarized in Table 3.8 by Tao et al., (2021).

Table 3.8.

Organisms	Substrates	Fermentation Modes	Titer (g/L)	Yield (g/g)
Bacteroides ruminicola	l-Rhamnose	Batch	0.070	0.0853
Salmonella typhimurium	l-Fucose	Flask	1.37	0.417
Salmonella typhimurium	l-Rhamnose	Flask	1.46	0.445
Klebsiella pneumoniae	l-Fucose	Flask	1.49	0.454
Klebsiella pneumoniae	l-Rhamnose	Flask	1.52	0.463
Clostridium strain AK-1	l-Rhamnose	Batch	1.68	0.375
Clostridium thermosaccharolyticum HG-8	Glucose	Batch	9.05	0.20
Clostridium thermosaccharolyticum HG-8	Glucose	Fed-batch	7.90	0.27
Clostridium thermosaccharolyticum HG-8	Xylose	Flask	3.23	0.123
Clostridium thermosaccharolyticum HG-8	Mannose	Flask	3.1	0.115
Clostridium thermosaccharolyticum HG-8	Cellobiose	Flask	3.1	0.108
Clostridium thermosaccharolyticum HG-8	Whey permeates	Flask	2.8	–
Lactobacillus buchneri	Lactic acid	Flask	0.571	0.384
Candida polymorpha	l-Rhamnose	Flask	–	–
Pichia robertsii	l-Rhamnose	Flask	–	–
Saccharomyces cerevisiae	Citrus stillage	Flask	0.211	–
Saccharomyces cerevisiae	Whey stillage	Flask	0.123	–

Saccharomyces cerevisiae	Corn stillage	Flask	0.105	–

However, the challenges of low efficiency and the requirement of petro-based compounds for production have led to researchers using metabolic engineering to facilitate other microbes to produce 1,2-propanediol. Several such examples have been reviewed by Tao et al., (2021) in the Table 3.9.

Table 3.9.

Host	Pathway involved	Substrates	Genetic modification	1,2-propanediol concentration
E. coli	Methylglyoxal pathway	Glucose	*gldA* (*E. coli*) ↑/*dhaD* (*K. pneumoniae*) ↑/*mgsA* (*E. coli*) ↑	0.7 g/L
E. coli	Methylglyoxal pathway	Glucose	*mgsA* (*B. subtilis*) ↑/*gldA* (*E. coli*) ↑/*fucO* (*E. coli*) ↑/Δ*zwf*/Δ*tpiA*/Δ*poxB*/Δ*ackA*/Δ*ubiC*	1.2 g/L
E. coli	Methylglyoxal pathway	Glucose	*fdh1* (*C. boidinii*) ↑/*mgsA* (*B. subtilis*) ↑/*gldA* (*E. coli*) ↑/*fucO* (*E. coli*) ↑/Δ*zwf*/Δ*tpiA*/Δ*ldhA*/Δ*gloA*/Δ*adhE*	5.13 g/L
E. coli	Methylglyoxal pathway	Glycerol	*mgsA* (*E. coli*) ↑/*yqhD* (*E. coli*) ↑/*gldA* (*E. coli*) ↑/Δ*ackA-pta*/Δ*ldhA*/*dhaK*::*dhaKL* (*Citrobacter freundii*)	5.6 g/L

E. coli	Methylglyoxal pathway	Glycerol	*mgsA* (*E. coli*)/*dhak* (*E. coli*)/*gldA* (*E. coli*)/*fucO* (*E. coli*)=>BMC	11.56 mM/OD_{600}
E. coli	Methylglyoxal pathway	Cellobiose	BGL Tfu0937 (*Thermobifida fusca* YX) ↑/*mgsA* (*B. subtilis*) ↑/ *gldA* (*E. coli*) ↑/*fucO* (*E. coli*) ↑/Δ*ptsG*/Δ*tpiA*/ Δ*gloA*/Δ*hchA*/Δ*ldhA*/ Δ*aldA*	0.11 g/L
E. coli	Lactate pathway	Glucose	*MavCAR* (*Mycobacterium avium*) ↑/*yahK* (*E. coli*) ↑/ Δ*dld*/Δ*lldD*/Δ*adhE*/Δ*frdA*/ Δ*pflB*/Δ*mgsA*/Δ*aldA* or *ldhA*:*lldh* (*Pediococcus acidilactici*)	0.73 g/L *S*-1,2-PDO 0.53 g/L *R*-1,2-PDO
E. coli	Lactate pathway	Glucose	*pct* (*M. elsdenii*) ↑/*pdup* (*S. enterica*) ↑/*yahK* (*E. coli*) ↑/Δ*dld*/Δ*lldD*/ Δ*adhE*/Δ*frdA*/Δ*pflB*/ Δ*mgsA*/Δ*aldA*/Δ*arcA* or *ldhA*::*lldh* (*Pediococcus acidilactici*)	9.3 g/L *S*-1,2-PDO 17.3 g/L *R*-1,2-PDO
E. coli	Lactate pathway	L-lactate	*pct* (*Clostridium propionicum DSM 1682*) ↑/*pdcD* (*Yersinia enterocolitica subsp. enterocolitica 8081*) ↑/ *mmsB* (*Bacillus cereus ATCC 14579*) ↑/*Lldh* (*Bacillus coagulans*)↑/ *Zppdc* (*Zymobacter palmae*)↑/mhpF (*E. coli MG1655*)↑/Δ*lldD*/Δ*ackA-pta*/Δ*adhE*/Δ*poxB*/Δ*ldhA*/ Δ*mgsA*/Δ*frdABCD*	1.04 g/L *S*-1,2-PDO
E. coli	Lactate pathway	l/d-lactate	*pct* (*M. elsdenii*) ↑/*pdup* (*S. enterica*) ↑/*yahK* (*E. coli*) ↑/Δ*lldD*/Δ*dld*/Δ*ldhA*/ Δ*adhE*	1.7 g/L *S*-1,2-PDO 1.5 g/L *R*-1,2-PDO
E. coli	Lactate pathway	Starch	*amyA* (*Streptococcus bovis*)↑/*pct* (*M. elsdenii*)↑/ *pdup* (*S. enterica*)↑/*yahK* (*E. coli*)↑	0.028 g/L
S. cerevisiae	Methylglyoxal pathway	Glucose	*mgsA* (*E. coli*) ↑/*gldA* (*E. coli*) ↑/Δ*tpi1*	1.11 g/L
S. cerevisiae	Methylglyoxal pathway	Glycerol	*mgsA* (*E. coli*) ↑/*gldA* (*E. coli*) ↑/*GUT1* (*S. cerevisiae*) ↑/*GUT2* (*S. cerevisiae*) ↑/*gdh* (*S. cerevisiae*) ↑/*GUP1* (*S. cerevisiae*) ↑	2.19 g/L

S. cerevisiae	Methylglyoxal pathway	Glycerol	*mgsA* (*E. coli*) ↑/ *yqhD* (*E. coli*) ↑/*gldA* (*E. coli*) ↑/*Opgdh* (*O. parapolymorpha*)↑/*DAK1* (*S. cerevisiae*)↑/TPI1 (*S. cerevisiae*)↓/Δ*GUT1*	>4 g/L
S. cerevisiae	Methylglyoxal pathway	Galactose	*mgsA* (*E. coli*) ↑/*dhaD* (*Citrobacter freundii*) ↑	0.45 g/L
C. glutamicum	Methylglyoxal pathway	Glucose	*mgsA* (*E. coli*) ↑/ *cgR_2242* (*C. glutamicum*)↑	1.83 g/L
C. glutamicum	Methylglyoxal pathway	Glucose	*mgsA* (*E. coli*) ↑/*yqhD* (*E. coli*) ↑/*gldA* (*E. coli*) ↑/ Δ*hdpA*/Δ*ldh*	0.343 mol/mol glucose
C. glutamicum	–	Pyrolysis water	*gldA* (*E. coli*) ↑/Δ*pqo*/ Δ*aceE*/Δ*ldhA*/Δ*mdh*	1.39 g/L
Synechococcus elongatus PCC 7942	Methylglyoxal pathway	CO_2	*mgsA* (*E. coli*) ↑/*yqhD* (*E. coli*) ↑/*adh* (*Clostridium beijerinckii*)↑	~0.15 g/L
Synechocystis sp. PCC6803	Methylglyoxal pathway	CO_2	*mgsA* (*E. coli*) ↑/*yqhD* (*E. coli*) ↑/*adh* (*Clostridium beijerinckii*) ↑/Δ*glgC*	~1 g/L

Source: Tao YM, Bu CY, Zou LH, Hu YL, Zheng ZJ, Ouyang J. A comprehensive review on microbial production of 1,2-propanediol: micro-organisms, metabolic pathways, and metabolic engineering. Biotechnol biofuels. 2021;14(1):216. Published 2021 Nov 18. doi:10.1186/s13068–021–02067-w:

3.2. FOOD PRODUCTS AND NUTRACEUTICALS

"To me, supplements are a necessity, not a luxury. I think of them as medicine: Instead of asking a doctor to prescribe me a drug when I'm ill, I'd rather take something that can help me avoid getting sick in the first place."

—Laird Hamilton

According to Wang & Zhang (2023), "cell factories" have become important with the development of the "big food view." The "big food view" aims to "start from better meeting people's needs for a better life," which includes expanding food sources and production approaches.

The amino acid L-aspartate serves as a precursor of amino acids and energy in our bodies. At a pharmaceutical level, it is used as a sweetener and in the treatment of certain conditions. According to Data Bridge Market Research, the market for this amino acid was $93.15 million in 2021, with a projected compound annual growth rate (CAGR) of 6.20% from 2022 to 2029. β-alanine, a non-essential amino acid produced by the liver, is the source of the antioxidant carnosine. It is used to regulate acidity and enhance flavor. Global Market Insights reported that the market for this molecule was $88.41 million in 2022. Table 3.10, compiled by Shi et al., (2023), review metabolic engineering approaches for both of these molecules (Table 3.11).

Table 3.10. *Metabolic Engineering Approaches for L-Aspartate Production in Literature*

Organism	Metabolic Engineering Strategies	Substrate	Titer (g/L)	Yield a (g/g)	Fermentation Strategy
E. coli XAR31	Introducing and overexpressing *CgaspC, Cgppc, Mspck, glk, bt-ca, acs, Cgasp, BsrocG* and *CR,* deleting genes involved in byproduct biosynthesis, developing a cofactor self-sufficient system, optimizing the fermentation conditions	glucose	33.1	0.39	fed-batch

C. glutamicum SLV. pEKEx3-aspB	Deleting genes involved in byproduct biosynthesis *(sdhCAB, ldhA* and *avtA)*	glucose	~5.72 a	0.75	Flask
B. flavum 70	Developing several mutations: a citrate synthase-defective glutamate auxotroph, S-(2-aminoethyl)-L-cysteine-resistant mutant, a methionine-insensitive revertant and phosphoenolpyruvate carboxylase, a supplement of biotin	glucose	22.6	0.22	Flask
E. coli pMA-RBS4-G27A/G171A	Co-overexpressing maleate cis-trans isomerase (MaiA) mutant and aspartase (AspA) on the plasmid and optimizing their activity ratio by ribosome binding site (RBS) regulation	maleate	419.8	0.72	5-L fermenter
E. coli JCL1258/pBAW2/pASP400	Overexpressing *aspC* and *tyrB* on plasmid pASP400 and overexpressing *parB* and *aspA* on plasmid pBAW2	fumarate	77.60 a	0.83	

Source: Shi A, Liu Y, Jia B, Zheng G, Yao Y. Metabolic engineering of microorganisms to produce l-aspartate and its derivatives. fermentation. 2023; 9(8):737. https://doi.org/10.3390/fermentation9080737:

Table 3.11. *Metabolic Engineering Approaches for B-Alanine Production in Literature*

Organism	Metabolic Engineering Strategies	Substrate	Titer (g/L)	Yield (g/g)	Productivity (g/(L.h))	Fermentation Strategy
E. coli XBR41	Introducing *Bs panD*, deleting genes involved in byproduct biosynthesis, developing a cofactor self-sufficient system, optimizing the fermentation conditions	glucose	37.7	0.75 a		Fed-batch
E. coli NL-A13	High-throughput method to screen L-aspartate-α-decarboxylase variant ADCK43Y, evaluation and elevation cells' tolerance to β-alanine, improving fumarate supply and strengthening the pathway of fumarate and OAA to L-aspartate, optimizing culture medium	Glucose glycerol	11.9			Fed-batch
E. coli W3110	Introducing *panD* from *Bacillus subtilis,* rerouting fluxes of the central carbon metabolism, relieving the inactivation of L-aspartate-α-decarboxylase, optimizing the fed-batch bioprocess	glucose	85.18	0.24	1.05	Fed-batch

E. coli ALA17/ pTrc99a-panDBS-aspBCG	Introducing *panD* from *B. subtilis* and *aspB* from *C. glutamicum,* inactivating the β-alanine uptake system, the aspartate kinase I and III, iclR, ptsG, aroG, galR, overexpressing ppc, aspC, aceB, aceA, glk, and gltBD operon	glucose	43.94	0.20		Fed-batch
E. coli W FZβA-10	Introducing an L-aspartate a-decarboxylase gene from *Bacillus tequilensis,* a L-aspartate dehydrogenase gene from *Pseudomonas aeruginosa* and a pyruvate decarboxylase from *Corynebacterium glutamicum,* overexpressing aspA, deleting three native L-aspartate kinase genes and genes for byproduct biosynthesis (ldhA, pflB, pta and adhE)	glucose	43.12		0.89	Fed-batch
E. coli eBA32	Biosensor-enabled high-throughput screening, cofactor balancing and pathway modification	glucose	34.8			Fed-batch

C. glutamicum XQ-5	Deleting byproduct biosynthesis pathway (lactate dehydrogenase and alanine/ valine aminotra-nsferases), replacing L-aspartate kinase (AK) with wild-type AK, introducing and overexpressing a mutated L-aspartate-α-decarboxylase ($BsADC^{E56S/I88M}$) from *B. subtilis*	glucose	56.5	39.5% b	0.79	Fed-batch
C. glutamicum BAL10 (pBA2_ tr18)	Introducing *panD* from *B. subtilis*, overexpressing PTS-independent glucose uptake system, ppc, pyc, aspB, rocG from *B. subtilis*, aspA from *E. coli* and β-alanine exporter (NCgl0580), replacing the native pck with that from *Mannheimia succinici* producens (encoded by Mspck), deleting odx and mdh	glucose	166.6	0.28	1.74	Fed-batch
B. megaterium BMDBPG	Introducing a codon-optimized *panD* from *B. subtilis*, overexpressing aspB, ppc and NADH-dependent glutamate dehydrogenase (gdh)	glucose	17.60	0.23	0.78	
Methyl-otrophic *Pichia pastoris* 2ADC-Spe	Overexpressing panD from *B. subtilis* and aspDH from *S. proteamaculans*	Methanol	5.6			Fed-batch

L-aspartate-α-decarb-oxylase (ADC) from *C. glutamicum*	Overexpressed in *Escherichia coli* BL21(DE3), optimal at 55 °C and pH 6 with excellent stability at 16–37°C and pH 4–7	L-aspartate	12.85	97.2% b		Purified enzyme
E. coli BTW	Co-expressing two different types of L-aspartate-α-decarboxylases: one was from *B. subtilis* and the other was from *Tribolium castaneum*	L-aspartate	271.5	92.4% b		Whole cell
E. coli BTEW	Co-expressing three enzymes: two types of L-aspartate-α-decarboxylase (one was from *B. subtilis* and the other was from *Tribolium castaneum)* and one type of L-aspartase (AspA) from *E. coli*	fumarate	200.3	90.0% b		Whole cell
B. megaterium BMDA-6	Balancing the expression of L-aspartate-1-decarboxylases (ADC) from *B. subtilis* and aspartate ammonia-lyase (AspA) from *B. megaterium,* optimizing the cultivation conditions and biocatalysis process parameters	fumarate	11.68	0.78		Whole cell

Source: Shi A, Liu Y, Jia B, Zheng G, Yao Y. Metabolic Engineering of Microorganisms to Produce L-Aspartate and Its Derivatives. Fermentation. 2023; 9(8):737. https://doi.org/10.3390/fermentation9080737:

Apart from its role in protection from thermal stress, ectoine finds application in cosmetics given its water-holding abilities. The market for this molecule was 20 million USD in 2021 according to Business Research Insights (Table 3.12).

Table 3.12. Table Summarizes Metabolic Engineering Approaches for Producing Ectoine

Organism	Metabolic Engineering Strategies	Substrate	Titer (g/L)	Yield (g/g)	Productivity (g/(L.h))	Fermentation Strategy
E. coli ET11 (ectA:ectB:ectC = 1:2:1)	Introducing the ectABC gene cluster from *Halomonas venusta* ZH, regulating the copy number of ectA, ectB and ectC, eliminating byproduct metabolic pathways, optimizing the culture medium	glucose	53.2	0.33	1.11	fed-batch
E. coli Ect05	Introducing ectABC gene cluster from *Halomonas elongata* and a feedback-resistant L-aspartate kinase (lysC) from *Corynebacterium glutamicum,* deleting thrA and iclR, improving ppc expression by promoter replacement	glucose	25.1	0.11	0.84	fed-batch
E. coli S16-ectBAC	Introducing ectABC gene cluster from *Aestuariispira* SWCN16T into *E. coli* BL21	sodium aspartate and glycerol	2.26			cell suspension bioconversion reactions in the optimum buffer
E. coli ET01	Introducing the ectABC operon from *Halomonas venusta* ZH, optimizing the fermentation process	glucose	47.8			fed-batch

E. coli BW25113	Introducing ectABC from *Halomonas elongata* and overexpressing these three genes with an arabinose-inducible promoter, optimizing the fermentation process	aspartate and glycerol	25.1	4.1 b	1.04	whole-cell catalysis
E. coli ECT2	Introducing the ectABC genes from *Halomonas elongata,* deleting lysA	glycerol and sodium aspartate	12.7	1.27	0.53	whole-cell catalysis
C. glutamicum ectABCopt	Introducing the ectABC genes from *Pseudomonas stutzeri* and regulating their expression with different promoters and three linker elements	glucose, sucrose and fructose	65	0.19	2.3 f	fed-batch
C. glutamicum ECT-2	Introducing a codon-optimized synthetic ectABCD gene cluster from *Pseudomonas stutzeri,* inactivating the L-lysine exporter, optimizing the fed-batch process	glucose	4.5	0.24 d	0.28 e	fed-batch
C. glutamicum CB5L6	Introducing the ectBAC cluster from *Pseudomonas stutzeri,* deleting pck, ldh and sugR, improving the precursor supply (overexpression of Ecasd and CglysCS301Y), constructing repressor libraries (BetI from *E. coli* and LmrA from *B. subtilis*)	glucose	45.52	0.25		fed-batch
H. hydrothermalis Y2/ΔectD/ΔdoeA	Identifying the pathways for ectoine synthesis and catabolism, deleting genes involved in ectoine catabolism (EctD and DoeA) and Na^+/ H^+ antiporter (Mrp), optimizing the culture medium	monosodium glutamate	10.5	0.21		fed-batch

H. bluephagenesis TD-ADEL-58	Overexpressing three clusters related to ectoine biosynthesis, including ectABC, lysC and asd, deleting byproduct biosynthetic pathways	glucose	28	0.21	1.0	fed-batch
Chromohalobacter salexigens	Optimizing the medium composition, especially the C/N ratio, to regulate the metabolic pattern	glucose	4.21			fed-batch
Chromohalobacter salexigens	Producing ectoine with two continuously operated bioreactors, regulating the hyperosmotic conditions and thermal stress	glucose	8.2		2.1	fed-batch
Brevibacterium sp. JCM 6894	Inducing ectoine biosynthesis with 2 M NaCl, fermentation with non-sterilized medium	polypepton and dried yeast extract	2.4			flask
Brevibacterium epidermis DSM20659	Optimizing the fermentation conditions and the extraction technology	monosodium glutamate	8	0.05	0.08 e	fed-batch
H. boliviensis LC1T	Optimizing NaCl concentrations and the medium for fed-batch cultivations	glucose, monosodium glutamate	4.3	0.07 c	0.12 e	fed-batch
H. boliviensis LC1T	Optimizing the nutrient parameters in the fed-batch fermenter	glucose, monosodium glutamate	9.2		0.26 e	fed-batch
H. salina BCRC17875	Optimizing the agitation speed and medium composition	yeast extract	13.94			fed-batch
Sinobaca sp. H24	Isolating an ectoine producer from soil, optimizing culture medium, identifying the genes involved in ectoine biosynthesis	yeast extract, glycerol	0.01			flask
H. salina DSM 5928	Optimizing the culture medium and NaCl concentration	monosodium glutamate	6.9		0.33 e	batch

H. elongata DSM2581	Two nanostructures, multiwalled carbon nanotube (MWCNT) and iron oxide nanoparticle (Fe_2O_3 NPs), to increase the availability of the substrate	glucose	14.25			batch
Marinococcus sp. MAR2	Optimizing the culture condition with response surface methodology (RSM) and a fed-batch strategy	yeast extract	5.6		0.16 e	fed-batch
H. salina DSM 5928T	Optimizing the two-step fermentation conditions: growing of cells and production of ectoine by resting cells	monosodium glutamate	14.86	0.14	0.32 e	batch
Marinococcus sp. ECT1	Developing semi-synthesized medium (YAMS medium), optimizing the yeast extract concentrations	yeast extract	2.5			batch
H. brevibacterium sp. JCM 6894	Optimizing the conditions for ectoine biosynthesis	polypepton, glucose, yeast extract	2.5			
H. campaniensis G8–52 (CCTCCM2019777)	Developing a higher ectoine producer by multiple rounds of UV mutation, identifying the key mutations (orf00723 and orf02403 (lipA)) related to ectoine biosynthesis	sodium L-glutamate	1.51			flask
H. elongata DSM 2581T	Testing NaCl influence on ectoine biosynthesis, revealing higher NaCl concentration activating genes involved in the pentose phosphate pathway, Entner–Doudoroff pathway, flagellar assembly pathway, ectoine metabolism, repressing genes involved in the tricarboxylic acid cycle and fatty acid metabolism		12.91			

H. boliviensis DSM 15516(T)	Optimizing the conditions for two-step fermentation and producing ectoine with milking process	glucose	8.9		0.38 e	fed-batch
P. halophilum DSM 102817T	Optimizing the culture medium, developing strategies for ectoine isolation	glucose	0.41 a			flask
H. elongate 1A01717	Optimizing the ectoine extraction and purification process	glucose	15.9			fed-batch

Source: Shi A, Liu Y, Jia B, Zheng G, Yao Y. Metabolic engineering of microorganisms to produce l-aspartate and its derivatives. fermentation. 2023; 9(8):737. https://doi.org/10.3390/fermentation9080737: © 2023 by the authors. Licensee MDPI, Basel, Switzerland. This article is an open access article distributed under the terms and conditions of the Creative Commons Attribution (CC BY) license (https://creativecommons.org/licenses/by/4.0/).

Nutraceuticals have attracted much interest given their ability to target "age-associated" diseases like diabetes, cancer, depression, etc. To illuminate the importance of these molecules sourced mainly from plants or microbes, Puri et al., (2022) summarized clinical data on the effects of nutraceuticals that have been shared in Table 3.13.

Table 3.13.

Nutraceuticals	**Bioactive Molecule**	**Dosage**	**Formulation**	**Duration of Study**	**Action**	**Population Size (Volunteers)**
Omega-3	Eicosapentaenoic acid (EPA) and Docosahexaenoic acid (DHA)	60 mg/kg/ day	Capsule	12 months	Cystic fibrosis	15
Omega-3	Eicosapentaenoic acid (EPA) and Docosahexaenoic acid (DHA)	900 mg in the omega-3 rich group and 250 mg in the omega-3 poor group	Capsule	15 days	Reduction of chemotherapy-related toxicities	61
Omega-3	Eicosapentaenoic acid (EPA) and Docosahexaenoic acid (DHA)	2000 mg/ day	Capsule	21 days	Appetite	72
Aloe vera	aloe-emodin, aloin, aloesin, emodin, and acemannan	-	Gel	2 months	Acute dermatitis	120
Aloe vera	aloe-emodin, aloin, aloesin, emodin, and acemannan	60 mL	Juice	5 months	Irritable bowel syndrome	110

Seaweed	polysaccharides, proteins, lipids and polyphenols	2000 mg/d	Extract	12 weeks	High-density lipoprotein (hdl) cholesterol	34
Wheatgrass	vitamins, (A, B, C and E), minerals such as iron, calcium, magnesium, benzo(a)pyrene, ferulic, gallic, caffeic, syringic and p-coumaric acid	-	Cream	12 weeks	Plantar fasciitis	134
Wheatgrass	vitamins, (A, B, C and E), minerals such as iron, calcium, magnesium, benzo(a)pyrene, ferulic, gallic, caffeic, syringic and p-coumaric acid	100 cc/day	Juice	1 month	Active distal ulcerative colitis	23
Ginseng	Ginsenosides	100 mg twice a day	Capsule	12 weeks	Psychomotor functions	16
Ginseng	Ginsenosides	-	Hydrogel	2 weeks	Skin homeostasis	20
Echinacea	polysaccharides, glycoproteins, alkamides, cichoric acid, caftaric acid and chlorogenic acids	20 mg or 40 mg twice a day	Powder	6 weeks	Anti-anxiety and anti-depressant	104
Echinacea	polysaccharides, glycoproteins, alkamides, cichoric acid, caftaric acid and chlorogenic acids	5 mL	Oral Suspension	6 months	Tonsillitis	300

Source: Puri V, Nagpal M, Singh I, Singh M, Dhingra GA, Huanbutta K, Dheer D, Sharma A, Sangnim T. A comprehensive review on nutraceuticals: therapy support and formulation challenges. Nutrients. 2022; 14(21):4637. https://doi.org/10.3390/nu14214637: © 2022 by the authors. Licensee MDPI, Basel, Switzerland. This article is an open access article distributed under the terms and conditions of the Creative Commons Attribution (CC BY) license (https://creativecommons.org/licenses/by/4.0/).

Given the challenges of complex production systems and low yield in plants, researchers have turned to metabolic engineering to produce these molecules in microbes. Figure 3.16 summarizes the scheme of producing nutraceuticals from carbon sources using metabolic engineering.

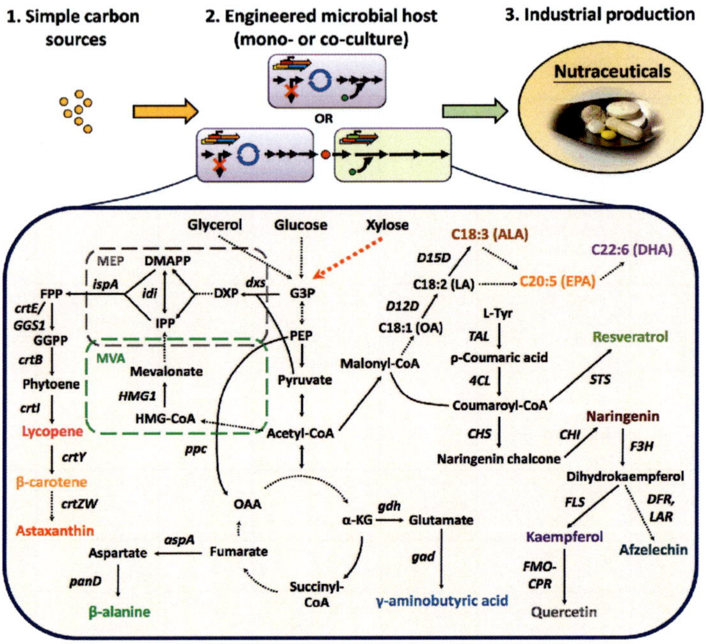

Figure 3.16

Source: Yuan, SF., Alper, H.S. Metabolic engineering of microbial cell factories for production of nutraceuticals. Microb Cell Fact 18, 46 (2019). https://doi.org/10.1186/s12934–019–1096-y: © The Author(s) 2019: *Open Access. This article is distributed under the terms of the Creative Commons Attribution 4.0 International License (http://creativecommons.org/licenses/by/4.0/), which permits unrestricted use, distribution, and reproduction in any medium, provided you give appropriate credit to the original author(s) and the source, provide a link to the Creative Commons license, and indicate if changes were made. The Creative Commons Public Domain Dedication waiver (http://creativecommons.org/publicdomain/zero/1.0/) applies to the data made available in this article, unless otherwise stated.*

Key: dxs, DXP synthase; idi, IPP isomerase; ispA, FPP synthase; HMG1, 3-hydroxy-3-methyl-glutaryl-coenzyme A reductase; crtE or GGS1, geranylgeranyl pyrophosphate synthase; crtB, phytoene synthase, crtI phytoene desaturase; crtY, lycopene cyc□ase; crtW, β-carotene keto□ase; crtZ, β-carotene hydroxy□ase; ppc, phosphoeno□-pyruvate carboxy□ase; aspA, aspartase; panD, □-aspartate-α-decarboxy□ase; gdh, g□utamate dehydrogenase; gad, g□utamate decarboxy□ase; D12D, Δ12-desaturase; D15D, Δ15-desaturase; TAL, tyrosine ammonia □yase; 4CL, 4-coumarate:CoA □igase; CHS, chalcone synthase; CHI, chalcone isomerase; STS, stilbene synthase; F3H, flavanone 3-hydroxy□ase; FLS, flavono□synthase; FMO, flavonoid 30-monooxygenase; CPR, cytochrome P450 reductase; DFR, dihydroflavona□4-reductase; LAR, leucoanthocyanidin reductase. Abbreviations of chemicals are G3P, glyceraldehyde 3-phosphate; DXP, 1-deoxy-D-xylulose-5-phosphate; IPP, isopentenyl pyrophosphate; DMAPP, dimethylallyl pyrophosphate, FPP, farnesyl pyrophosphate; L-Tyr, L-tyrosine; PEP, phosphoenolpyruvate; α-KG, 2-oxoglutarate; OAA, oxaloacetate; OA, oleic acid; LA, linoleic acid; ALA, -linolenic acid; EPA, Eicosapentaenoic acid

Table 3.14 reviewed and presented by Yuan & Alper (2019) summarizes how metabolic engineering approaches were used for producing nutraceuticals ([a]DCW represents EPA content at dry cell weight).

Table 3.14.

Product	Titer (Mg/L)	Carbon Source	Platform Organism	Medium, Fermentation Type and Parameters
α-Linolenic acid	1400	Glucose	*Y. lipolytica* L36DGA1	YSC medium contained 80 g/L glucose; a pulse of 80 g glucose was added at 72 h/fed-batch 2 L bioreactor, 20°C
EPA	56.6% in total lipids with 15% DCWa 50% in total lipids with 25% DCWa	Glucose	*Y. lipolytica* ATCC 20362 *Y. lipolytica* Y4305	Nitrogen-rich medium contained 20 g/L glucose for the first stage fermentation, and nitrogen-limited medium contained 80 g/L glucose for the second stage fermentation/two-stage flask, 30°C. Two-stage 2 L bioreactor, 30°C
DHA	2.4 (5% of total fatty acids)	Complex media	*E. coli* DH5α	LB medium was supplemented with 1 mg/L cerulenin/flask, 15°C after 1 mM IPTG induction
DHA	5.6% in total lipids	Glucose	*Y. lipolytica* Y4305	Cells were grown in MM medium containing 20 g/L glucose for 48 h, then transferred to HGM medium containing 80 g/L glucose for additional 72 h fermentation/two-stage fermentation, 30°C
Naringenin	100.64	Glucose	*E. coli* BL21 (DE3)	MOPS medium contained 5 g/L glucose and 4 g/L NH_4Cl/flask, 30°C
Naringenin	54.4/112.9	Glucose	*S. cerevisiae* CEN. PK	Synthetic medium contained 20 g/L glucose and 10 g/L (NH4)2SO4/ flask or batch 2 L bioreactor, 30°C
Naringenin	21	Xylose	*E. coli* BW25113 and *S. cerevisiae* CEN.PK2–1C	Synthetic fermented medium contained 40 g/L xylose, 5 g/L yeast extract and inorganic salt/ flask, 30°C
Resveratrol	416 (glucose)/531 (ethanol)	Glucose or ethanol	*S. cerevisiae* CEN. PK102–5B	Medium contained 40 g/L glucose (for batch phase), trace metals and vitamin solutions; 16 g/L glucose or 17 g/L ethanol for feeding/fed-batch 1 L bioreactor, 30°C
Resveratrol	812 (glucose)/755 (ethanol)	Glucose or ethanol	*S. cerevisiae* CEN. PK102–5B	Medium contained 40 g/L glucose (for batch phase), trace metals and vitamin solutions; 88 g/L glucose or 79 g/L ethanol for feeding/fed-batch 1 L bioreactor, 30°C

Resveratrol	22.6	Glycerol	*E. coli* W3110 coculture	M9 medium contained 0.3 mM l-phenylalanine and 10 g/L glycerol/batch 1 L bioreactor, 30°C
Kaempferol	27	Glucose	*S. cerevisiae* CEN. PK102–5B	Synthetic feed-in-time medium contained vitamins, dextrose polymer and enzyme minx/96-deep well plate, 30°C
Quercetin	20			
Afzelechin	41	Glycerol	*E. coli* BL21star™(DE3) *ΔsucC ΔfumC* (upstream strain) and BL21star™(DE3) (downstream strain)	Initial AMM medium contained 20 g/L glycerol; feed solution contained 2 × MOPS mix with 250 g/L glycerol/fed-batch bioreactor, 30°C
Lycopene	1230	Glycerol	*E. coli* BL21(DE3)	M9 medium contained 40 g/L glycerol; engineered strain consumed around 130 g/L glycerol/ fed-batch 150 L bioreactor, 30°C
Lycopene	2370	Glucose or ethanol	*S. cerevisiae* CEN. PK2–1D	Feeding solution contained 500 g/L glucose and 15 g/L yeast extract for the first stage fermentation, and ethanol was used for the second stage fermentation/Two-stage fed-batch 7 L bioreactor, 30°C
β-Carotene	3200	Glycerol	*E. coli* BL21(DE3)	Optimized medium contained 20 g/L glycerol; 400 g/L glycerol was fed at a rate of 3 g/L/h/fed-batch 5 L bioreactor, 34°C after IPTG induction
β-Carotene	2100	Glycerol	*E. coli* ATCC 8739	Synthetic medium contained 10 g/L glycerol; 500 g/L glycerol was fed at a rate of 20 mL/h/fed-batch 7 L bioreactor, 37°C
β-Carotene	4000	Glucose	*Y. lipolytica* MYA2613	Optimized medium with the C/N ratio at 3:1.5 for the first stage fermentation, and 600 g/L glucose was used for the second stage fermentation/Two-stage fed-batch 2 L bioreactor, 30°C
β-Carotene	6500	Glucose	*Y. lipolytica* Po1d	YPD medium contained 20 g/L yeast extract, 40 g/L peptone and 5 g/L glucose; additional glucose was added after 6 h at a rate of 6 g/h/ fed-batch 5 L bioreactor, 28°C
Astaxanthin	432.8 (7.12 mg/g DCW)	Glycerol	*E. coli* W3110	Modified medium contained 30 g/L glucose and 5 g/L yeast extract (for batch phase); glycerol concentration was maintained at 0–2 g/L for feeding; 0.5 mM IPTG was added when OD600 reached 30–40/fed-batch 5 L bioreactor, 30°C
Astaxanthin	217.9 (13.8 mg/g DCW)	Glucose	*S. cerevisiae* BY4742	YPD medium contained 20 g/L glucose; glucose feeding was controlled below 2 g/L and 30 g yeast extract was added every 12 h/ fed-batch 5 L bioreactor, 30°C

Astaxanthin	54.6 (3.5 mg/g DCW)	Glucose	*Y. lipolytica* GB20	YPD medium contained 80 g/L glucose/microtiter plate, 30°C
β-Alanine	32300	Glucose	*E. coli* W3110	Synthetic medium contained 20 g/L glucose and 9 g/L $(NH_4)_2SO_4$; 240 g/L glucose was consumed/fed-batch 6.6 L bioreactor, 37°C
GABA	4800	Glucose	*E. coli* BW25113	M9 medium contained 20 g/L glucose/flask, 37°C
GABA	39000	Glucose	*C. glutamicum* ATCC 13032	GP1 medium contained 100 g/L glucose and 50 μg/L biotin; 498 g glucose was consumed/fed-batch 5 L bioreactor, 30°C

Source: Yuan, SF., Alper, H.S. Metabolic engineering of microbial cell factories for production of nutraceuticals. Microb cell fact 18, 46 (2019). https://doi.org/10.1186/s12934–019–1096-y: © The Author(s) 2019: *Open AccessThis article is distributed under the terms of the Creative Commons Attribution 4.0 International License (http://creativecommons.org/licenses/by/4.0/), which permits unrestricted use, distribution, and reproduction in any medium, provided you give appropriate credit to the original author(s) and the source, provide a link to the Creative Commons license, and indicate if changes were made. The Creative Commons Public Domain Dedication waiver (http://creativecommons.org/publicdomain/zero/1.0/) applies to the data made available in this article, unless otherwise stated.*

3.3. FINAL WORDS

"If we knew what we were doing, it would not be called research, would it?"

— *Albert Einstein*

The aim is to get to zero carbon emissions according to the COP26 UN climate change summit. This has led to a shift towards sustainable and green resources and technologies, especially using microbial cells. This makes the entry of next-generation feedstocks (NGFs), C1 substrates (carbon dioxide, carbon monoxide, methane, methanol, and formate) and C2 substrates (predominantly ethanol and acetate). Zhang et al. (2022) reviewed the advantages of such NGF feedstocks and substrates, which are shown in Figure 3.17.

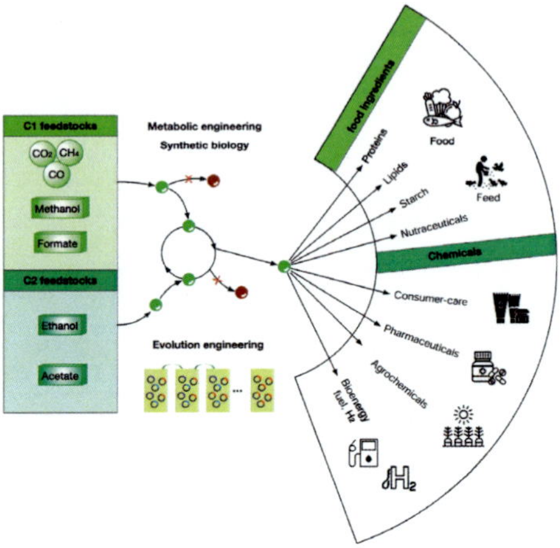

Feedstocks	Chemical formula	Water solubility [g/L][a]	Price ($/ton) [b]	Sources	Advantages and uniqueness	Disadvantages
Carbon dioxide	CO_2	1.69	0–80[c]	Earth's atmosphere; human activities, e.g., burning fossil fuels for electricity, heat, and transportation; Arctic permafrost thawing	-Naturally abundant and free and even get carbon credit by reducing CO_2 release -Non-toxic and non-flammable -Tremendous efforts, and technological breakthrough from academy and industry -Breakthrough in CO_2-fixing biotechnology in synthetic microbes -Well explored in gas fermentation using anaerobic acetogens (Liew et al., 2016)	-Most oxidized, and zero reducing power (**Figure 1B**), requiring large amount of reducing power supply and hydrogen source from water, methanol (Lebloas et al., 1996) or H_2 -Very low solubility in water that limits mass transfer and microbial productivity -Difficulty in storage and transportation
Carbon monoxide	CO	0.028	27–298[d]	Industrial waste and electrosynthesis of CO_2	-Can provide reducing equivalent in the Wood-Ljungdahl Pathway for CO_2 assimilation -A diverse group of bacteria and archaea, referred to as carboxydotrophs, can use CO as a primary carbon and energy source -Well explored in gas fermentation using anaerobic acetogens (Liew et al., 2016)	-Lower reducing power and requiring additional reducing power supply -Very low solubility in water that limits mass transfer and microbial productivity -Difficulty in storage and transportation -Toxic, flammable and explosive
Methane	CH_4	0.023	200–320[e]	Natural and shale gas; syngas and human activities e.g., landfills, agricultural activities, coal mining, wastewater treatment	-Naturally abundant and low price -Highest degree of reduction, energy intensive -Naturally used by methanotrophs -Can be used as sole feedstock to supply both carbon and energy	-Very low solubility in water that limits mass transfer and microbial productivity -Difficulty in storage and transportation -Flammable and explosive -Challenges in heterologous expression of methane monooxygenases (MMOs) (Jiang et al., 2021) -Challenges in engineering methanotrophs
Methanol	CH_3OH	Miscible	150–300[f]	Synthesis from natural gas, syngas and hydrogenation of CO_2	-A bulk chemical and relatively cheap -Higher degree of reduction and electron rich -Completely water miscible and higher mass transfer and supports higher microbial productivities -Easy transportation and storage	-Formation of the very toxic intermediate formaldehyde so that methanol concentration must be kept low (~5 g/L) -Low productivity by wildtype methylotrophs and engineered biotechnological microbes while used as sole feedstock. Currently, the shortest reported doubling time is 8.5 h in an evolved *E. coli* (Chen et al., 2020b) -Flammable -More research efforts are required for faster assimilation of methanol in microbes

(Continued on following page)

Feedstocks	Chemical formula	Water solubility [g/L][a]	Price ($/ton)[b]	Sources	Advantages and uniqueness	Disadvantages
					-Can be used as sole feedstock to supply both carbon and energy -Higher energetic efficiency as compared to H_2/CO_2 or CO when used by acetogens (Claassens et al., 2019)	-High fermentation cost required to neutralize the heat generated by methanol oxidation -High oxygen demand
Formic acid	HCOOH	972	450–500[g]	Electrochemical, photoreduction of CO_2, or hydrogenation of CO_2	-High solubility in water and other polar solvents, higher mass transfer and supports higher microbial productivities -Inflammable and higher degree of reduction than CO_2 and CO. -Easy transportation and storage -Higher energetic efficiency as compared to H_2/CO_2 or CO when used by acetogens (Claassens et al., 2019)	-Relatively higher price than methanol -Formation of the toxic intermediate formaldehyde when assimilated by microbes -Less studied as compared to methanol as microbial feedstock -More oxidized than methanol and thus less reducing power -Low productivity while used as main feedstock, doubling time is 65.9 h in a highly engineered *E. coli* (Bang et al., 2020) -More research efforts are required for faster assimilation of formate in microbes -Alkali is required to neutralize the acidity as in aerobic fermentation, weak acids are a powerful respiratory uncoupler so have to be used under strict carbon limiting conditions and relatively high pH to limit the amount of free formic acid
Ethanol	C_2H_5OH	Miscible	250–350[h]	fermentation from starch based raw materials and lignocellulose	-A bulk chemical and relatively cheap -Great advance in bio-ethanol technology -High solubility in water, higher mass transfer and supports higher microbial productivities -Easy assimilation by industrial workhorse microorganisms -Can be fed into bioreactor in pure form -Produce acetyl-CoA, a key precursor for several value-added bioproducts (e.g., lipids, terpenoids, polyketides, PHB) (Kiefer et al., 2020)	-Relatively more expensive than methanol -The technology of cellulosic ethanol should be further improved -Further boosting the productivity of microbes growing on ethanol -Metabolic engineering efforts are required to further boost the conversion yield of ethanol to high-value bioproducts -Flammable -High fermentation cost required to neutralize the heat generated by ethanol oxidation -High oxygen demand
Acetic acid	CH_3COOH	1,233	300–450[i]	Methanol carbonylation, sugar fermentation, depolymerization of lignocellulose and acetogen fermentation from C1 gas	-Natural product found in animal metabolism and food -Lower toxicity than C1 chemicals -A bulk chemical and with increasing global market, and bio-	-More expensive than methanol and ethanol -Technology of bio-acetate should be further improved

(Continued on following page)

Feedstocks	Chemical formula	Water solubility [g/L][a]	Price ($/ton)[b]	Sources	Advantages and uniqueness	Disadvantages
					acetic acid market is growing rapidly -High solubility in water and other polar solvents, higher mass transfer and supports higher microbial productivities -Easy assimilation by industrial workhorse microorganisms -Can be fed into bioreactor in pure form -Acetate is the direct precursor for acetyl-CoA that is used for the biosynthesis of numerous products (e.g., lipids, terpenoids, polyketides, PHB) (Kiefer et al., 2020)	-Further boosting the growth rate and productivity of microbes growing on acetate -Metabolic engineering efforts are required to further boost the conversion yield of acetate to high-value bioproducts -As a respiratory uncoupler, alkali is required to neutralize the acidity of acetate and minimise substrate toxicity -Central metabolism topology needs fine tuning to improve growth efficiency
Glucose	$C_6H_{12}O_6$	909	300–400	Hydrolysis of starch from corn, potato, wheat, and cassava	-Well established metabolic systems in microorganisms -Extensive knowledge on metabolism -High efficiency for microbial fermentation -Non-toxic and non-flammable and easier to transport as solids	-Competing with food source -Releasing high-amount CO_2 during fermentation process as compared to NGFs

[a]Solubility of next generation feedstocks in water at 1 atm pressure and 293 K (Kaye and Laby, 1986).
[b]The current prices for methane, methanol, ethanol and acetate are considerably higher than 1–2 years ago due to the global supply chain disruption, which is caused by COVID pandemic and political tensions, here we use the median price of pre-COVID period.
[c]The price is adjusted by factoring in carbon credit, https://www.reuters.com/business/cop/carbon-needs-cost-least-100tonne-now-reach-net-zero-by-2050-2021-10-25/
[d]Price is based on syngas, refer to Table 1, Biotechnol Biofuels. 2017; 10: 150.
[e]Gas lower heating value (lhv) is assumed for ship fuel based on 1 $/mmBTU (lhv)= 46.76 $/ton, https://www.dnv.com/maritime/insights/topics/lng-as-marine-fuel/current-price-development-oil-and-gas.html
[f]https://www.methanol.org/methanol-price-supply-demand/
[g]https://www.echemi.com/produce/pr2106011005-formic-acid-99-powder-saa6598-saa.html
[h]https://tradingeconomics.com/commodity/ethanol
[i]https://www.echemi.com/produca/pr2104271858-glacial-acetic-acid.html

Figure 3.17.

Source: Zhang C, Ottenheim C, Weingarten M & Ji L (2022) Microbial utilization of next-generation feedstocks for the biomanufacturing of value-added chemi-

Given that microbes can produce and use such feedstocks, the application of metabolic engineering becomes relevant here. To get a glimpse of how microbes and enzymes have assisted this goal, Zhang et al., (2022) reviewed select processes shown in Table 3.15.

Table 3.15

Feedstock	Bioproduct	Application	Microorganism/enzyme	Cultivation strategy	Titre (g/L)	Yield (g/g)[a]	References
CO_2 and H_2	Amylose	Food	Enzymes	Cell-free system	1.64	—[b]	Cai et al. (2021)
CO_2 and pyruvate	Acetate	Bulk chemicals	Acetobacterium woodii DSM 1030	Continuous gas fermentation	59.2	—[b]	Kantzow et al. (2015)
CO_2 and H_2	Ethanol	Bulk Chemical	*Clostridium ljungdahlii*	Continuous gas fermentation	10 g/L/d	—	Gaddy et al. (2007)
CO_2 and H_2	Acetone	Bulk Chemical	*Clostridium autoethanogenum*	Continuous gas fermentation	3 g/L/d	—	Liew et al. (2022)
CO_2 and H_2	Isopropanol	Bulk Chemical	*C.autoethanogenum*	Continuous gas fermentation	3 g/L/d	—	Liew et al. (2022)
Syngas	n-Butanol	Fuel	*C. ljungdahlii*	Batch	0.148	—[b]	Köpke et al. (2010)
Methane	Methanol	Bulk chemicals	*Methylosinus trichosporium*	Fed-batch	1.34	—[b]	Hur et al. (2017)
Methane	Astaxanthin	Nutraceuticals	*Methylomonas* sp.	Batch	2.4 mg/g DCW	—[b]	Ye et al. (2007)
Methane	α-bisabolene	Consumer-care	*Methylotuvimicrobium alcaliphilum*	Batch	24.55	—[b]	Nguyen et al. (2021)
Methanol	L-glutamate	Food	*Bacillus methanolicus*	Fed-batch	60	—[b]	Heggeset et al. (2012)
Methanol	α-humulene	Consumer-care	*M. extorquens*	Fed-batch	1.65	0.03	Sonntag et al. (2015)
Methanol	Cadaverine	Precursor to polymers	*Bacillus methanolicus*	Fed-batch	11.3	—[b]	Nærdal et al. (2015)
Methanol	PHB	Biopolymers	*M. extorquens*	Fed-batch	52.9	0.12	Bourque et al. (1995)
Methanol	Pyruvate	Precursor to food and pharmaceuticals	*S. ceravisiae*	Batch	0.26	0.25	Dai et al. (2017)
Methanol and glycerol	Lovastatin	Pharmaceuticals	*P. pastoris*	Fed-batch	250.8	—[b]	Liu et al. (2018)
Acetate	Mevalonic acid	Precursor for pharmaceuticals/nutraceuticals	*E. coli*	Fed-batch	7.85	0.27	Xu et al. (2017b)
Acetate	β-Caryophyllene	Consumer-care	*E. coli*	Fed-batch	1.05	0.02	Yang and Nie, (2016)
Acetate	MNEI protein	Food	*E. coli*	Fed-batch	0.18	0.02	Leone et al. (2015)
Acetate	Lipids	Food or feed	*R. toruloides*	Batch	2.1	0.11	Huang et al. (2016)
Acetate	Lipids	Food or feed	*Y. lipolytica*	semicontinuous	115	0.16	Xu et al. (2017a)
Acetate	PHB	Biopolymers	*C. necator*	Fed-batch	43	—[b]	Garcia-Gonzalez and De Wever, (2018)
Acetate	PHB	Biopolymers	*E. coli*	Batch	1.27	0.25	Chen et al. (2018)
Acetate	Acetone	Bulk chemical	*E. coli*	Fed-batch	6.57	0.29	Yang et al. (2019)
Ethanol	PHB	Biopolymers	*E. coli*	Fed-batch	35.67	0.27	Sun et al. (2020)
Ethanol	PHB	Biopolymers	*E. coli*	Batch	1.1	0.11	Liang et al. (2021)
Ethanol and glucose	Amorphadiene	Drug precursor	*S. cerevisiae*	Fed-batch	40	—[b]	Westfall et al. (2012)

[a]*Here, yield refers to mass of product per mass of substrate (g/g).*
[b]*Not determined or no data available.*

Source: Zhang C, Ottenheim C, Weingarten M & Ji L (2022) Microbial Utilization of Next-Generation Feedstocks for the Biomanufacturing of Value-Added Chemicals and Food Ingredients. Front. Bioeng. Biotechnol. 10:874612. doi: 10.3389/fbioe.2022.874612:

A slew of approaches has been tested on other "non-conventional organisms" for producing biofuels like ethanol, paving the way for further tweaking of pathways to address fuel needs. Table 3.16 summarizes a few reports of how some microbes have been engineered.

Table 3.16

Non-conventional yeast species	Glucose	Ethanol	Temperature (°C)	Salt	Acetic acid	Available gene manipulation tools	Reference
Kluyveromyces marxianus			52			NHEJ mediated Transformation Complete genome is available	Hoshida et al. (2014)
Zygosaccharomyces bailii	60% (W/V)				24 g/L	Homologous recombination Complete genome is available	Branduardi et al. (2014)
Dekkera bruxellensis		10–16%				Well-established transformation protocols Complete genome is available	Miklenic et al. (2013)
Ogataea polymorpha			50			Recombinant vectors are available with promoters	Saraya et al. (2012)
Hansenula polymorpha		9.8 g/L (xylose as carbon source)	50			Tightly regulated native promoters Complete genome is available	Wagner and Alper (2016); Kurylenko et al. (2014)
Kluyveromyces lactis		4 mol of ethanol from 1 mol of lactose				Tightly regulated native promoters More similar to *S. cerevisiae* Complete genome is available	Wagner and Alper (2016); Gonzalez-Siso et al. (2015)
Pichia pastoris						Tightly regulated native promoters Complete genome is available	Wagner and Alper (2016)
Yarrowia lipolytica						Strong constitutive hybrid promoters	Wagner and Alper (2016)
Zygosaccharomyces rouxii	90% (W/V)			3M		Homologous recombination Modified plasmids	Leandro et al. (2011)

Source: Madhavan A, Jose AA, Binod P, Sindhu R, Sukumaran RK, Pandey A and Castro GE (2017) Synthetic Biology and Metabolic Engineering Approaches and Its Impact on Non-Conventional Yeast and Biofuel Production. Front. Energy Res. 5:8. doi: 10.3389/fenrg.2017.00008:

As an attractive field of research, the application of cell factories to produce molecules of interest is immense. Given the current scenario of overexploited resources and endangered sources of molecules for human life and health, the concept warrants the need for transparent research and process optimization.

Figure 3.18 shows the potential of cell factories for producing plant natural products (PNPs).

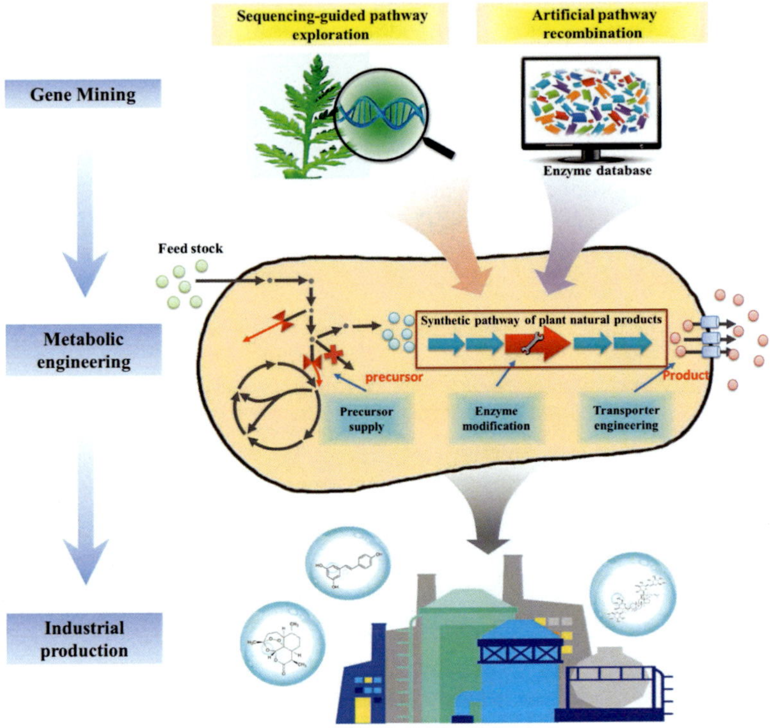

Figure 3.18

Source: Liu X, Ding W, Jiang H. Engineering microbial cell factories for the production of plant natural products: from design principles to industrial-scale production. Microb Cell Fact. 2017;16(1):125. Published 2017 Jul 19. doi:10.1186/s12934–017–0732–7:

"Research is seeing what everybody else has seen and thinking what nobody else has thought."

— Albert Szent-Györgyi

REFERENCES

1. Chen, L., Zhang, J., & Chen, W. N. (2014). Engineering the *Saccharomyces cerevisiae* β-oxidation pathway to increase medium chain fatty acid production as potential biofuel. *PLoS ONE, 9*(1), Article e84853. https://doi.org/10.1371/journal.pone.0084853.
2. Ji, R.-Y., Ding, Y., Shi, T.-Q., Lin, L., Huang, H., Gao, Z., & Ji, X.-J. (2018). Metabolic engineering of yeast for the production of 3-hydroxypropionic acid. *Frontiers in Microbiology, 9*, Article 2185. https://doi.org/10.3389/fmicb.2018.02185.
3. Jiang, W., Li, C., Li, Y., & Peng, H. (2022). Metabolic engineering strategies for improved lipid production and cellular physiological responses in yeast Saccharomyces cerevisiae. *Journal of Fungi, 8*(5), 427. https://doi.org/10.3390/jof8050427.
4. Liu, S., Sun, Y., Wei, T., Gong, D., Wang, Q., Zhan, Z., & Song, J. (2023). Engineering 3-hydroxypropionic acid production from glucose in *Yarrowia lipolytica* through Malonyl-CoA pathway. *Journal of Fungi, 9*(5), 573. https://doi.org/10.3390/jof9050573.
5. Liu, X., Ding, W., & Jiang, H. (2017). Engineering microbial cell factories for the production of plant natural products: From design principles to industrial-scale production. *Microbial Cell Factories, 16*(1), Article 125. https://doi.org/10.1186/s12934-017-0732-7.
6. Madhavan, A., Jose, A. A., Binod, P., Sindhu, R., Sukumaran, R. K., Pandey, A., & Castro, G. E. (2017). Synthetic biology and metabolic engineering approaches and its impact on non-conventional yeast and biofuel production. *Frontiers in Energy Research, 5*, Article 8. https://doi.org/10.3389/fenrg.2017.00008.
7. Nawab, S., Wang, N., Ma, X., et al. (2020). Genetic engineering of non-native hosts for 1-butanol production and its challenges: A review. *Microbial Cell Factories, 19*, Article 79. https://doi.org/10.1186/s12934-020-01337-w.
8. Puri, V., Nagpal, M., Singh, I., Singh, M., Dhingra, G. A., Huanbutta, K., Dheer, D., & Sharma, A., et al. (2022). A comprehensive review on nutraceuticals: Therapy support and formulation challenges. *Nutrients, 14*(21), Article 4637. https://doi.org/10.3390/nu14214637.
9. Shi, A., Liu, Y., Jia, B., Zheng, G., & Yao, Y. (2023). Metabolic engineering of microorganisms to produce L-aspartate and its derivatives. *Fermentation, 9*(8), 737.
10. Tao, Y. M., Bu, C. Y., Zou, L. H., Hu, Y. L., Zheng, Z. J., & Ouyang,

J. (2021). A comprehensive review on microbial production of 1,2-propanediol: Micro-organisms, metabolic pathways, and metabolic engineering. *Biotechnology for Biofuels, 14*(1), Article 216. https://doi.org/10.1186/s13068-021-02067-w.

11. Topaloğlu, A., Esen, Ö., Turanlı-Yıldız, B., Arslan, M., & Çakar, Z. P. (2023). From *Saccharomyces cerevisiae* to ethanol: Unlocking the power of evolutionary engineering in metabolic engineering applications. *Journal of Fungi, 9*(10), 984. https://doi.org/10.3390/jof9100984.
12. Wang, J., & Zhang, X. (2023). The big food view and human health from the prospect of bio-manufacturing and future food. *Frontiers in Nutrition, 10*, Article 1160743. https://doi.org/10.3389/fnut.2023.1160743.
13. Yuan, S. F., & Alper, H. S. (2019). Metabolic engineering of microbial cell factories for production of nutraceuticals. *Microbial Cell Factories, 18*, Article 46. https://doi.org/10.1186/s12934-019-1096-y.
14. Zhang, C., Ottenheim, C., Weingarten, M., & Ji, L. (2022). Microbial utilization of next-generation feedstocks for the biomanufacturing of value-added chemicals and food ingredients. *Frontiers in Bioengineering and Biotechnology, 10*, Article 874612. https://doi.org/10.3389/fbioe.2022.874612.

Chapter 4

MACHINE LEARNING FOR PREDICTING METABOLIC PATHWAYS

Shiv Sanjeevi Sripathi

"Critical investments in infrastructure are bearing fruit and, as described below, synthetic biology is increasingly becoming, at least part of, the solution to many of our present and future needs in medicine, food and energy production, remediation, manufacturing, and national security."

—— El Karoui M, Hoyos-Flight M and Fletcher L (2019)
Future Trends in Synthetic Biology—A Report.
Front. Bioeng. Biotechnol. 7:175. doi: 10.3389/fbioe.2019.00175

4.1. INTRODUCTION

This chapter briefly summarizes the role of machine learning (ML) in predicting pathways given the growing use of artificial intelligence (AI) in all spheres. The development of metabolic engineering and synthetic biology is also fueled by data generated from OMICS tools. This has set the stage for ML to be applied in this arena to augment the results from traditional approaches as depicted in Figure 4.1.

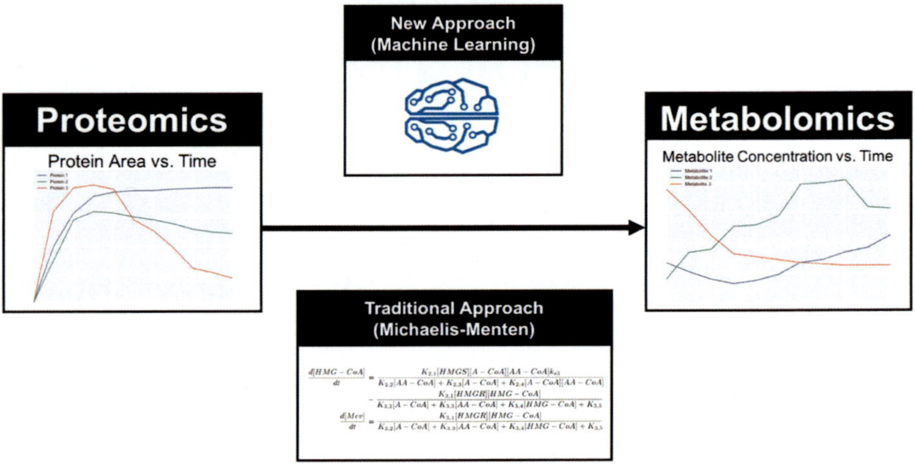

Figure 4.1

Source: Costello, Z., Martin, H.G. A machine learning approach to predict metabolic pathway dynamics from time-series multiomics data. npj Syst Biol Appl 4, 19 (2018). https://doi.org/10.1038/s41540–018–0054–3:

The process of building a biological pathway entails mining information based on a data-driven objective (DDO) or a knowledge-driven objective (KDO). In the former, a set of genes and/or proteins is obtained by high-throughput experiments such as microarray and mass spectrometry (MS), and in the latter, a broad topic of interest is selected, and the relevant knowledge is mined from resources like literature and knowledge bases. The assembled information is then put together (Step 2) using pathway building tools to create a pathway, followed by iterations of feedback from domain experts (Step 3) and refinement (Step 4) (Viswanathan et al., 2008). Figure 4.2 summarizes these steps.

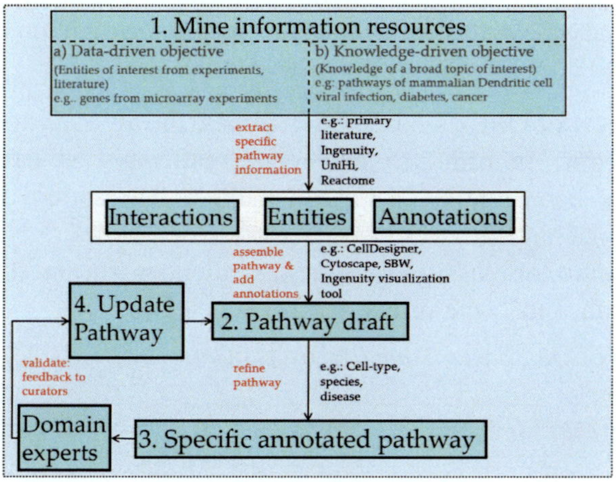

Figure 4.2

Source: Viswanathan GA, Seto J, Patil S, Nudelman G, Sealfon SC. Getting started in biological pathway construction and analysis. PLoS Comput biol. 2008;4(2): e16. doi: 10.1371/journal.pcbi.0040016: © 2008 Viswanathan et al. This is an open-access article distributed under the terms of the Creative Commons Attribution License, which permits unrestricted use, distribution, and reproduction in any medium, provided the original author and source are properly credited.

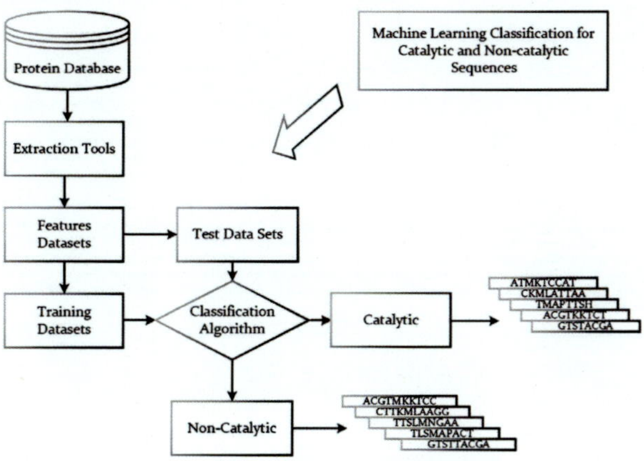

Figure 4.3

Source: Shah HA, Liu J, Yang Z & Feng J (2021) Review of Machine Learning Methods for the Prediction and Reconstruction of Metabolic Pathways. Front. mol.

Given the slew of OMICs-based approaches, the amount of data generated becomes enormous. In order to categorize pathways as catalytic, several approaches have been employed. These include the random forest method, support vector machine (SVM), and K-nearest neighbors (KNN). Figure 4.3 represents a scheme for classifying protein sequences as catalytic ones (Tables 4.1 and 4.2) (Shah, Liu, Yang & Feng, 2021).

Table 4.1. *A List of Databases Compiled by Viswanathan et al., (2008)*

Database		Description
Protein–Protein Interaction Databases: Organize experimental and/or in silico interactions	BIND	200,000 documented biomolecular interactions and complexes
	MINT	Experimentally verified interactions
	HPRD	Elegant and comprehensive presentation of the interactions, entities, and evidences
	MPact	Yeast interactions. A part of MIPS
	DIP	Experimentally determined interactions
	IntAct	Database and analysis system of binary and multiprotein interactions
	PDZBase	PDZ Domain containing proteins
	GNPV	Based on specific experiments and literature
	BioGrid	Physical and genetic interactions
	UniHi	Comprehensive human protein interactions
	OPHID	Combines PPI from BIND, HPRD, and MINT
Metabolic Pathways Databases: Compendium of pathways describing metabolic and physical processes (Primary source for metabolic information initiated by Stanford Research Initiative)	EcoCyc	Entire genome and biochemical machinery of *E. coli*
	MetaCyc	Pathways of more than 165 species
	HumanCyc	Human metabolic pathways and the human genome
	BioCyc	Collection of databases for several organism
Signaling Pathways Databases: Pathways pertaining to signal transduction	KEGG	Comprehensive. Links to several useful databases
	PANTHER	Compendium of pathways built using CellDesigner
	Reactome	Hierarchical layout. Extensive links to relevant databases
	Biomodels	Domain experts curated pathways and associated mathematical models
	STKE	Repository of canonical pathways
	Ingenuity Systems	Commercial mammalian biological knowledgebase
	PID	Compendium of several assembled signaling pathways
	BioPP	Repository of biological pathways built using CellDesigner

Most databases have a graphics viewer for displaying entities and interactions. Refer to Table S1 for a more detailed description and URLs of these databases.
BIND, Biomolecular Interaction Network Database; BioPP, Biological Pathway Publisher; DIP, Database of Interacting Proteins; EcoCyc, Encyclopaedia of *E. coli* Genes and Metabolism; GNPV, Genome Network Platform Viewer; HPRD, Human Protein Reference Database; KEGG, Kyoto Encyclopedia of Genes and Genomes; MetaCyc, a Metabolic Pathway database; MINT, Molecular INTeration database; MIPS, Munich Information center for Protein Sequences; OPHID, Online Predicted Human Interaction Database; PANTHER, Protein Analysis through Evolutionary Relationship database; PID, The Pathway Interaction Database; STKE, Signal Transduction Knowledge Environment; UNIHI, Unified Human Interactome.
doi:10.1371/journal.pcbi.0040016.t001

Table 4.2

Database	Curation type	GO Annotation (Y/N)	Description	URL
1. Protein-Protein Interactions Databases				
BIND	M	N	200,000 documented biomolecular interactions and complexes	http://www.bind.ca/
MINT	M	N	Experimentally verified interactions	http://mint.bio.uniroma2.it/mint/
HPRD	M	N	Elegant and comprehensive presentation of the interactions, entities and evidences	http://www.hprd.org/
MPact	B	N	Yeast interactions. A part of MIPS	http://mips.gsf.de/genre/proj/mpact/
DIP	B	Y	Experimentally determined interactions	http://dip.doe-mci.ucla.edu/
IntAct	M	Y	Database and analysis system of binary and multi-protein interactions	http://www.ebi.ac.uk/intact
PDZBase	M	N	PDZ Domain containing proteins	http://icb.med.cornell.edu/services/pdz/start/
GNPV	B	Y	Based on specific experiments and literature	http://genomenetwork/nig.ac.jp/
BioGrid	M	Y	Physical and genetic interactions	http://thebiogrid.org/
UniHi	B	Y	Comprehensive human protein interactions	http://theoderich.fb3.mdc-berlin.de:8080/unihi/
OPHID	B	Y	Combines PPI from BIND, HPRD, and MINT	http://ophid.utoronto.ca/ophid/
2. Metabolic Pathway Databases				

EcoCyc	B	Y	Entire genome and biochemical machinery of *E. coli*	http://ecocyc.org
MetaCyc	M	N	Pathways of over 165 species	http://metacyc.org
HumanCyc	B	N	Human metabolic pathways and the human genome	http://humancyc.org
BioCyc	B	N	Collection of databases for several organism	http://biocyc.org
3. Signaling Pathway Databases				
KEGG	M	Y	Comprehensive collection of pathways such as human disease, signaling, genetic information processing pathways. Links to several useful databases.	http://www.genome.ad.jp/kegg/
PANTHER	M	N	Compendium of metabolic and signaling pathways built using Cell Designer. Pathways can be downloaded in SBML format.	http://panther.appliedbiosystems.com/
Reactome	M	Y	Hierarchical layout. Extensive links to relevant databases such as NCBI, ENSEMBL, UNIPROT, HAPMAP, KEGG, CHEBI, PubMed, GO. Follows PSI-MI standards.	http://www.reactome.org/
Biomodels	M	Y	Domain experts curated biological connection maps and associated mathematical models	http://www.ebi.ac.uk/biomodels/
STKE	M	N	Repository of canonical pathways.	http://stke.sciencemag.org/cm/
Ingenuity Systems	M	Y	Commercial mammalian biological knowledgebase about genes, drugs, chemical, cellular and disease processes, and signaling and metabolic pathways.	http://ingenuity.com/

PID	M	Y	Compendium of several highly structured, assembled signaling pathways	http://pid.nic.nih.gov/PID/
BioPP	B	Y	Repository of biological pathways built using Cell Designer	http://tsb.mssm.edu/ pathwayPublisher/boradcast/

Key: M – Manual curation, A – Automated curation, B- Both manual and automated curation; Y – Yes, N – No; BIND – Biomolecular Interaction Network Database, DIP – Database of Interacting Proteins, GNPV – Genome Network Platform Viewer, HPRD = Human Protein Reference Database, MINT – Molecular INTeration database, MIPS – Munich Information Center for Protein Sequences, UNIHI – Unified Human Interactome, OPHID – Online Predicted Human Interaction Database, EcoCyc – Encyclopaedia of E. coli Genes and Metabolism, MetaCyc – aMetabolic Pathway database, KEGG – Kyoto Encyclopedia of Genes and Genomes, PANTHER – Protein Analysis Through Evolutionary Relationship database, STKE – Signal Transduction Knowledge Environment, PID – The Pathway Interaction Database, BioPP – Biological Pathway Publisher.

Source: Viswanathan GA, Seto J, Patil S, Nudelman G, Sealfon SC. Getting started in biological pathway construction and analysis. PLoS Comput Biol. 2008;4(2): e16. doi: 10.1371/journal.pcbi.0040016:

The conventional approach for identifying a metabolite entails the comparison of its mass spectrometry (MS) data with references. However, given the incomplete data present in reference databases, ML approaches enter the arena to fill the gaps. This can entail the use of programs to search the MS fingerprint or MS spectrum with the closest match to the compound/metabolite being studied. A scheme made by Shah et al., (2021) depicting the identification of metabolites has been presented in Figure 4.4.

The prediction of reactions becomes challenging given the complexity and resonance of molecular structures. This is where ML-based methods can assist in predicting reactions. This can entail predicting precursors or products for a compound (compound-based approach) or predicting whether a pair of compounds can react (compound pair method) (Shah, Liu, Yang & Feng, 2021).

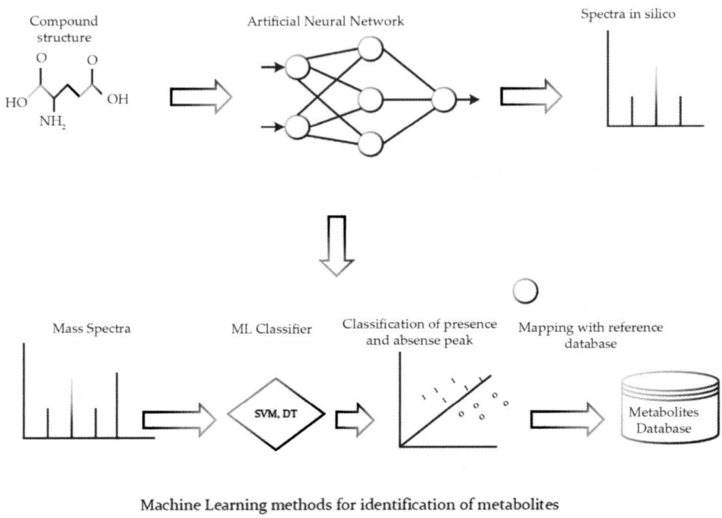

Figure 4.4

Source: Shah HA, Liu J, Yang Z & Feng J (2021) Review of Machine Learning Methods for the Prediction and Reconstruction of Metabolic Pathways. Front. Mol. Biosci. 8:634141. doi: 10.3389/fmolb.2021.634141:

4.2. CASE STUDY 1

A 2011-published study by Grimaldi et al., reported the construction of "a novel inference method called Reverse Engineering Gene Networks with Artificial Neural Networks (RegnANN)." It entails Multilayer Perceptrons (MLPs), which builds a multi-variable regressor for each gene once it is trained by microarray data. This approach as illustrated by the researchers is produced in Figure 4.5.

Subsequent evaluation of the accuracy and comparison with three reference algorithms, ARACNE, CLR, and KELLER (that are employed in research) revealed favorable outcomes. The performances in terms of the AUC (the area under the curve) and the Matthews Correlation Coefficient (MCC). Select AUC curves of the developed tool vs. those of the references have been presented in Figure 4.6.

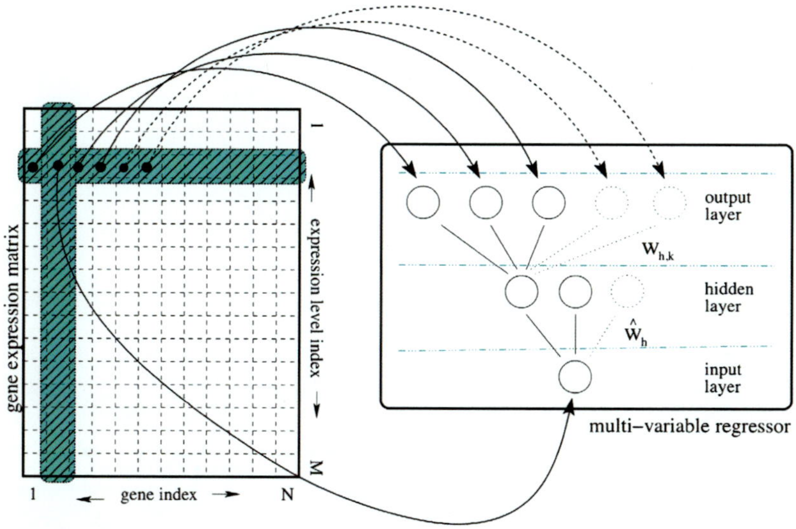

Figure 4.5

Source: Grimaldi M, Visintainer R, Jurman G. RegnANN: Reverse engineering gene networks using artificia□neura□networks. *plos one.* 2011;6(12): e28646. doi: 10.1371/journal.pone.0028646: *This is an open-access article distributed under the terms of the Creative Commons Attribution License, which permits unrestricted use, distribution, and reproduction in any medium, provided the original author and source are properly credited.*

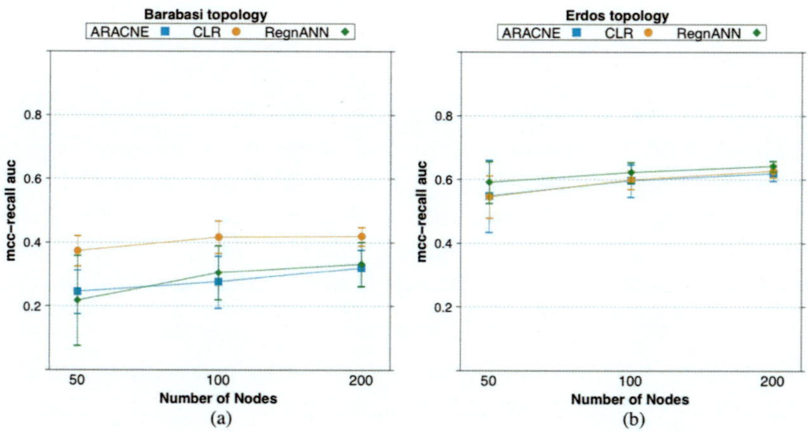

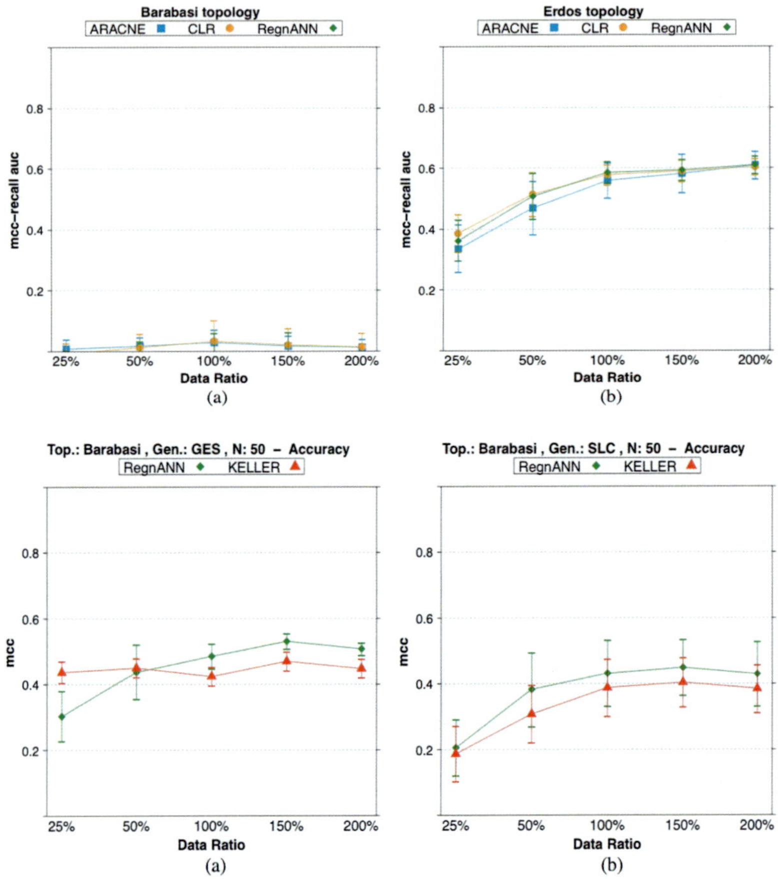

Figure 4.6

Source: Grimaldi M, Visintainer R, Jurman G. Regn ANN: Reverse Engineering Gene Networks using Artificia□Neura□Networks. *PLoS One.* 2011;6(12): e28646. doi: 10.1371/journal.pone.0028646*: This is an open-access article distributed under the terms of the Creative Commons Attribution License, which permits unrestricted use, distribution, and reproduction in any medium, provided the original author and source are properly credited.*

Overall, Regn ANN could discover second-order correlations and showed robust agreement with synthetic and biological data.

Chen et al. (2010) constructed a computational approach to predict the network of substrate-enzyme-product triads. The reference benchmark included 4,592 networking triads and 729,600 non-networking triads (the number of negative triads was about 50 times that of the positive triads). The curves as charted by the researchers are given in Figure 4.7.

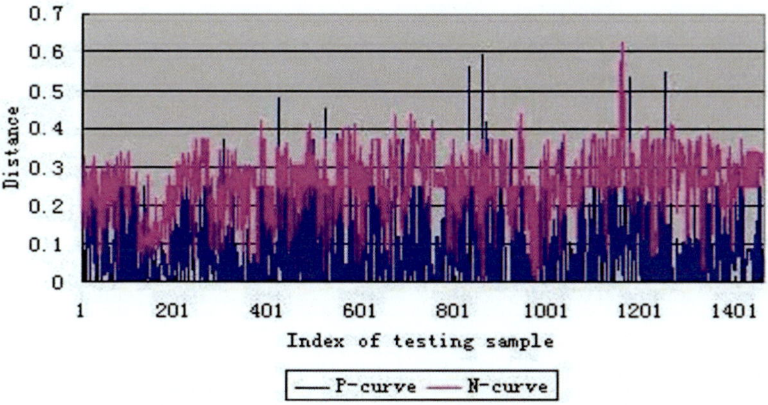

Interval	Frequency		Interval	Frequency	
	Positive triads	Negative triads		Positive triads	Negative triads
0.00~0.05	750	16	0.05~0.10	208	50
0.10~0.15	146	108	0.15~0.20	95	124
0.20~0.25	51	153	0.25~0.30	186	556
0.30~0.35	5	254	0.35~0.40	12	181
0.40~0.45	1	14	0.45~0.50	2	2
0.50~0.55	1	0	0.55~0.60	3	1
0.60~0.65	0	1			

Figure 4.7

Source: Chen L, Feng KY, Cai YD, Chou KC, Li HP. Predicting the network of substrate-enzyme-product triads by combining compound similarity and functional domain composition. BMC Bioinformatics. 2010; 11:293. Published 2010 May 31. doi:10.1186/1471–2105–11–293:

The overall success rate was 98.71%, with a success rate of 95.41% for the 1,460 testing networking triads and 98.77% for the 72,960 testing non-networking triads. Although there were 67 misclassified triad samples, 70.15% of them had a distance of less than 0.1, i.e., close to the margin. The researchers recommended using the software to analyze substrate-enzyme-product networks. The distances of these misclassified triads, as reported by the authors, are presented in Table 4.3.

Table 4.3

Substrates	Enzymes	Products	Distance		Differences
			Positive Triads	Negative Triads	
C00002	YIL139C	C06397	0.24	0.19125	0.04875
C00002	YPL271W	C00008	0.25	0.22125	0.02875
C00002	YPR033C	C00020	0.1	0.03375	0.06625
C00003	YKR066C	C00004	0.25	0.225	0.025
C00003	YPR167C	C00004	0.25	0.177831	0.072169
C00010	YER090W	C00024	0.25	0.1125	0.1375
C00010	YER178W	C00024	0.189188	0.1425	0.046688
C00024	YAL054C	C00033	0.21	0.199626	0.010374
C00024	YCL030C	C06548	0.25	0.0975	0.1525
C00024	YLR153C	C00033	0.21	0.199626	0.010374
C00025	YHR037W	C03912	0.375	0.202643	0.172357
C00026	YIR034C	C00449	0.271688	0.25	0.021688
C00035	YGL047W	C00096	0.1875	0.165	0.0225
C00037	YOL049W	C00051	0.48375	0.25	0.23375
C00047	YPL096W	C12989	0.25	0.225	0.025
C00055	YBL013W	C04121	0.177831	0.12375	0.054081

C00055	YDR410C	C04121	0.25	0.22125	0.02875
C00055	YKR069W	C04121	0.25	0.19125	0.05875
C00065	YBR263W	C00143	0.375	0.25	0.125
C00065	YLR058C	C00143	0.375	0.25	0.125
C00083	YPL231W	C12647	0.073223	0.02625	0.046973
C00085	YKL104C	C00352	0.25	0.2325	0.0175
C00086	YIR029W	C00499	0.4525	0.375	0.0775
C00096	YBR252W	C00144	0.25	0.12	0.13
C00096	YGR036C	C00636	0.25	0.24375	0.00625
C00108	YDR354W	C04302	0.375	0.2325	0.1425
C00109	YCL018W	C06032	0.383376	0.36625	0.017126
C00118	YGL026C	C03506	0.375	0.37375	0.00125
C00143	YGL125W	C00440	0.3025	0.28375	0.01875
C00155	YNL256W	C01118	0.25	0.22125	0.02875
C00167	YJR131W	C00191	0.25	0.21	0.04
C00191	YOR065W	C05787	0.25	0.19875	0.05125
C00223	YDR062W	C12096	0.0825	0.082244	0.000256
C00223	YMR296C	C12096	0.0825	0.04875	0.03375
C00234	YDR408C	C04376	0.36625	0.32125	0.045

C00333	YJR153W	C00470	0.375	0.12375	0.25125
C00448	YDL205C	C16144	0.225	0.19125	0.03375
C00582	YHL003C	C05598	0.25	0.1875	0.0625
C00582	YKL008C	C05598	0.25	0.1875	0.0625
C00632	YDR120C	C05831	0.25	0.15	0.1
C00652	YML086C	C06316	0.565	0.36625	0.19875
C00842	YDR127W	C06017	0.1125	0.09	0.0225
C00864	YDR531W	C03492	0.25125	0.25	0.00125
C00931	YDL205C	C01024	0.59625	0.375	0.22125
C01063	YBL015W	C09813	0.1275	0.1125	0.015
C01079	YDR044W	C03263	0.41875	0.25	0.16875
C01096	YCL030C	C02888	0.25	0.22875	0.02125
C01100	YIL116W	C01267	0.375	0.25	0.125
C01902	YML008C	C08830	0.375	0.25	0.125
C02411	YGR155W	C03058	0.09	0.075	0.015
C02909	YHR007C	C14098	0.25	0.195	0.055
C03012	YDR402C	C11713	0.36375	0.2575	0.10625
C03598	YPR167C	C04297	0.25	0.1875	0.0625

C04751	YAR015W	C04823	0.34	0.32875	0.01125
C04874	YDR452W	C05925	0.16875	0.125	0.04375
C06102	YLR231C	C06105	0.535	0.25	0.285
C06397	YBR029C	C07838	0.18375	0.17625	0.0075
C06599	YNL202W	C06600	0.147938	0.113376	0.034562
C06714	YDR127W	C06723	0.0975	0.08625	0.01125
C07649	YDR402C	C12673	0.55	0.3625	0.1875
C07732	YGR234W	C07733	0.3075	0.25	0.0575
C09811	YGL063W	C09812	0.1125	0.10125	0.01125
C11907	YPR118W	C11908	0.25	0.22875	0.02125
C11923	YFR015C	C12384	0.25	0.03	0.22
C11923	YLR258W	C12384	0.25	0.03	0.22
C14082	YHR007C	C14089	0.25	0.195	0.055
C15786	YGR060W	C15797	0.09375	0.08625	0.0075

Source: Chen L, Feng KY, Cai YD, Chou KC, Li HP. Predicting the network of substrate-enzyme-product triads by combining compound similarity and functional domain composition. BMC Bioinformatics. 2010; 11:293. Published 2010 May 31. doi:10.1186/1471–2105–11–293:

4.3. CASE STUDY 2

In another study, Nagao et al. (2014) reported the use of random forests to annotate the functions of enzymes using a tool called the enzyme function prediction method by random forests (EFPrf). The developed EFPrf had a set of binary predictors for enzymes in each CATH superfamily where the random forests-derived specificity determining residues (SDRs) (rf-SDRs) were experimentally predicted residue positions/specificities. The overview of the approach as depicted by the authors is presented in Figures 4.8 and 4.9.

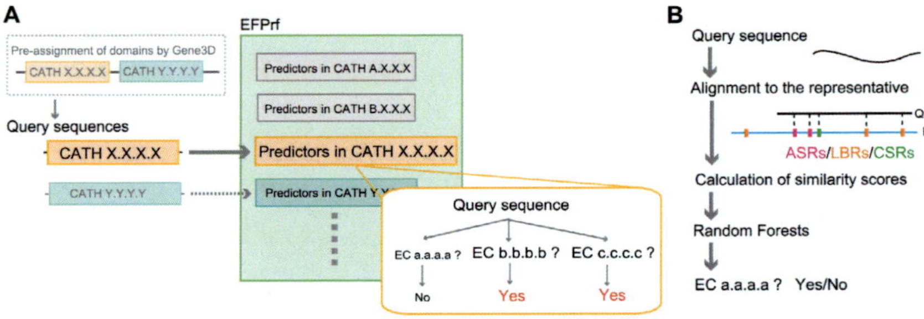

Figure 4.8

Source: Nagao C, Nagano N, Mizuguchi K. Prediction of detailed enzyme functions and identification of specificity determining residues by random forests. PLoS One. 2014;9(1): e84623. Published 2014 Jan 8. doi: 10.1371/journal.pone.0084623:

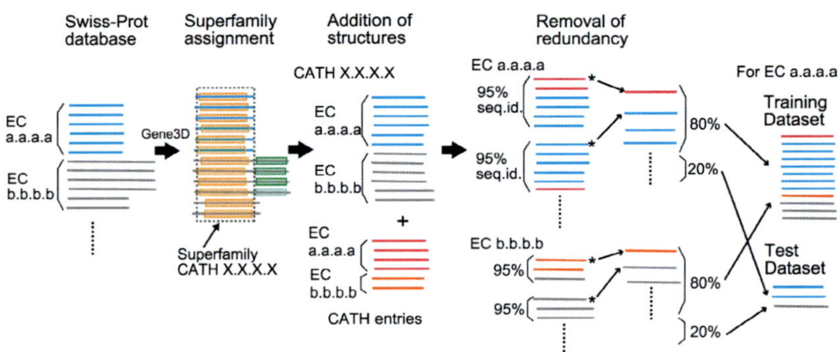

Figure 4.9

Source: Nagao C, Nagano N, Mizuguchi K. Prediction of detailed enzyme functions and identification of specificity determining residues by random forests. PLoS

One. 2014;9(1):e84623. Published 2014 Jan 8. doi: 10.1371/journal.pone.0084623:

Subsequent cross-validated benchmark assessment revealed the approach to have a precision of 0.98 and a recall of 0.89.

The utility of protein-protein interaction (PPI) studies becomes relevant in analyzing metabolic pathway prediction, phylogeny, and sequence analysis, especially for non-model organisms where there is a paucity of data. In this context, Kludas et al. (2016) used multiple kernel learning (MKL), pairwise kernels, and kernelized structured output prediction to predict PPIs in *S. cerevisiae*. The Receiver Operating Characteristic curves (ROC) with associated area-under-curve (AUC) statistics and precision-recall (PR) curves for PPI prediction of the baker's yeast as obtained by the authors are given in Figure 4.10.

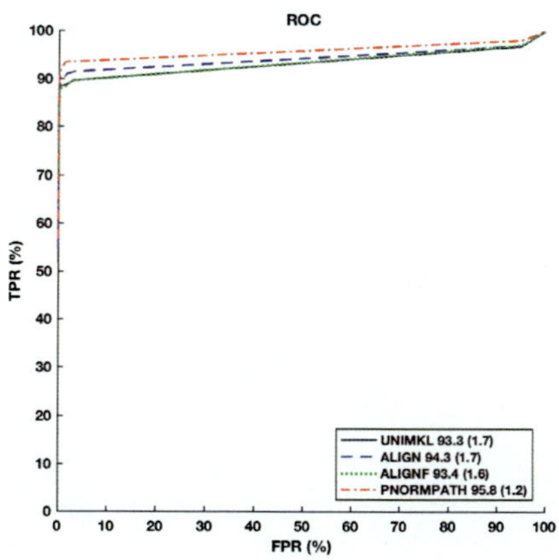

Figure 4.10

Source: Kludas J, Arvas M, Castillo S, Pakula T, Oja M, Brouard C, et al. (2016) Machine learning of protein interactions in fungal secretory Pathways. Plos one 11(7): e0159302. https://doi.org/10.1371/journal.pone.0159302:

*Subsequent application of this training (*the training data from *S. cerevisiae* was applied to the target species *T. reesei) in the related T. reesei* revealed unique pathways. The predicted secretory (242) and unknown (14) proteins have been shown in Figure 4.11 by the authors- thick edges are negative (red) and positive (green); Pink nodes have no interactions in STRING. B) Pie chart of the functional aspects of the *T. reesei* secretion network.

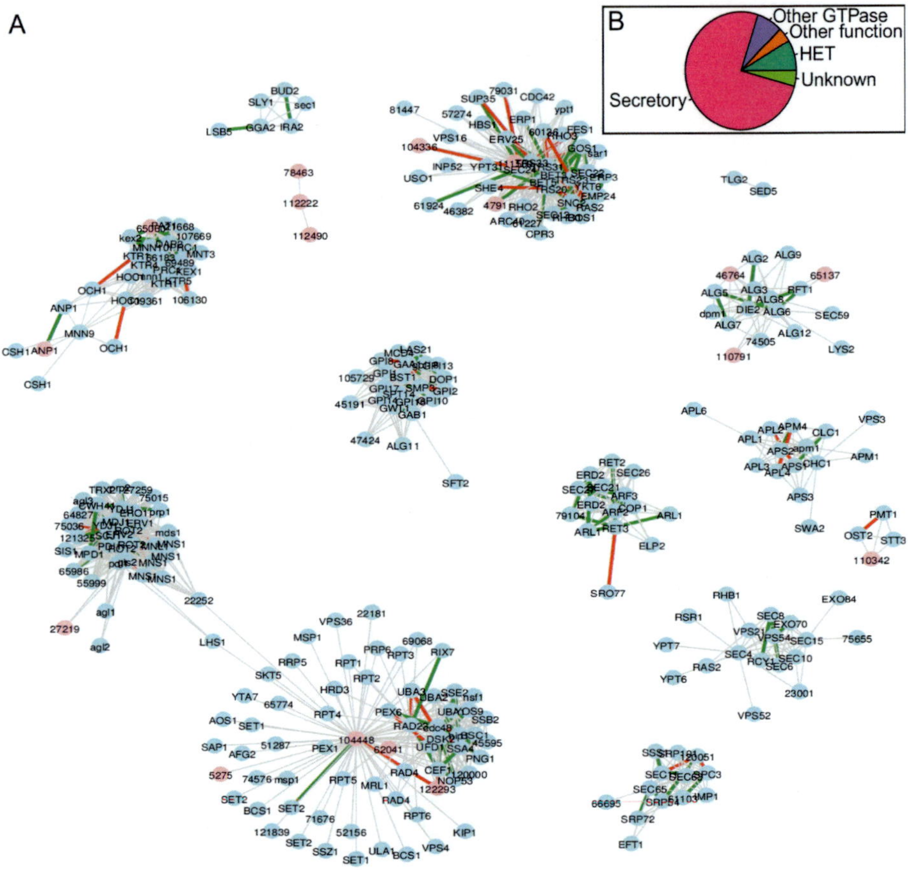

Figure 4.11

Source: Kludas J, Arvas M, Castillo S, Pakula T, Oja M, Brouard C, et al. (2016) Machine learning of protein interactions in fungal secretory pathways. Plos ONE 11(7): e0159302. https://doi.org/10.1371/journal.pone.0159302: © 2016 Kludas et al. This is an open access article distributed under the terms of the Creative Commons Attribution License, which permits unrestricted use, distribution, and reproduction in any medium, provided the original author and source are credited.

Costello and Martin (2018) proposed an ML approach as an alternative to traditional kinetic modeling. The approach entailed the use of training data to compute the changes in protein and metabolite levels. The association between time-series proteomics and metabolomics data and the metabolite change is derived from a supervised ML algorithm to simulate the dynamics of a specific strain for use in industrial production. The approach as proposed by the researchers is presented in Figure 4.12.

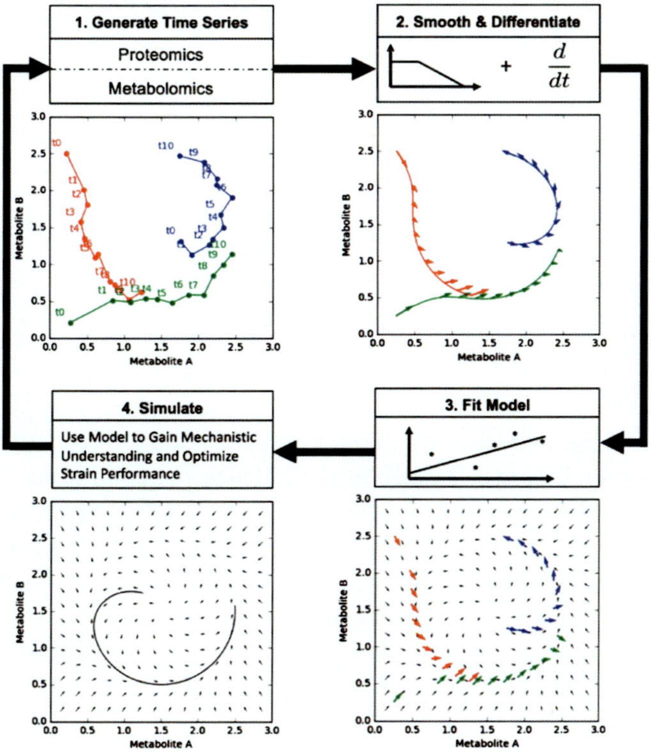

Figure 4.12

The model was then tested for metabolic engineering of the limonene and isopentenol metabolic pathways, which are depicted in Figures 4.13 and 4.14.

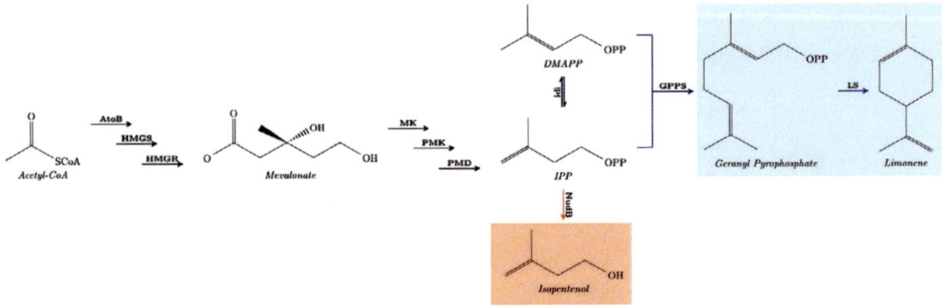

Most metabolites could be predicted using a two time-series (strains) with the predictions for limonene and isopentenol (the final products) reaching accuracy on a quantitative level.

*Given that the method was not based on an*y interaction, it was proposed to be employed for guiding metabolic engineering.

In what was presented as "the first metabolic model of an organism in the *phylum Rhodothermaeota,"* Bagheri et al. (2019) used genomic and biochemical data available for the strain *Salinibacter ruber* DSM13855 to develop a genome-scale metabolic network reconstruction. The model was called MB631, which had 1459 reactions, 1363 metabolites, and 631 genes. Subsequent lab-based assays revealed that iMB631 could correctly predict growth in 91% of cases where growth was detected experimentally and 83% of conditions in which the organism did not grow.

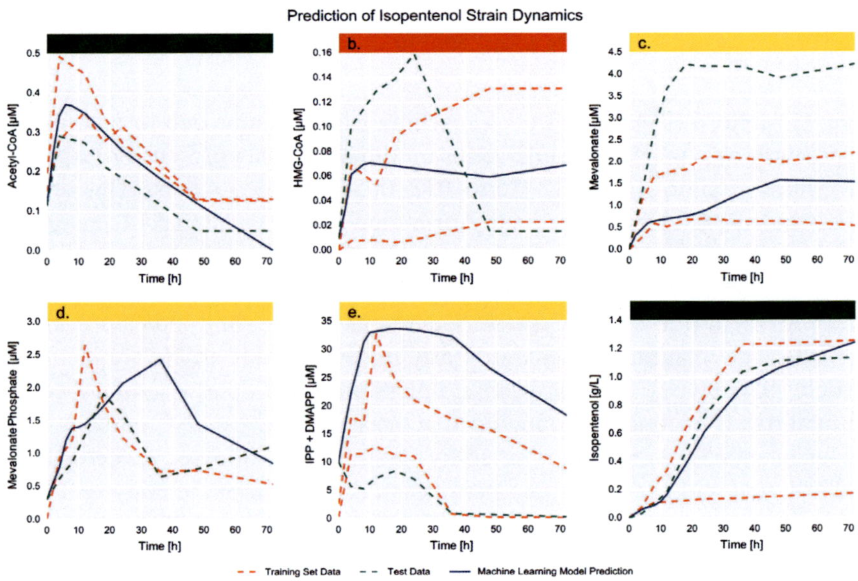

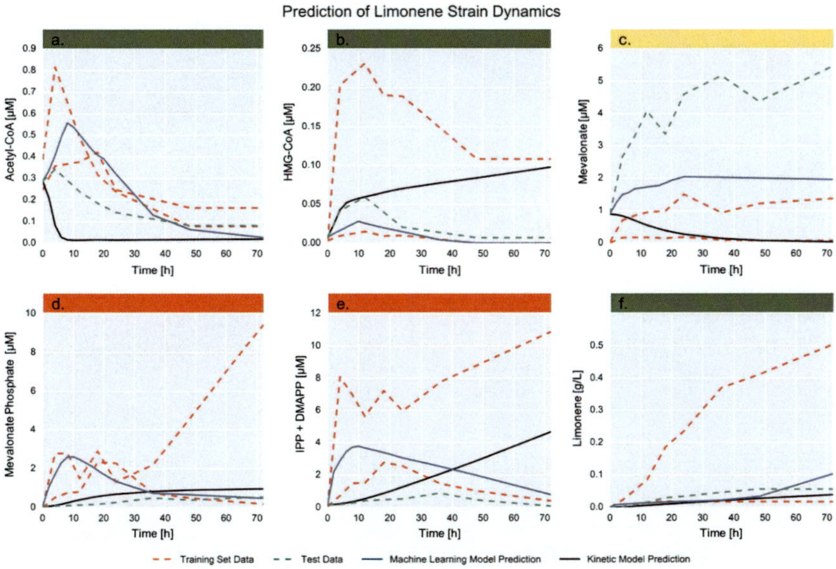

Figure 4.13

Source: Costello, Z., Martin, H.G. A machine learning approach to predict metabolic pathway dynamics from time-series multiomics data. npj Syst Biol Appl 4, 19 (2018). https://doi.org/10.1038/s41540–018–0054–3: Open Access This article is licensed under a Creative Commons Attribution 4.0 International License, which permits use, sharing, adaptation, distribution and reproduction in any medium or format, as long as you give appropriate credit to the original author(s) and the source, provide a link to the Creative Commons license, and indicate if changes were made. The images or other third-party material in this article are included in the article's Creative Commons license, unless indicated otherwise in a credit line to the material. If material is not included in the article's Creative Commons license and your intended use is not permitted by statutory regulation or exceeds the permitted use, you will need to obtain permission directly from the copyright holder. To view a copy of this license, visit http://creativecommons.org/licenses/by/4.0/.

Further, as the training sets were increased in numbers, the predictions were improved.

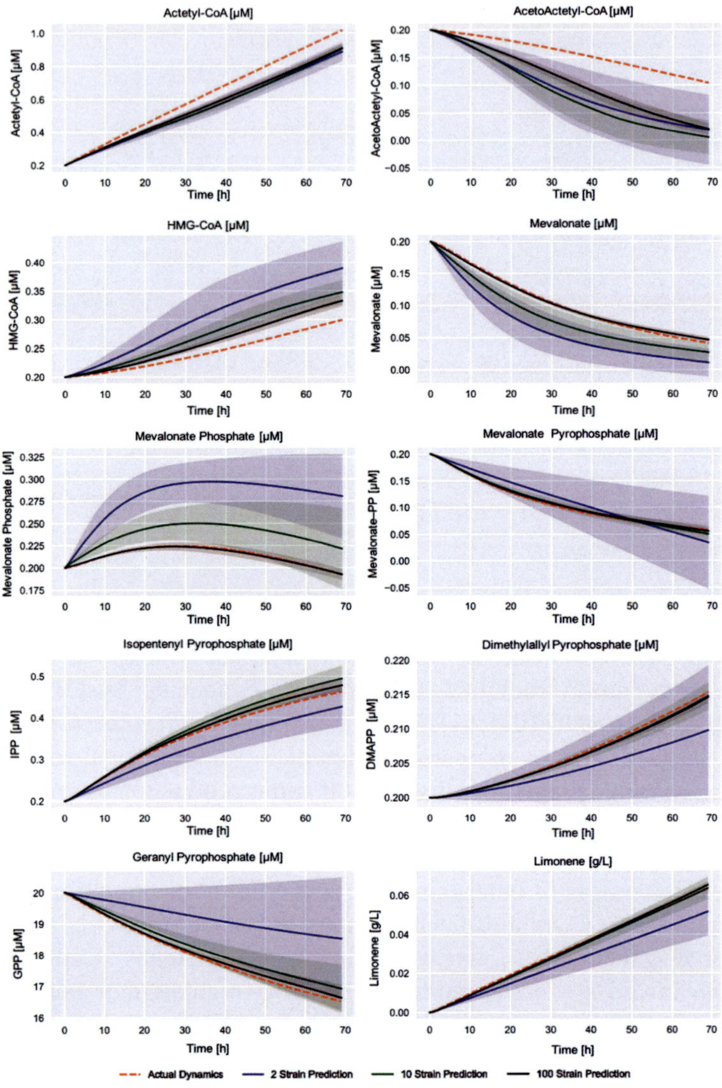

Figure 4.14

Source: Costello, Z., Martin, H.G. A machine learning approach to predict metabolic pathway dynamics from time-series multiomics data. npj Syst Biol Appl 4, 19 (2018). https://doi.org/10.1038/s41540–018–0054–3: Open Access This article is licensed under a Creative Commons Attribution 4.0 International License, which permits use, sharing, adaptation, distribution and reproduction in any medium or format, as long as you give appropriate credit to the original author(s) and the source, provide a link to the Creative Commons license, and indicate if changes were made. The images or other third-party material in this article are included in the article's Creative Commons license, unless indicated otherwise in a credit line

Table 4.4, made by the authors, presents the analysis of the model prediction and earlier results obtained for various activities of *Salinibacter ruber* DSM13855.

Table 4.4.

Utilization Source	Experimental Results	*In Silico* Results	Results
Catalase activity	+	+	✓
Starch hydrolysis	+	+	✓
DNA hydrolysis	+	+	✓
Gelatin hydrolysis	+	+	✓
Indole production	+	+	✓
glycerol	+	+	✓
glucose	+	+	✓
maltose	+	+	✓
fructose	-	+	×
galactose	-	-	✓
arabinose	-	-	✓
ribose	-	-	✓
xylose	-	-	✓
sucrose	-	+	×
lactose	-	-	✓
cellobiose	-	+	×
raffinose	-	-	✓
L-tryptophan	+	-	×
L-asparagine	-	-	✓
L-arginine	+	+	✓
L-methionine	+	+	✓

L-tyrosine	+	+	✓
L-cysteine	+	+	✓
L-aspartate	-	+	×

Source: Bagheri M, Marashi SA, Amoozegar MA. A genome-scale metabolic network reconstruction of extremely halophilic bacterium *Salinibacter ruber*. *PLoS One.* 2019;14(5): e0216336. Published 2019 May 9. doi: 10.1371/journal.pone.0216336.

The reductive tricarboxylic acid cycle (rTCA) was revealed to support autotrophic behavior under specific conditions and the Sa[illegible]ni xanthin biosynthesis pathway was also modeled. The rTCA pathway outlined has been reproduced in Figure 4.15

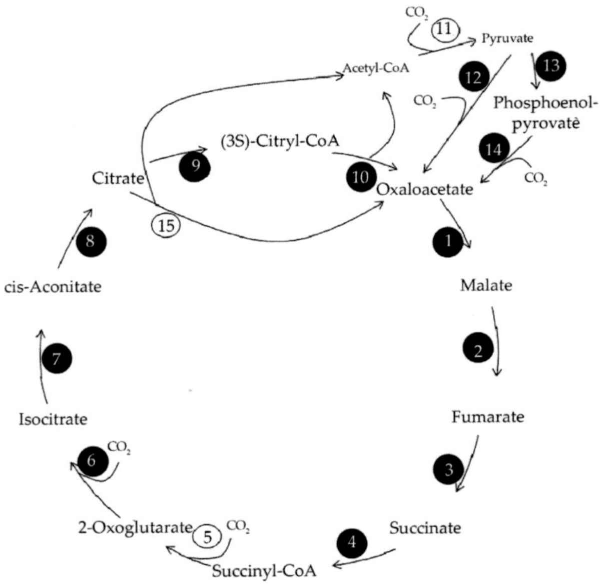

Figure 4.15

Source: Bagheri M, Marashi SA, Amoozegar MA. A genome-scale metabolic network reconstruction of extremely halophilic bacterium Salinibacter ruber. PLoS One. 2019;14(5): e0216336. Published 2019 May 9. doi: 10.1371/journal.pone.0216336.

Flux balance analysis by iMB631 predicted the pentose phosphate pathway as opposed to glycolysis as the active glucose consumption method in *S. ruber*. This becomes significant as this source was a subject of ambiguity. This pathway identified by modelling is presented in Figure 4.16 as reported by the researchers.

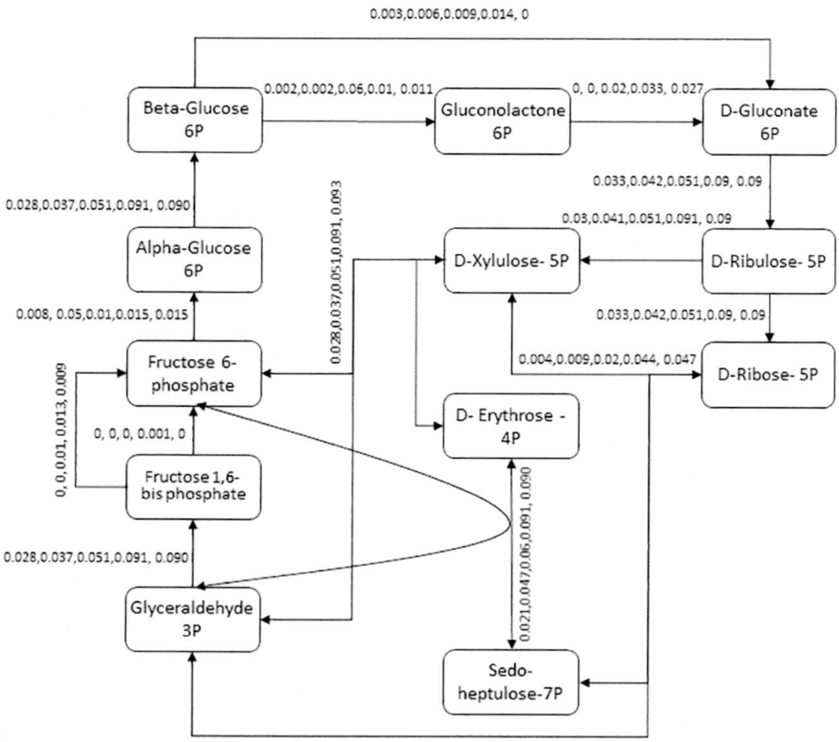

Figure 4.16

Source: Bagheri M, Marashi SA, Amoozegar MA. A genome-scale metabolic network reconstruction of extremely halophilic bacterium salinibacter ruber. PLoS One. 2019;14(5): e0216336. Published 2019 May 9. doi: 10.1371/journal.pone.0216336.

A tool called Automated Recommendation Tool (ART) was developed to assist the Design-Build-Test-Learn (DBTL) cycle. This cycle is vital in metabolic engineering, consisting of design (D) of a biological system, after which the system is built (B) from DNA into cells using synthetic biology tools. The next step is testing (T) whether this constructed biological system functions as required by the design. Given that success in the first attempt is rare, multiple steps need the Learn (L) step, which uses previously generated data to feed the next D step.

ART was designed as a tailor-made system to improve microbial strain bioengineering. Based on the input profiling data (OMICs), it predicts the response of a variable (target product). The researchers made a scheme of its working, which has been shown in Figure 4.17.

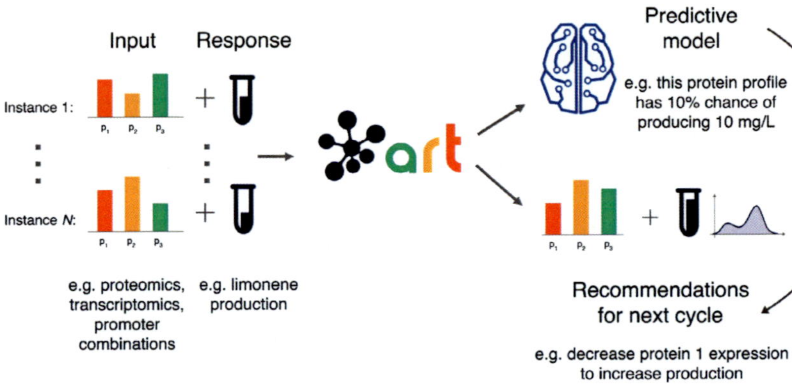

Figure 4.17

Source: Radivojević, T., Costello, Z., Workman, K. et al. A machine learning automated recommendation tool for synthetic biology. Nat commun 11, 4879 (2020). https://doi.org/10.1038/s41467–020–18008–4:

The testing of ART was done on limonene production (which is a biofuel) from real-time data from the mevalonate pathway designed from bacteria like *E. coli, S. cerevisiae*, and *S. aureus* along with plant-encoded genes – geranyl-diphosphate synthase from *A. grandis* and limonene synthase from *M. spicata.* Training of ART revealed that it could provide recommendations to guide bioengineering at par with the algorithm Principal Component Analysis of Proteomics (PCAP). Figure 4.18 shows the results obtained by the authors:

(a) training ART and predicted production (c). The ART recommendations that caused a 40% increase in production, the second DBTL cycle strains (d), close to the experimentally measured values (three blue points in b). Adding these three points from DBTL cycle 2 yielded 30 strains for training (e) with recommendations for higher production.

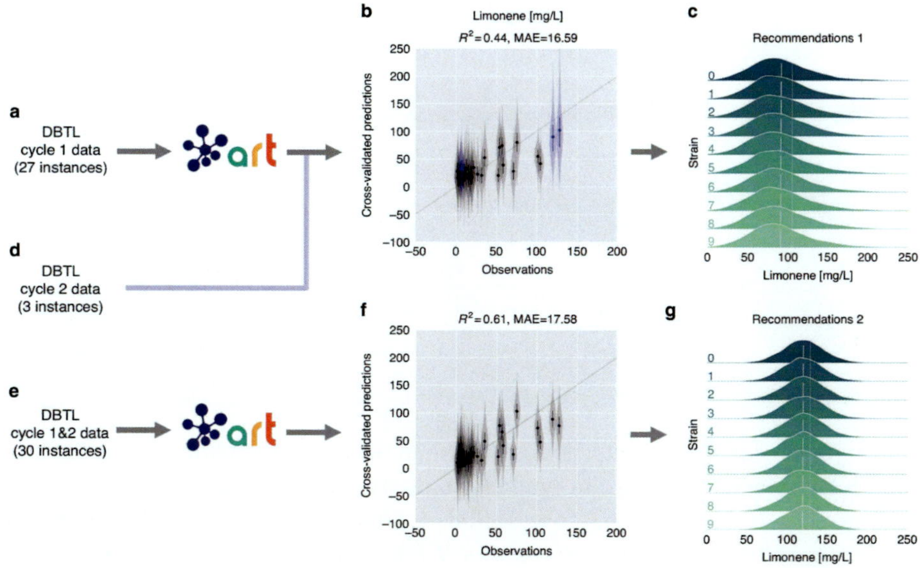

Figure 4.18

Source: Radivojević, T., Costello, Z., Workman, K. et al. A machine learning automated recommendation tool for synthetic biology. Nat commun 11, 4879 (2020). https://doi.org/10.1038/s41467–020–18008–4: *Open Access This article is licensed under a Creative Commons Attribution 4.0 International License, which permits use, sharing, adaptation, distribution and reproduction in any medium or format, as long as you give appropriate credit to the original author(s) and the source, provide a link to the Creative Commons license, and indicate if changes were made. The images or other third-party material in this article are included in the article's Creative Commons license, unless indicated otherwise in a credit line to the material. If material is not included in the article's Creative Commons license and your intended use is not permitted by statutory regulation or exceeds the permitted use, you will need to obtain permission directly from the copyright holder. To view a copy of this license, visit http://creativecommons.org/licenses/by/4.0/.*

The second test was ART in the production of hop beer from yeast (*S. cerevisiae*) without hops. ART could recommend protein profiles predicted to reach the flavor profile of Pale Ale (PA) flavor was also close to the metabolite profile of Torpedo (T) (Figure 4.19).

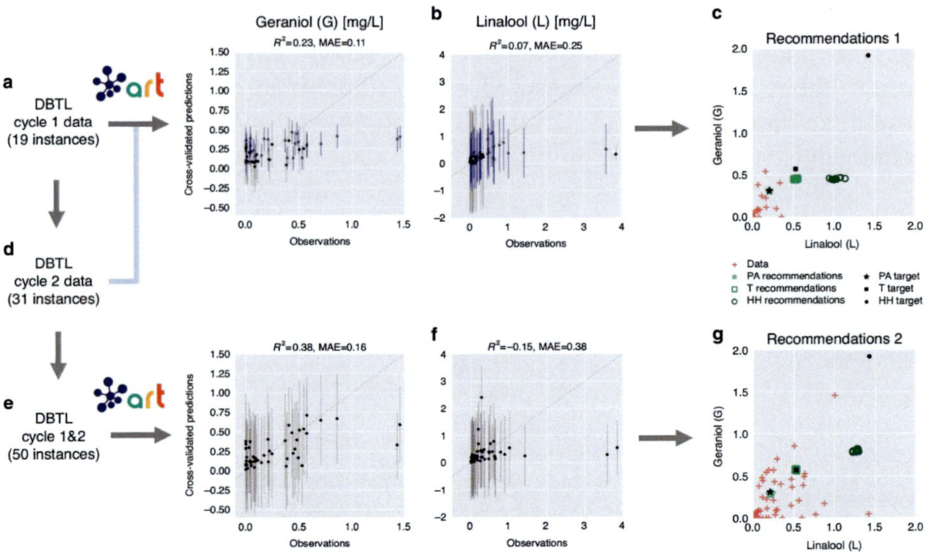

Figure 4.19

Source: Radivojević, T., Costello, Z., Workman, K. et al. A machine learning automated recommendation tool for synthetic biology. Nat commun 11, 4879 (2020). https://doi.org/10.1038/s41467–020–18008–4: *Open Access This article is licensed under a Creative Commons Attribution 4.0 International License, which permits use, sharing, adaptation, distribution and reproduction in any medium or format, as long as you give appropriate credit to the original author(s) and the source, provide a link to the Creative Commons license, and indicate if changes were made. The images or other third-party material in this article are included in the article's Creative Commons license, unless indicated otherwise in a credit line to the material. If material is not included in the article's Creative Commons license and your intended use is not permitted by statutory regulation or exceeds the permitted use, you will need to obtain permission directly from the copyright holder. To view a copy of this license, visit http://creativecommons.org/licenses/by/4.0/.*

Notwithstanding the limitations, such tools can be subjected to further research and fine-tuning for potential use. In another study, Zhang et al. (2020) combined mechanistic and machine learning models to facilitate the prediction of genotype-to-phenotype. The authors targeted the aromatic amino acid (AAA) pathway that produces tryptophan in *Saccharomyces cerevisiae*. A 7776-member combinatorial library design space was made of five genes from genome-scale models (GSM) simulations, each controlled by six promoters obtained from transcriptomics data mining. Figure 4.20 shows the genes and promoters for tryptophan engineering in yeast by the researchers.

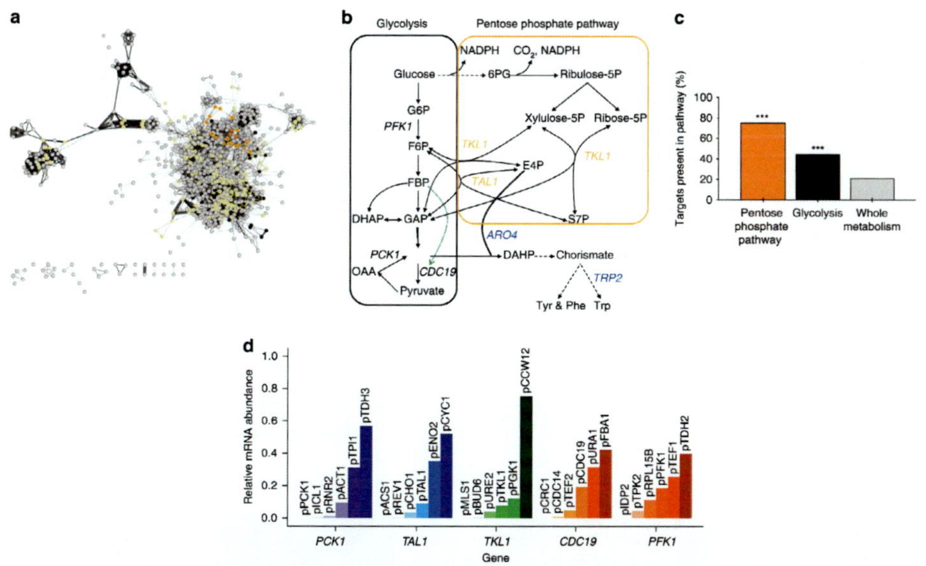

Figure 4.20

Source: Zhang, J., Petersen, S.D., Radivojevic, T. et al. Combining mechanistic and machine learning models for predictive engineering and optimization of tryptophan metabolism. Nat Commun 11, 4880 (2020). https://doi.org/10.1038/s41467–020–17910–1:

Fluorescent read-outs of an engineered tryptophan biosensor encoded into over 500 different strain designs were the source of more than 124,000 experimental time series data points for training tryptophan biosynthesis rate models. Subsequent training of machine learning systems revealed that these models could identify patterns yielding tryptophan more than the best designs employed for training by 74%.

Figure 4.21 shows the results obtained by the authors: (a, b) Learning curves for ART and EVOLVE algorithms, respectively. Mean absolute error (MAE) of training is shown as blue curves; Promoter distributions for the 30 recommendations of the exploitative (ART) (c), and explorative (EVOLVE) approach; (d; e, f) Cross-validated predictions vs average GFP synthesis-MFI mean fluorescence intensity.

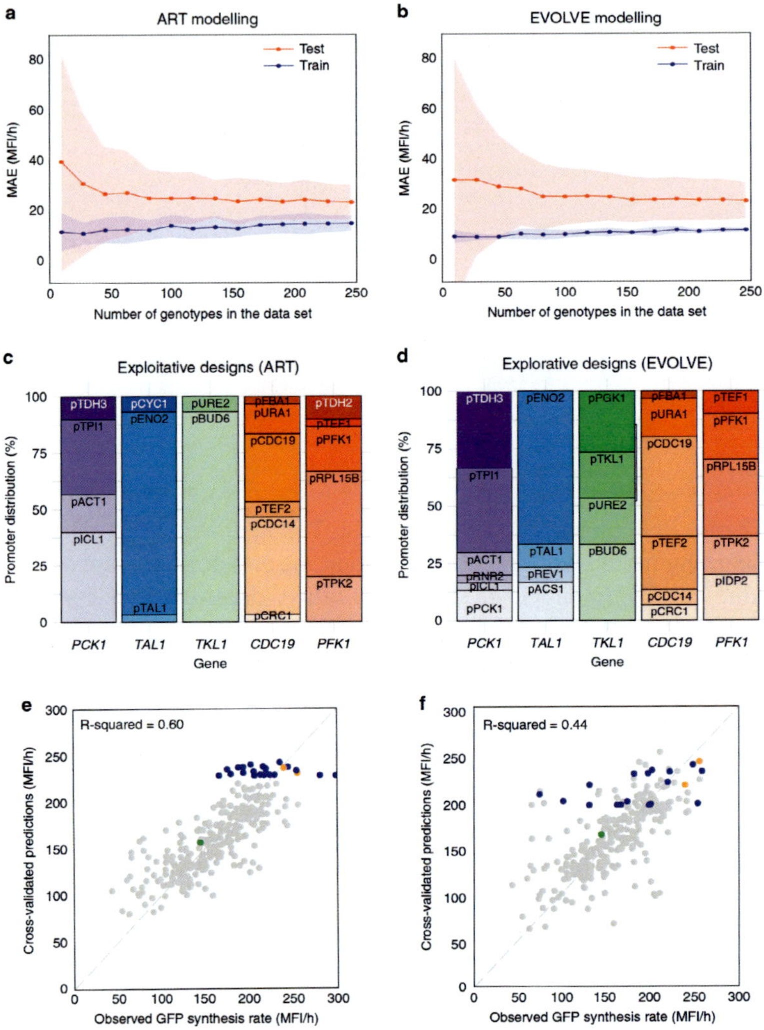

Figure 4.21

Source: Zhang, J., Petersen, S.D., Radivojevic, T. et al. Combining mechanistic and machine learning models for predictive engineering and optimization of tryptophan metabolism. Nat Commun 11, 4880 (2020). https://doi.org/10.1038/s41467–020–17910–1: Open Access This article is licensed under a Creative Commons Attribution 4.0 International License, which permits use, sharing, adaptation, distribution and reproduction in any medium or format, as long as you give appropriate credit to the original author(s) and the source, provide a link to the Creative Commons license, and indicate if changes were made. The images or other third party material in this article are included in the article's Creative Commons license, unless indicated otherwise in a credit line to the material. If material is not included in the article's Creative Commons license and your intended use is not permitted by statutory regulation or exceeds the permitted use, you will need to obtain permission directly from the copyright holder. To view a copy of this license, visit http://creativecommons.org/licenses/by/4.0/.

An ML workflow tool called METIS (Machine-Learning guided Experimental Trials for Improvement of Systems, named after the ancient goddess of wisdom and crafts) was developed by Pandi et al. (2022) for data-driven optimization of pathways. The scheme of this tool as represented by the authors is given in Figure 4.22.

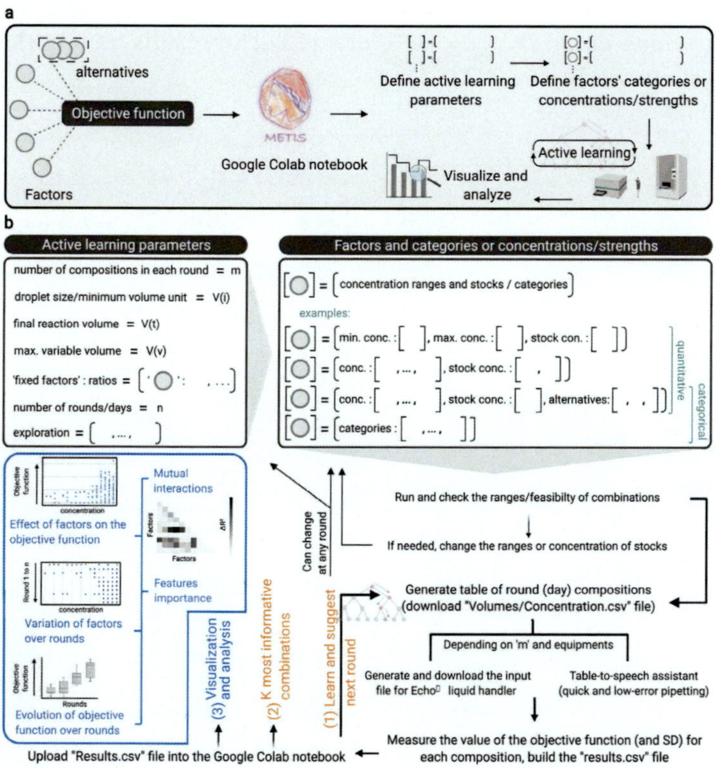

Figure 4.22

Source: Pandi, A., Diehl, C., Yazdizadeh Kharrazi, A. et al. A versatile active learning workflow for optimization of genetic and metabolic networks. Nat Commun

The workflow was first applied to augment the applicability (activity) of a cell-free in vitro gene circuit. The activity of a previously reported *LacI*-based multi-level controller was increased by a factor of two where the workflow identified resource competition as a challenge in the limited applicability of the system, which could thereby be addressed. The results as reported by the authors are shown in Figure 4.23.

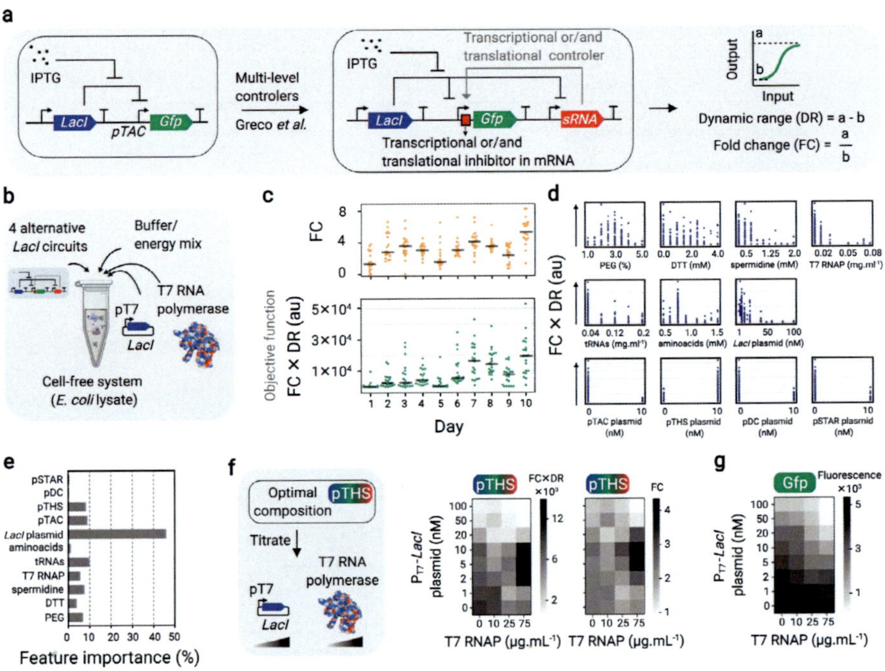

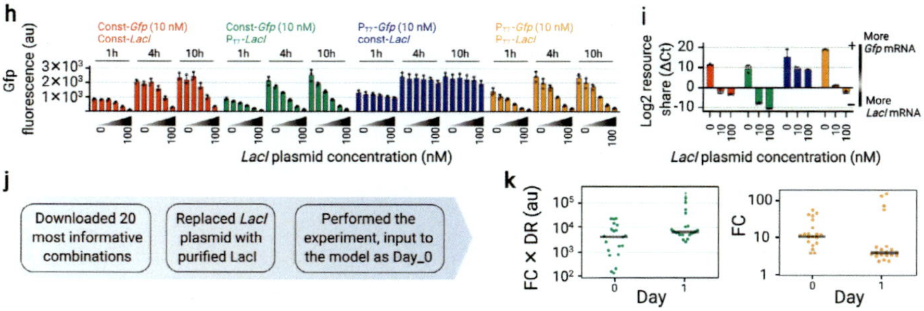

Figure 4.23

The protein produced from an optimized system of a T7 promoter, ribosome binding site (RBS), N-terminal amino acids 3, 4, and 5, and the last two C-terminal amino acids was increased by a factor of ten. Subsequently, a synthetic cycle called the crotonyl-CoA/ethylmalonyl-CoA/hydroxybutyryl-CoA (CETCH) based on a highly efficient CO_2-fixing enzyme called crotonyl-CoA carboxylase/reductase (Ccr) was assessed. Despite showing increased efficiency over the Calvin-Benson-Bassham (CBB) cycle of photosynthesis, further optimization was required, which limited the application of this cycle. The application of METIS facilitated the productivity of this cycle composed of 17 enzymes and 10 factors to increase by a factor of ten in 1,000 experiments, which resulted according to the authors as "in the most efficient CO_2-fixing in vitro system described to date."

The results obtained have been presented here: (a) The CETCH cycle (pco: propionyl-CoA oxidase, ccr: crotonyl-CoA carboxylase/reductase, epi: ethylmalonyl-CoA/methylmalonyl-CoA epimerase, mcm: methylmalonyl-CoA mutase, scr: succinyl-CoA reductase, ssr: succinic semialdehyde reductase, hbs: 4-hydroxybutyryl-CoA synthetase, hbd: 4-hydroxybutyryl-CoA dehydratase, ecm: ethylmalonyl-CoA mutase, mco: methylsuccinyl-CoA oxidase, mch: mesaconyl-CoA hydratase, mcl: β-methylmalyl-CoA lyase, gor: glyoxylate reductase, kat: catalase, fdh: formate dehydrogenase, ck: creatine phosphokinase); (b) Active learning and LC-MS for the glycolate content; (c) Optimizing the CETCH cycle with the glycolate yield.; (d) Summary of the optimization; (e) Transformed data; (f, g) Factors; (h–l) Manual experiment plots of the glycolate production over 8 h (h) and its first 15 min with slopes (i), initial production rate versus the final glycolate yield (j), total amount of measured CoA esters after 8 h versus the final glycolate yield (k), and quantified CoA esters over 8 h (l) (Figure 4.24).

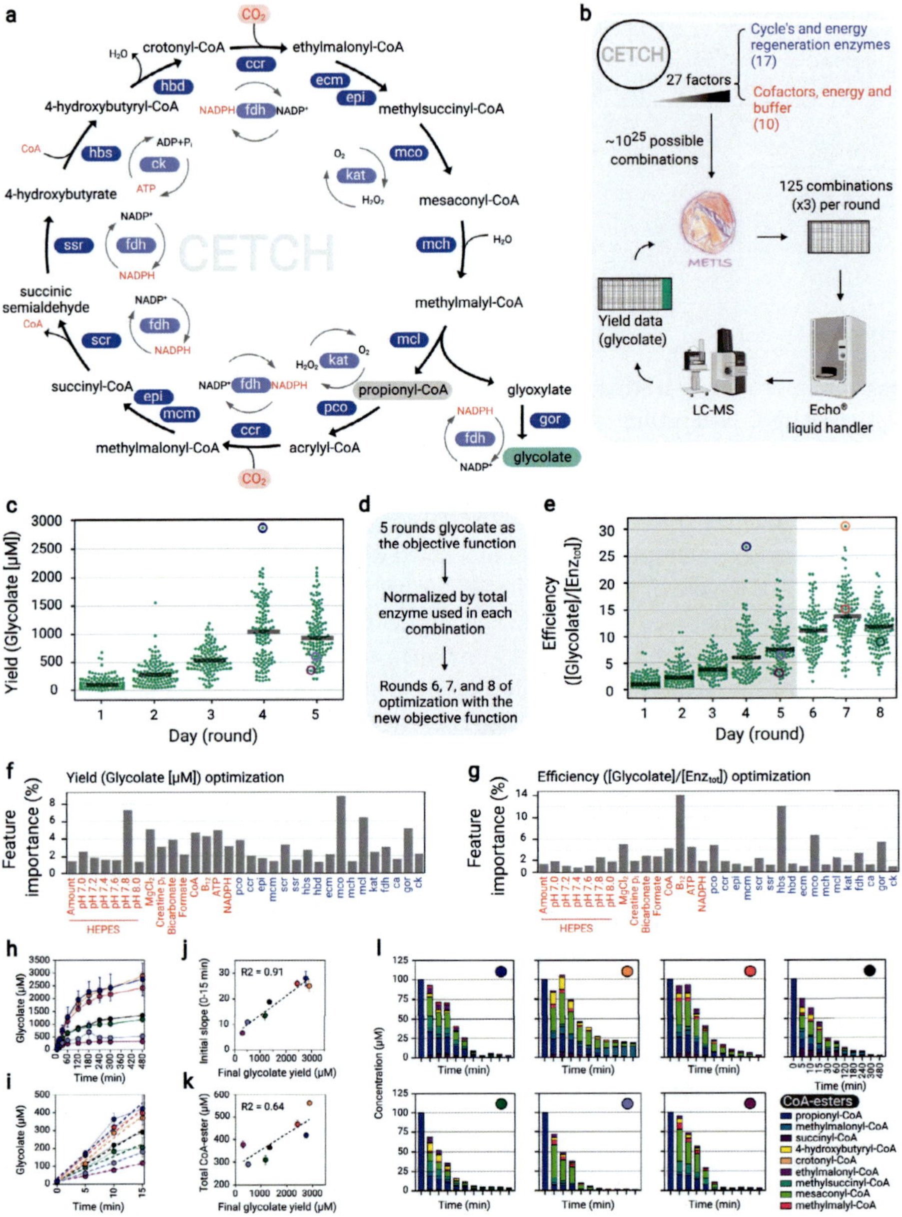

Figure 4.24

Source: Pandi, A., Diehl, C., Yazdizadeh Kharrazi, A. et al. A versatile active learning workflow for optimization of genetic and metabolic networks. Nat Commun 13, 3876 (2022). https://doi.org/10.1038/s41467–022–31245-z: Open Access This article is licensed under a Creative Commons Attribution 4.0 International License, which permits use, sharing, adaptation, distribution and reproduction in any me-

In a study by Tack et al. (2023), an ML-based approach was used for predictive genetic sensor engineering using a large-scale dataset of more than 60,000 variants of a protein-based genetic sensor, LacI (the lac repressor). The main aim was the need to specifically and precisely engineer synthetic circuits based on genetic sensors for dynamic gene regulation control. The researchers used the tool LANTERN to train the ML model with the dataset to predict EC_{50} and G_{∞} (these factors are from the Hill equation; G_0 is the basal output level; G_{∞} is the saturating output level and EC_{50} is the input level required to give an output midway between G_0 and G_{∞}). The dose-response effects were assessed by selecting 16 mutations in the LacI core domain, which had different effects for the forward engineering. After the synthesis of LacI sequences for the variants, cytometry was done for the dose response and compared with the LANTERN model predictions. The prediction accuracy of the latter was 1.93-fold for EC_{50} and 1.19-fold accuracy for G_{∞}. The analyzes are shown in Figure 4.25.

Given that most variants had a high EC_{50}, the researchers used the LANTERN approach with biophysics for engineering improved inverted LacI variants. An improvement in EC_{50} and G_{∞} values was reported for one mutation (V80L) where the EC_{50} was reduced by a factor of 5 or 6, and G_{∞} by a factor of about 1.3. The results of improving the genetic sensor as reported by the authors are given in Figure 4.26: (A) The parent variant with A87P, V301M, and E357G missense mutations. (B) The parent variant with V96E, T154I, S158R, V238D, M254I, and V264I missense mutations.

Top: (A) LacI structure with mutations as colored spheres. (B) Dose-response of single-mutant LacI variants obtained by ML-based forward engineering: blue (for silent mutations), orange (for non-silent mutations), and the wild-type dose (gray).

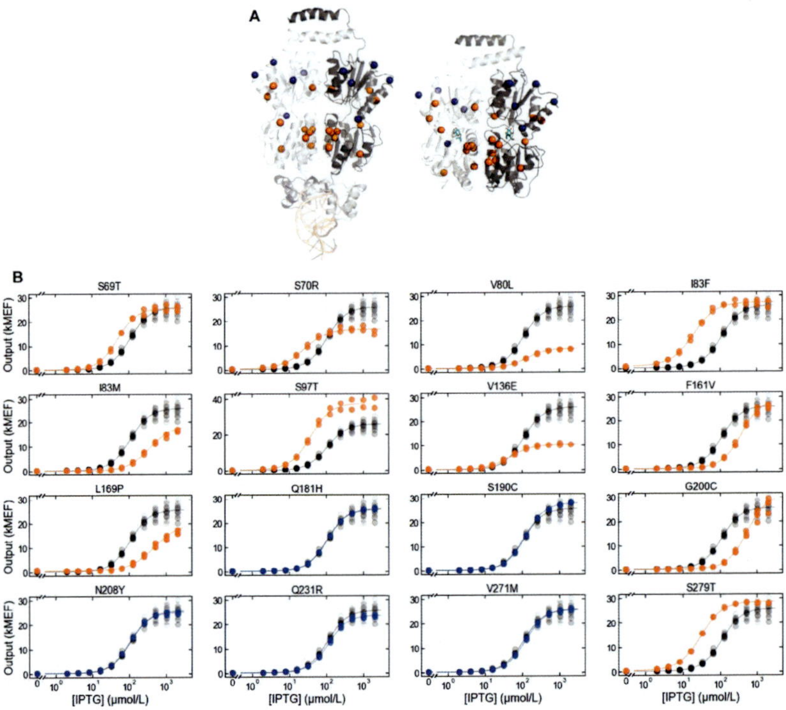

Bottom: (A) EC50 from the flow cytometry vs. EC50 predicted by the LANTERN ML mode□ (B) $G\infty$ from the flow cytometry vs. $G\infty$ predicted by the LANTERN ML model.

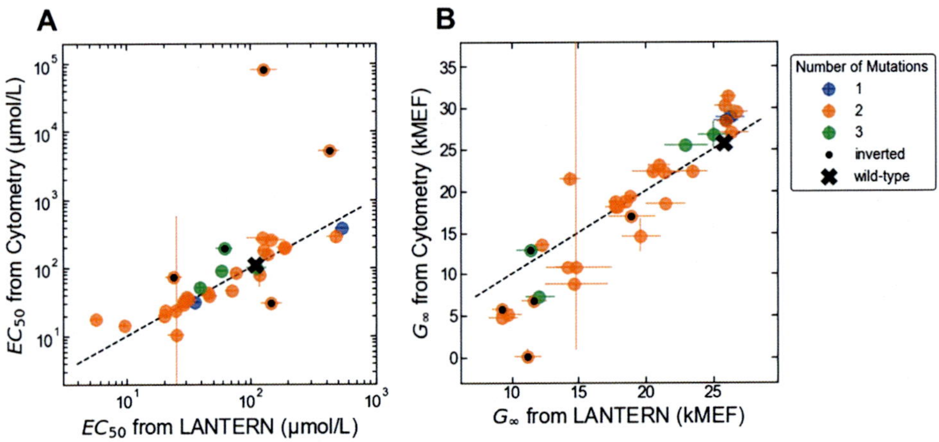

Figure 4.25

Source: Tack DS, Tonner PD, Pressman A, et al. Precision engineering of biological function with large-scale measurements and machine learning. PLoS One. 2023;18(3): e0283548. Published 2023 Mar 29. doi: 10.1371/journal.pone.0283548:

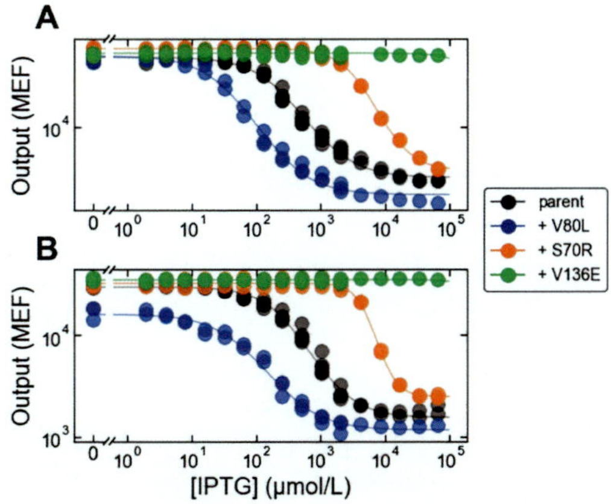

Figure 4.26

Source: Tack DS, Tonner PD, Pressman A, et al. Precision engineering of biological function with large-scale measurements and machine learning. *PLoS* One. 2023;18(3): e0283548. Published 2023 Mar 29. doi: 10.1371/journal.pone.0283548:

A summary of systems biology and AI to circumvent the challenges of plant metabolic engineering, including instability, post-translational modifications, and misfolding, was reviewed by Parthiban et al. (2023). This becomes pertinent given that plant systems are safe in humans, minimally contaminated, suitable for scale-up, are 99% pure, rapid, and cost-effective. For instance, McNulty et al. (2020) stated that plants could produce antimicrobial proteins with an accumulation of 3 g/kg fresh weight and the unit production cost of such plant-based proteins commercially is \$3.00–6.88/g, which can facilitate competitive selling (Table 4.5).

Table 4.5. Several Applications of Ml in this Field

Component	Name of the program	Type of ML algorithm	Architecture	Function/ Parameter	Model system/ training dataset	References
mRNA	APARENT (APA REgressionNeT)	CNN	One-hot encoded matrix system with two convolutional layers	• mRNA isoform prediction and polyadenylation within +10 to +35 nt downstream of 6-base central sequence element (CSE) • cleavage site prediction across polyA signal	HEK293	Bogard et al. (2019)
	6-mer Logistic Regression Baseline	Linear logistic regression	One-hot encoded matrix system with 6-mer counts	• mRNA isoform prediction and polyadenylation • cleavage site prediction		
mRNA, gene enhancers and protein	DEN (Deep Exploration Network)	Deep Convolutional Generative Adversarial Networks (DC-GANs)	One-hot encoded matrix Latent Seed Sequence Tensor	• polyadenylation signals conformed to mRNA isoforms and 3' cleavage sites • differential splicing • maximum transcriptional activation of gene enhancers • functional variants of GFP (Green Fluorescent Protein)	HEK293 HeLa MCF7 CHO	Linder et al. (2020)
	APARENT	CNN				
	-	GP regression				
	APA VAE (Variational Autoencoder)	Residual Neural Network (ResNet)				
	KL-bounded DEN	CNN				
Gene interaction and expression	scCapsNet	DNN	Capsule Neural Network	Discovery of gene interactions; closely related in function but presenting differential gene expression pattern in single cell types (based on transcriptome analysis)	scRNA-seq dataset including mouse retinal bipolar (mRBC) cells and human peripheral blood mononuclear cells (hPBMC)	Wang et al. (2020)
Transcription factor	Independent Component Analysis (ICA)	Unsupervised ML	-	Gene expression and transcriptional regulation in *E. coli* through transcriptome analysis	*E. coli* K12 RNA-seq expression profiles	Sastry et al. (2019)
Transcription factor binding	FactorNet	Convolutional RNN	One hot encoded 4-row bit matrix, LSTM	Transcription Factor (TF) cell type specific binding site prediction. (Eg.TF E2F1 binding to GM12878 and HeLa-S3)	DNase-seq, ChIp-seq and RNA-seq data of chromosomes X and 1-22 from ENCODE-DREAM challenge	Quang and Xie (2019)
Promoters	Hybrid biophysical-ML approach	Ridge regression model	-	• Synthetic promoter designing • Identification of -35 and -10 motifs and optimal spacer length	*E. coli*	LaFleur et al. (2022)
Synthetic promoter	DL model	Deep CNN	Transformer model with BiLSTM	Design regulatory sequences including orthologous promoters	RNA-seq data from *S. cerevisiae* and 10 other Ascomycota species	Vaishnav et al. (2022)
Protein	DeepRHD	DNN	CNN based bidirectional GRU (Gated Recurrent Units)	Remote homology prediction of protein sequences using physico chemical properties and evolutionary information	SCOP1.67 dataset	Routray et al. (2022)
Protein	ProtT5	pLM (protein language models) Logistic Regression	Attention based deep dilated residual networks consisting of convolution layers (ResNet CNN)	Protein (transmembrane beta barrel proteins – OmpX and variants) structure prediction from sequences	High resolution protein 3D structure dataset from ProteinNet12	Weissenow et al. (2022)
Protein	ML model	Linear regression models including glmnet, partial least squares, averaged neural network, SVM with radial basis function kernel, stochastic gradient boosting, boosted generalized linear model, random forest, cubist and naïve Bayes models	Caret package in R	Factors influencing recombinant protein stability including Molecular weight, cysteine residues and N-linked glycosylation	CHO cells expressing human secretome	Masson et al. (2022)
Protein	ASPIRER	DL model	XGBoost and N-terminal sequence-based CNN	Prediction of Non-classical secreted proteins (NCSPs)	Gram positive bacteria NCSPs dataset from UniProt	Wang et al. (2022)
Protein	eUniRep	DL NN	UniRep multiplicative LSTM	Protein, avGFP and TEM-1 β-lactamase, engineering (Low-*N* engineering) using small number of functional variants	*E. coli* DH5α	Biswas et al. (2021)

Protein	UniRep	SVM LR Random Forest (RF) ANN	RNN	Prediction of recombinant gene expression and protein solubility	*B. subtilis*	Martiny et al. (2021)
Protein	ECNet	RNN	BiLSTM, Transformer architecture with TAPE integration	Protein fitness prediction based on evolutionary context, engineered TEM-1 β-lactamase variants showing enhanced ampicillin resistance	*E. coli* DH5α Diverse large-scale deep mutational scanning (DMS) datasets and random mutagenesis datasets	Luo et al. (2021)
Protein	EPSOL	Keras based DL model	Multidimensional Embedding, multi-convolutional-pooling module and a Multi-layer Perceptron (MLP)	Protein solubility prediction	Heterologous expressed *E. coli* soluble and insoluble protein dataset compiled by Smialowski et al. (2012)	Wu and Yu (2021)
Protein	DEEPred	Multi-layered perceptrons (MLPs)	Feed-forward multitask DNN	Sequence/Gene Ontology (GO) based functional definition prediction of proteins	*Pseudomonas aeruginosa* strain reference genome and UniProtKB/Swiss-Prot dataset	Sureyya Rifaioglu et al. (2019)
Protein	ML models	GANs Logistic regression Decision Tree SVM Naïve Bayes Cforest XGboost ANNs	Generator Neural Network and Discriminator Neural Network	Prediction of Protein solubility	eSol database dataset	Han et al. (2019)
Protein	DeepSol	DL model	CNN, non-linear high-dimensional k-mer vector spaces, deep feed-forward neural network (FFNN)	Protein solubility prediction	Heterologous expressed *E. coli* soluble and insoluble protein dataset compiled by Smialowski et al. (2012)	Khurana et al. (2018)
Protein	ML	RNN	BiLSTM, One-hot encoded matrix	Identification and function prediction of protein homologs including iron sequestering proteins, cytochrome P450, serine and cysteine proteases and G-Protein coupled receptors, detection through fluorescence (GFP)	*E. coli*	Liu (2017)
Protein	SPIDER2	Deep learning neural network	Stacked sparse autoencoder	Protein secondary structure, solvent accessible surface area, main chain torsion angle prediction	Non-redundant high resolution protein structures dataset	Yang et al. (2017)
Amyloidogenic proteins	AbsoluRATE	SVM	Sequence-based regression	Aggregation kinetics prediction of amyloidogenic proteins	CPAD 2.0 database dataset	Rawat et al. (2021)
Antibody	DeepAb	Deep residual convolutional network (Deep RCN) with Rosetta-based protocol	RNN, BiLSTM, LSTM	Antibody Fv structure prediction from sequence	Observed Antibody Space (OAS) database, SAbDab database	Ruffolo et al. (2022)
Antibody	DeepH3	Deep residual network	One dimensional and two dimensional convolutions	Prediction of *de novo* CDR H3 loop structures	Rosetta and SAbDab dataset	Ruffolo et al. (2020)
mAbs	solPredict	ESM1b-based Multilayer perceptron (MLP2Layer)	Pretrained protein language model EMS1b embedding	• Rapid, large-scale high throughput screening of mAb sequences (IgG1, IgG2 and IgG4) and quantitative solubility prediction	HEK293/CHO	Feng et al. (2022a)

			module and a Multi-layer Perceptron (MLP)		protein dataset compiled by Smialowski et al. (2012)	
Protein	DEEPred	Multi-layered perceptrons (MLPs)	Feed-forward multitask DNN	Sequence/Gene Ontology (GO) based functional definition prediction of proteins	*Pseudomonas aeruginosa* strain reference genome and UniProtKB/Swiss-Prot dataset	Sureyya Rifaioglu et al. (2019)

Protein	ML models	GANs Logistic regression Decision Tree SVM Naïve Bayes Cforest XGboost ANNs	Generator Neural Network and Discriminator Neural Network	Prediction of Protein solubility	eSol database dataset	Han et al. (2019)
Protein	DeepSol	DL model	CNN, non-linear high-dimensional k-mer vector spaces, deep feed-forward neural network (FFNN)	Protein solubility prediction	Heterologous expressed *E. coli* soluble and insoluble protein dataset compiled by Smialowski et al. (2012)	Khurana et al. (2018)
Protein	ML	RNN	BiLSTM, One-hot encoded matrix	Identification and function prediction of protein homologs including iron sequestering proteins, cytochrome P450, serine and cysteine proteases and G-Protein coupled receptors, detection through fluorescence (GFP)	*E. coli*	Liu (2017)
Protein	SPIDER2	Deep learning neural network	Stacked sparse autoencoder	Protein secondary structure, solvent accessible surface area, main chain torsion angle prediction	Non-redundant high resolution protein structures dataset	Yang et al. (2017)
Amyloidogenic proteins	AbsoluRATE	SVM	Sequence-based regression	Aggregation kinetics prediction of amyloidogenic proteins	CPAD 2.0 database dataset	Rawat et al. (2021)
Antibody	DeepAb	Deep residual convolutional network (Deep RCN) with Rosetta-based protocol	RNN, BiLSTM, LSTM	Antibody Fv structure prediction from sequence	Observed Antibody Space (OAS) database, SAbDab database	Ruffolo et al. (2022)
Antibody	DeepH3	Deep residual network	One dimensional and two dimensional convolutions	Prediction of *de novo* CDR H3 loop structures	Rosetta and SAbDab dataset	Ruffolo et al. (2020)
mAbs	solPredict	ESM1b-based Multilayer perceptron (MLP2Layer)	Pretrained protein language model EMS1b embedding	• Rapid, large-scale high throughput screening of mAb sequences (IgG1, IgG2 and IgG4) and quantitative solubility prediction	HEK293/CHO	Feng et al. (2022a)
MHC Class II Antigen Presentation	NNAlign_MAC	ANN	NNAlign_MA ML framework	• CD4 T cell epitope prediction • MHC class II antigen presentation prediction • Prediction of protein-drug immunogenicity	Single allele and Multiple allele dataset & IEDB dataset	Barra et al. (2020)
Signal Peptide	XGBoost	Regression model	-	Increasing the protein translocation rates to ER by optimizing synthetic signal peptide-protein (mAb/ScFv) complex formation	CHO-K1 cells	O'Neill et al. (2023)
Signal peptide	Sequence-to-sequence model	Attention-based neural network	Transformer model	Signal peptide prediction from Amylase, lipase, protease and xylanase enzymes	*B. subtilis*	Wu et al. (2020)
Signal peptide	SignalP 5.0	DL model	Non-linear PSSMs (position specific scoring matrix), BiLSTM and a conditional random field	Peptide identification (three classes including Sec/SPI, Sec/SPII, Tat/SPI) in prokaryotes	Reference proteomes of *E. coli* K12 and *S. cerevisiae*	Almagro Armenteros et al. (2019)
Toxic motifs	ToxDL Domain2Vec	Deep CNN	Bidirectional GRU, one-hot encoded matrix Skip-gram model	Toxicity assessment of genetically engineered organisms by highlighting toxic motifs and alteration of toxicity	Toxic/venom protein dataset from Animal Toxin Annotation Project in UniProt	Pan et al. (2020)
NSAID	Ensemble Decision Tree (DT)	Extremely Random Tree (ET) Random Forest (RF) Gradient Boosting	Multiple base trees with bagging strategy Sequence of base predictors	Non-steroidal anti-inflammatory drug, Oxaprozin, solubility in supercritical CO_2 fluid	Oxaprozin solubility dataset from Khoshmaram et al. (2021)	Alshehri et al. (2022)

Source: Parthiban S, Vijeesh T, Gayathri T, Shanmugaraj B, Sharma A and Sathishkumar R (2023) Artificial intelligence-driven systems engineering for next-generation plant-derived biopharmaceuticals. Front. Plant Sci. 14:1252166. doi: 10.3389/fpls.2023.1252166: © 2023 Parthiban, Vijeesh, Gayathri, Shanmugaraj, Sharma

4.4. FINAL WORDS

Given that gene regulation, especially in eukaryotes, is complex, control in metabolic engineering and synthetic biology has been a challenge. Research studies are documenting the construction of models for predicting genes and expression levels using increased datasets. These models can facilitate the designing of regulatory sequences to obtain a specific gene expression level. Research on ML models, algorithms, and guided interpretation can pave the way for ML to be used for the rational designing of complex pathways (de Jongh et al., 2020).

"Synthetic biology is giving us insights into the criteria and processes that underpin all living systems; in turn, we can take this insight and design and use it to build a 'better biology.' However, we need to take the public with us on this journey, create meaningful and considered dialog about the work we may do and the impact it might have on our world."

—El Karoui M, Hoyos-Flight M and Fletcher L (2019) Future Trends in Synthetic Biology—A Report. Front. Bioeng. Biotechnol. 7:175. doi: 10.3389/fbioe.2019.00175

REFERENCES

1. Bagheri, M., Marashi, S. A., & Amoozegar, M. A. (2019). A genome-scale metabolic network reconstruction of extremely halophilic bacterium *Salinibacter ruber*. *PLoS ONE, 14*(5), e0216336. https://doi.org/10.1371/journal.pone.0216336.
2. Chen, L., Feng, K. Y., Cai, Y. D., Chou, K. C., & Li, H. P. (2010). Predicting the network of substrate-enzyme-product triads by combining compound similarity and functional domain composition. *BMC Bioinformatics, 11*, Article 293. https://doi.org/10.1186/1471-2105-11-293.
3. Costello, Z., & Martin, H. G. (2018). A machine learning approach to predict metabolic pathway dynamics from time-series multiomics data. *npj Systems Biology and Applications, 4*(1), 19. https://doi.org/10.1038/s41540-018-0054-3.
4. de Jongh, R. P. H., van Dijk, A. D. J., Julsing, M. K., Schaap, P. J., & de Ridder, D. (2020). Designing eukaryotic gene expression regulation using machine learning. *Trends in Biotechnology, 38*(2), 191–201. https://doi.org/10.1016/j.tibtech.2019.07.007.
5. Grimaldi, M., Visintainer, R., & Jurman, G. (2011). RegnANN: Reverse Engineering Gene Networks using Artificial Neural Networks. *PLoS ONE, 6*(12), e28646. https://doi.org/10.1371/journal.pone.0028646.
6. Kludas, J., Arvas, M., Castillo, S., Pakula, T., Oja, M., Brouard, C., & Penttilä, M. (2016). Machine learning of protein interactions in fungal secretory pathways. *PLoS ONE, 11*(7), e0159302. https://doi.org/10.1371/journal.pone.0159302.
7. McNulty, M. J., Gleba, Y., Tusé, D., Hahn-Löbmann, S., Giritch, A., Nandi, S., & McDonald, K. A. (2020). Techno-economic analysis of a plant-based platform for manufacturing antimicrobial proteins for food safety. *Biotechnology Progress, 36*(1), e2896. https://doi.org/10.1002/btpr.2896.
8. Nagao, C., Nagano, N., & Mizuguchi, K. (2014). Prediction of detailed enzyme functions and identification of specificity determining residues by random forests. *PLoS ONE, 9*(1), e84623. https://doi.org/10.1371/journal.pone.0084623.
9. Pandi, A., Diehl, C., Yazdizadeh Kharrazi, A., et al. (2022). A versatile active learning workflow for optimization of genetic and metabolic networks. *Nature Communications, 13*, Article 3876. https://doi.org/10.1038/s41467-022-31245-z.
10. Parthiban, S., Vijeesh, T., Gayathri, T., Shanmugaraj, B., Sharma, A., &

Sathishkumar, R. (2023). Artificial intelligence-driven systems engineering for next-generation plant-derived biopharmaceuticals. *Frontiers in Plant Science, 14*, Article 1252166. https://doi.org/10.3389/fpls.2023.1252166.

11. Radivojević, T., Costello, Z., Workman, K., et al. (2020). A machine learning Automated Recommendation Tool for synthetic biology. *Nature Communications, 11*, Article 4879. https://doi.org/10.1038/s41467-020-18008-4.
12. Shah, H. A., Liu, J., Yang, Z., & Feng, J. (2021). Review of machine learning methods for the prediction and reconstruction of metabolic pathways. *Frontiers in Molecular Biosciences, 8*, Article 634141. https://doi.org/10.3389/fmolb.2021.634141.
13. Tack, D. S., Tonner, P. D., Pressman, A., et al. (2023). Precision engineering of biological function with large-scale measurements and machine learning. *PLoS ONE, 18*(3), e0283548. https://doi.org/10.1371/journal.pone.0283548.
14. Viswanathan, G. A., Seto, J., Patil, S., Nudelman, G., & Sealfon, S. C. (2008). Getting started in biological pathway construction and analysis. *PLoS Computational Biology, 4*(2), e16. https://doi.org/10.1371/journal.pcbi.0040016.
15. Zhang, J., Petersen, S. D., Radivojevic, T., et al. (2020). Combining mechanistic and machine learning models for predictive engineering and optimization of tryptophan metabolism. *Nature Communications, 11*, Article 4880. https://doi.org/10.1038/s41467-020-17910-1.

Chapter 5

ENGINEERING CELLS FOR DRUG SYNTHESIS AND DELIVERY

Shiv Sanjeevi Sripathi

5.1. INTRODUCTION

Cell therapy shows great potential for treating a variety of diseases, including cancer. In fact, using genetically modified T cells from patients, called chimeric antigen receptors (CARs), has enabled a new approach to cancer therapy by changing the specificity of T cells. The U.S. Food and Drug Administration has recently approved CD19-specific CAR T cells for treating children and young adults with B-cell acute lymphoblastic leukemia, marking an exciting development in cancer therapy. In addition, research is currently being conducted on the use of cell transplantation to address cell function inadequacies through clinical studies (Fiorina et al., 2008). New approaches are presently being devised at different stages to utilize stem cells for tissue engineering (Bianco and Robey, 2001). The main challenge with these methods is meeting the clinical demand for donor cells, which can be difficult to obtain, as well as the need for immunosuppressive or other drugs to enhance cell performance, which have adverse effects and must be taken for life (Stephan et al., 2010). In order to address the issue of rejection by the host immune system, cell engineering methods that involve immunoisolation of cells using microcapsules or three-dimensional scaffolds have been proposed (Vegas et al., 2016). While the implant materials show promise, challenges arise from the foreign-body response they induce, which include the recruitment of immune cells, the formation of fibrous deposits, the obstruction of nutrient passage, and ultimately, the failure of the cells (Vegas et al., 2016).

In addition to this, extra work is necessary to evaluate the precision of transplantation and the degree of incorporated cells, in order to enhance the therapeutic results (Barnett et al., 2007). In light of this, there is a great need for innovative technologies in cell engineering to advance biomedical applications.

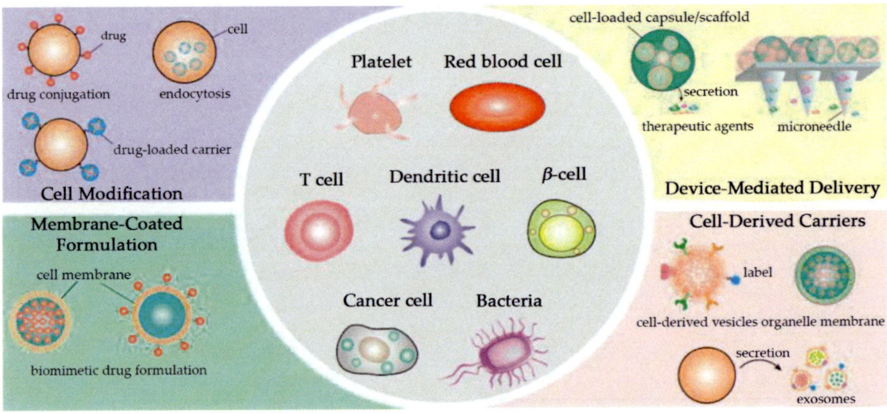

Figure 5.1. *Visual representation of the latest breakthroughs in cell engineering techniques from various viewpoints, specifically for drug delivery purposes.*

Source: Chen et al., (2018).

The capacity of cells within the body system to communicate and interact with each other and their surroundings enables them to carry out intricate functions and form intricate biological structures. One instance of this is the manner in which the numerous proteins found on the surface of red blood cells (RBCs) allow for information exchange with phagocytes, which in turn prevents an attack by the complement system and ultimately results in a longer lifespan for the red blood cells. Furthermore, research has shown that the targeted recruitment of certain cells in response to chemokines plays a role in immune responses, disease progression, and tissue formation (Yoo et al., 2011). Drawing inspiration from these processes, efforts have been initiated to incorporate therapeutic concepts into natural cells in order to develop the cutting-edge delivery platform of the future (Miller et al., 2015). In addition, "top-down" strategies that involve disguising synthetic nanomaterials with cellular membranes have been suggested to replicate the essential functions of the original cells and increase the usefulness of nanomaterials (Hu et al., 2015). The use of cell engineering techniques not only reduces the host's immune response but also preserves the intricate structures and crucial biological functions of natural cells, resulting in enhanced therapeutic effectiveness. This chapter will concentrate on recent research that utilizes cutting-edge cell engineering technologies for various applications, including cancer immunotherapy, targeted drug delivery, and diabetes treatment (as depicted in Figure 5.1).

5.2. CELL SURFACE DRUG MODIFICATION

The process of modifying the surface of cells can be categorized into two groups: physical modification and chemical modification.

5.2.1. Physical Modification

The physical modification process is relatively straightforward, but its stability is not very high. Physical modification techniques primarily involve noncovalent surface modification, such as intermolecular hydrogen bonds, van der Waals forces, hydrophobic interactions, and other forces to bind drugs to the cell surface. Some specific methods include electrostatic force adsorption and photolithographic deposition. For instance, intact cells were used to decorate the surface of RBCs with chlorin e6 (Ce6), a photosensitizer, by simply mixing them together. This allowed the RBC membrane to remain integrated and stable in a dark environment. However, when exposed to light, the cell membrane disintegrated, and singlet oxygen was generated by Ce6, enabling this system to provide light-controlled therapy by rapidly releasing DOX (Gao et al., 2017).

5.2.2. Chemical Modification

The chemical modification of materials to the cell surface is strongly associated. In reality, this method of chemical modification is one of the most effective techniques in membrane engineering (Yao et al., 2017). Cells can be altered through various methods, such as antigen-antibody interaction, biotin-avidin binding, and glycoengineering of the cell surface.

5.2.2.1. Antigen-Antibody Interaction

There are numerous antigens present on the surface of cells that can be directly linked to drugs that have been modified with antibodies. This connection can be established through the modification of antigens on the cell surface and the binding of antibody-modified drugs to them. Additionally, the direct modification of antibodies on cells can also be utilized for the purpose of disease detection or treatment (Villa et al., 2016). The use of antibody-modified living cells, particularly RBCs, has shown promise in preventing and treating autoimmune diseases. By covalently attaching disease-associated autoantigens to RBCs through transpeptidase sortase, unmodified RBCs can be modified to have prophylactic and therapeutic effects on a mouse model of multiple sclerosis in autoimmune encephalomyelitis. Additionally, RBCs modified with insulin-derived peptide have been found to be effective in preventing the onset of type 1 diabetes in a mouse model, allowing the animals to remain normoglycemic. These findings suggest that the use of transpeptidase sortase-deposited RBCs by disease-associated autoantigens could be a promising strategy for the prevention and treatment of certain autoimmune diseases (Pishesha et al., 2017). Various antibodies, including CD22-PE, were specifically designed to be attached to red blood cells (RBCs). Fluorescein isothiocyanate (FITC) was used to join the anti-human antibody (Ab), and 4′,6-diamidino-2-phenylindole (DAPI) was also inserted into the surface of RBCs. The fluorescence microscopy signals demonstrated that the PE- and FITC-labeled Abs were clearly bound to the activated conserved RBCs. Additionally, the fluorescence signals showed a high RBC loading capacity (Al-Essa et al., 2018). The use of dye-labeled antibodies to modify the surface of RBCs was helpful in conducting fluorescence-activated cell sorting (FACS) of target cells. The binding of fluorochrome-conjugated antibodies against CD45 and CD22 to RBCs can be assessed microscopically (Al-Essa et al., 2019).

5.2.2.2. Biotin-Avidin Linkage

One way to modify materials on the surface of living cells is through biotin-avidin linkage, which has strong non-covalent affinities and is at least 10,000 times more stable than the interaction between antibody and antigen. This means that drugs are less likely to leak out. For example, mesenchymal stem cells (MSCs) can be modified with DOX by using biotin conjugation, where avidin acts as a bridge between the DOX conjugates and the MSCs. The Biotinylated PEG-PAMAM-cis-aconityl-DOX (BPCD) conjugates are anchored to the MSCs via avidin (Yao et al., 2017). Nanoparticles may be attached to the surface of cells through biotin-avidin modification, such as by biotinylating the surface of neural stem cells (NSCs) and then combining them with streptavidin-conjugated polystyrene NPs, which can enhance the retention and targeting capabilities of NSC-NPs (Mooney et al., 2014). The use of NPS on the surface of biotinylated human NSCs resulted in a marked improvement in the delivery of docetaxel (DTX) to tumors. This delivery system was also found to be effective in a mouse model of triple-negative breast cancer (TNBC) when administered (Mooney et al., 2014).

5.2.2.3. Glycoengineering

Glycoengineering is a biocompatible approach that involves modifying the surface of cells. By using glycoengineering, it is possible to express non-natural azide groups on the surface of MSCs. These azide groups can be linked with dibenzyl cyclooctyne through click chemistry, which enables in vivo tumor tissue targeting. This technique has been tested in mouse subcutaneous and orthotopic tumor models and has proven effective as an imaging agent and drug delivery system (Layek et al., 2016). This method enables the targeted accumulation of drugs within tumors. By administering azide-labeled MSCs to mouse models, the concentration of functional nanoparticles linked through glycoengineering on the surface of MSCs at the tumor site can be significantly increased, leading to effective inhibition of ovarian cancer growth (Layek et al., 2020). In addition, cell surface glycoengineering enhances the selectin-mediated adhesion of mesenchymal stem cells (MSCs) and cardiosphere-derived cells (CDCs) by utilizing physiological selectin-ligands, which can improve stem cell engraftment (Lo et al., 2016).

5.2.2.4. Cell Backpack

Cell backpacks, consisting of multiple nanometer-thick layers of thin films, have the potential to serve as drug delivery vehicles at the cellular level (Polak et al., 2015). The use of cell backpacks on the surface of living cells has been widely adopted for modification, either through physical adsorption or chemical modification. Due to their unique shape, size, and composition, cell backpacks are resistant to being engulfed by macrophages, thereby ensuring the effective separation of drugs from cells, preserving drug integrity, and maintaining cell function (Klyachko et al., 2017). Typically, backpacks are composed of multiple layers that possess distinctive structures and textures (Shum & Heslop, 2018). Backpacks made with microcontact printing have the ability to effectively bind to macrophages and offer an antigen-agonist advantage. These backpacks are constructed from biodegradable polymers and feature multiple layers, including a cell-adhesive layer, a poly(lactic-co-glycolic) acid (PLGA) layer, a polyvinyl alcohol layer, and a second PLGA layer (Shields et al., 2020). Monocytes are the ideal cellular vehicles for transporting cell backpacks to inflamed tissues due to their ability to traverse strict biological barriers, such as the blood-brain barrier (BBB). To establish a connection between BP and monocytes, scientists typically utilize streptavidin and biotinylated antibodies. By undergoing a sequence of treatments, cells equipped with cell backpacks exhibit enhanced penetration capabilities and target two separate inflamed organs. This approach not only minimizes the likelihood of phagocytosis but also extends the cells' circulation time within the living organism. Experimental mouse models have demonstrated that these cells maintain their regular functions, such as differentiation and endocortical infiltration (Anselmo et al., 2015). Backpacks, which are loaded with catalase, can potentially be transported across the blood-brain barrier (BBB) in mice with lipopolysaccharide-induced encephalitis by attaching to macrophages through a biotin-avidin connection. This method could potentially be used as a treatment for a variety of neurodegenerative diseases (Klyachko et al., 2017). The use of a disulfide-containing bis-N-hydroxy succinimide (NHS) cross-linker (NHS-SS-NHS) to incorporate the CD45-targeting antibody can result in the formation of protein nanogels (NGs) that bind to the surface of T cells. By using these NGs to load T cells with large quantities of protein drugs, and then selectively releasing the cargo to activate the T cell receptor, this approach greatly increases the number of T cells in the tumor. This allows for the administration of a high dose of the cytokine without causing toxicity, resulting in improved therapeutic efficacy (Tang et al., 2018).

5.3. CELLS AS DELIVERY DEVICES

The use of nanoparticles to improve the effectiveness of medications has been widespread. However, some experts believe that they may not be suitable for universal drug delivery (Park, 2019). Based on the observations made, it has been noted that only a small fraction of injected nanoparticles, roughly 1%, actually reach their intended target sites. Additionally, there are concerns regarding the toxic effects of nanoparticle-cell interactions, as well as the challenge of achieving cell selectivity in diseased tissues that are heterogeneous, such as chronic wounds and cancer. Furthermore, phagocytic clearance leads to a decrease in the concentration of nanoparticles, and clinical success has been limited (Wilhelm et al., 2016). Although future endeavors may still incorporate this technology, the aforementioned obstacles have prompted the exploration of alternative strategies for administering drugs.

Delivering drugs using cells is another alternative approach. Cells have been referred to as "living drugs" due to their ability to persist in the body and produce a desired effect for a longer period of time than natural or synthetic pharmaceutical drugs. This is because when free proteins and therapeutics are injected into the body, the immune system often responds by removing them from the bloodstream quickly (Fliervoet & Mastrobattista, 2017). Cells possess the innate capacity to perceive, assimilate, and react to shifting surroundings in vivo, making them a promising means for the delivery of various medicinal compounds. Consequently, the utilization of live cells to treat or eradicate ailments has garnered considerable attention in the field of drug delivery, as they exhibit remarkable biocompatibility, possess an inherent aptitude for targeting tissues, demonstrate biodegradability, exhibit high drug-loading capacity, and possess the ability to traverse biological barriers (Agrahari et al., 2017). Utilizing the cellular half-life and tissue-specific properties in conjunction with the delivery of active drugs maximizes the benefits of both systems, resulting in distinct, tailored therapies that more effectively combat disease. In this section, we explore the potential of employing biological cells as delivery vehicles for therapeutic molecules and bioactive proteins, while also evaluating their advantages and drawbacks (Figure 5.2).

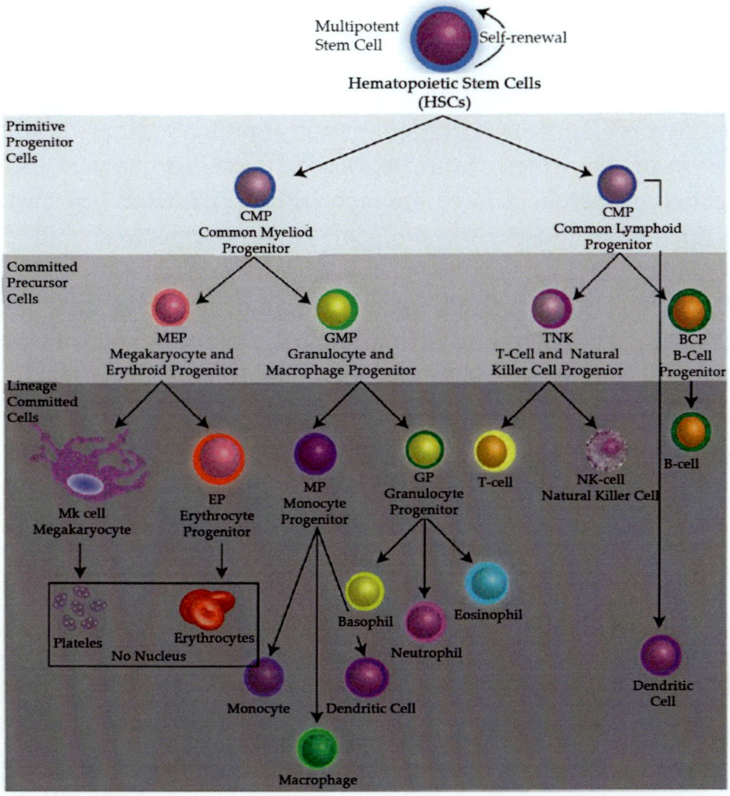

Figure 5.2. *Biological cells as therapeutic delivery vehicles.*

The choice of cell type for therapeutic delivery is crucial for different applications. For instance, anucleate cells such as red blood cells and platelets have a wide biodistribution but cannot divide in vitro, which may limit their effectiveness. On the other hand, mesenchymal stem cells are naturally regenerative and can target injured tissue, but they are also quickly cleared from the body and can be hijacked by cancer cells. Bacteria are easily manipulated and can penetrate tumors, but they may also disrupt the native microbiome and elicit immune responses. Finally, neutrophils are natural responders to sites of inflammation, but they have a short lifespan and poor survival after isolation, which may limit their usefulness in therapeutic delivery (Bush et al., 2021).

5.4. ANUCLEATE CELLS AS DELIVERY SYSTEMS

Red blood cells (RBCs) and platelets are unique components of the blood circulatory system, originating from hematopoietic stem cells. These cell types lack nuclei and, therefore, do not contain hereditary genomic information, which eliminates their inherent risk of abnormal growth or tumorigenic transformation. Their ability to circulate unimpeded in the bloodstream for several days makes them advantageous for drug delivery, especially when considering that the injection of free proteins often triggers a swift immune response leading to the rapid removal of the protein from the bloodstream. To enhance the delivery of therapeutics, researchers have engineered anucleate cells as custom drug delivery vehicles for transporting molecules and proteins to specific locations in the body and carrying designer payloads for treating injuries and diseases (Sun et al., 2019). The therapeutic molecules can be delivered to cells either by encapsulating them within the cells (intracellular encapsulation) or by attaching them to the cell surface (surface modification) to facilitate their circulation throughout the body.

5.5. ENGINEERING RBCS FOR DRUG DELIVERY

5.5.1. Intracellular Encapsulation

To enhance the circulation time of drugs, incorporating active drug molecules within the intracellular space of red blood cells (RBCs) creates an immune-privileged environment for the encapsulated drugs. This can be accomplished by placing RBCs in a hypotonic solution to cause the opening of resealable pores in the cell membrane, allowing molecules to enter the cell. Afterward, restoring the isotonicity of the solution seals the pores, trapping the molecules inside the RBCs (Coker et al., 2018).

This technique has been employed for the purpose of introducing prodrugs into the cytoplasm of red blood cells, which are unable to permeate through the red blood cell membrane (Biagiotti et al., 2011). The natural mechanisms of red blood cells (RBCs) convert the prodrug into its active form gradually, allowing it to diffuse through the cell membrane and exert its effects. This process was illustrated through the use of dexamethasone sodium phosphate, a prodrug with anti-inflammatory and immunosuppressive properties, which has proven effective in treating conditions such as inflammatory bowel disease, cystic fibrosis, and ataxia telangiectasia (Coker et al., 2018). Inside red blood cells (RBCs), the prodrug undergoes dephosphorylation by resident enzymes and transforms into its active, diffusible form, dexamethasone. Dexamethasone, in its active form, can cause side effects such as osteoporosis, glaucoma, and skin atrophy when administered at high and prolonged doses *in vivo* (Santiago & da Silva JA, 2014). However, by encapsulating the prodrug within RBCs, a slow release of the active molecule is achieved, diffusing into the bloodstream through the RBC membrane. This approach has demonstrated therapeutic concentrations of the active drug for several days. Enzymatic proteins, such as L-Asparaginase, have been loaded into red blood cells (RBCs) to decrease abnormal levels of the amino acid asparagine, which have been shown to impede the growth of tumor cells that rely on asparagine for their development (Hammel et al., 2020).

5.5.2. Surface Modification

Red blood cells (RBCs) are shaped like concave discs, which allow for optimal oxygen uptake and the ability to navigate through capillaries and release oxygen molecules. Additionally, these cells naturally display surface molecules, including receptors, proteins, and functional groups on their membrane surface, which provide binding sites for antibodies and drugs (Du & Chen, 2019). Scientists have effectively employed membrane surface molecules to modify the red blood cell (RBC) membrane with various antibodies and drugs, resulting in improved delivery efficiency of therapeutic molecules to sites of injury and disease (Agrahari et al., 2017).

In most cases, such tethering methods enhance the pharmacokinetics and biodistribution of the drug compared to the free drug, and in some cases, localization to the RBC surface increases the drug's specificity, leading to functional benefits that cannot be achieved with the free drug alone. For instance, researchers have genetically engineered RBCs in vitro to introduce non-native genes that encode enzyme recognition sites (ERS) on the RBC surface, allowing therapeutic proteins or drugs to be covalently attached to the RBC membrane while preserving the integrity of the membrane (Shi et al., 2014). This study initially used biotin to demonstrate that the modified RBCs could circulate in vivo for at least 28 days.

Then, antibodies were attached to the RBCs, enabling the engineered cells to specifically bind to target cells that express the antibody target protein, showing that RBCs can be engineered to target specific locations in the body. This same tethering technique was also used to display disease-associated autoantigens on the surface of RBCs for inducing antigen-specific tolerance in autoimmune diseases (Pishesha et al., 2017).

5.6. RBC BIOMIMETICS

The development of RBC biomimetics involves incorporating at least one native RBC element into the design of nanoparticles to improve their performance. For instance, coating nanoparticles with red blood cell membranes can extend their circulation time by evading the immune system's macrophage uptake and avoiding rapid clearance (Hu et al., 2013). Additionally, RBC membrane components such as lipids and associated membrane proteins are incorporated into the membrane coating, lending the nanoparticles the same bound proteins as RBCs (Figure 5.2, left panel). For example, iron oxide magnetic clusters have recently been coated with RBC membranes to enhance their blood retention time when injected into mice, allowing them to accumulate in tumors. When exposed to near-infrared light within the tumor, the magnetic clusters resulted in enhanced tumor cell killing (Ren et al., 2016). The use of RBC membrane coatings on various nanoparticles has also been explored for pathogen targeting (Escajadillo et al., 2017) and solid tumor penetration (Luk & Zhang, 2015).

Various efforts have been made to develop artificial red blood cells (RBCs) that possess characteristics similar to natural RBCs, such as size, shape, and deformability, allowing them to navigate through the vasculature, carry oxygen to tissues, and display biomarkers on their cell membranes, including self-recognition markers to prevent immune attack and rapid clearance from circulation (Merkel et al., 2011). Recently, biomimetic synthetic rebuilt RBCs (RRBCs) have been engineered from polymers with etched silica cores, incorporating all the features of natural RBCs and additional non-native functionalities, such as drug delivery, magnetism for imaging, and toxin sensing and sequestration (Guo et al., 2020). The functionality of these synthetic cells was demonstrated to be similar to that of natural RBCs, and they were loaded with MRI contrast agents for imaging and various anticancer drugs with different release profiles for use in cancer treatment.

5.7. PLATELETS AS DELIVERY VEHICLES

5.7.1. Intracellular Encapsulation

Activated platelets undergo significant structural changes and cause the release of their intracellular contents. Natural sites of platelet aggregation and activation have been utilized to achieve targeted therapeutic release upon activation. For instance, free doxorubicin (DOX), an anti-cancer drug, has poor circulation times and causes cardiotoxic effects (Shi et al., 2011). By passively loading DOX onto platelets through diffusion, these platelets home to tumor cells and release DOX upon activation (see Figure 5.2, right panel). This approach eliminates the negative effects of free DOX and results in the preferential killing of tumor cells in both in vitro and in vivo lymphoma models (Xu et al., 2017). Another strategy is to genetically engineer platelet progenitor stem cells to produce therapeutic proteins in platelets. This method has been used to treat hemophilia A by expressing platelet factor VIII (pFVIII) in induced pluripotent stem (iPS) cell-derived megakaryocytes (see Box 1). Platelets derived from these megakaryocytes survive in circulation and release pFVIII upon activation, improving hemostatic function in mouse models of hemophilia A (see Figure 5.2, right panel) (Lyde et al., 2019).

5.7.2. Surface Modification

The advantage of platelets being able to circulate unobstructed in the bloodstream, similar to red blood cells, is a significant advantage for drug delivery. Additionally, platelets naturally aggregate with circulating tumor cells (CTCs) through specific interactions between surface ligands, which promotes metastasis (Mammadova et al., 2020). To capitalize on this, hematopoietic stem cells were genetically engineered to produce the tumor necrosis factor-related apoptosis-inducing ligand (TRAIL), a ligand that triggers apoptosis in tumor cells. In this study, when the genetically modified HSCs differentiated into platelets, the TRAIL-expressing platelets were proven to kill cancer cells and significantly inhibit metastasis in a prostate cancer mouse model (Li et al., 2016). In another instance, megakaryocytes (MKs), which are the precursor cells of platelets, were genetically modified to express the programmed cell death protein 1 (PD-1) on their cell membrane. Tumor cells expressing PD-L1 suppress the immune system by depleting the T cells that would typically attack and eliminate the tumor. In this study, platelets expressing the PD-1 protein on their membrane accumulated in tumors and were shown to reinvigorate the exhausted T cells, allowing the T cells to significantly reduce the size of the tumor in mice (Zhang et al., 2018).

5.8. PLATELET BIOMIMETICS

The use of platelet membrane components to coat nanoparticles has been explored as a means to address the issue of short circulation time caused by interactions with the blood circulation. While platelet transfusions are currently the primary method for increasing a patient's platelet count, there is a need for synthetic biomimetic platelets due to their short storage life and potential for contamination. Initial attempts to create synthetic platelets involved sonication of platelets in the presence of nanoparticles, which resulted in complete platelet membrane coverage of the nanoparticle surface without affecting the nanoparticle composition (Figure 5.3, right panel). This method has been used to coat nanoparticles for the treatment of thrombocytopenia, multiple myeloma, and thrombus (Hu et al., 2016). Additionally, ongoing research is focused on engineering synthetic platelets from various biomaterials, such as liposomes, albumin microparticles, and latex particles, to match platelet shape and mechanics and enable similar clot capabilities as natural platelets (Nandi & Brown, 2016). These synthetic platelets are designed to target wounds and promote healing by interacting with surrounding cells and tissue. They can also be functionalized with various molecules, such as peptides, antibodies, and growth factors, to provide targeting and therapeutic capabilities. For instance, adding the Von Willebrand factor (VWF-A1) domain to synthetic platelets demonstrated their ability to adhere to injured endothelial vascular walls under high shear stress and form a clot at the injury site, while also maintaining hemostasis (Doshi et al., 2012).

5.9. MESENCHYMAL STEM CELLS AS DELIVERY SYSTEMS

Adult multipotent stem cells called mesenchymal stem cells (MSCs) are present in bone marrow and exhibit the capacity for self-renewal and differentiation into specialized tissues, such as bone, fat, and cartilage (Deans & Elisseeff, 2009). Additionally, MSCs have been shown to stimulate tissue regeneration and wound healing by secreting paracrine factors (Berglund et al., 2017).

5.9.1. Paracrine Function of MSCs

Mesenchymal Stem Cells (MSCs) possess tissue repair properties, as well as anti-apoptotic, anti-inflammatory, immunomodulatory, and pro-angiogenic qualities that are mediated through paracrine signaling, making them desirable delivery systems (Baraniak & McDevitt, 2010). Since MSCs possess inherent immunomodulatory properties, they can exhibit both pro-inflammatory and anti-inflammatory effects. The majority of therapies utilizing MSCs focus on their anti-inflammatory and immunomodulatory properties. For instance, osteoarthritis (OA) is a condition characterized by inflammation within joints that lacks effective treatments. Preclinical data indicates that injecting MSCs into affected joints can reduce pain and inflammation, believed to result from MSCs secreting anti-inflammatory cytokines such as interleukin (IL)-6 and IL-10 (Zhou et al., 2019). Additionally, emerging evidence suggests that MSCs play an essential immunomodulatory role in regulating the immune system by significantly influencing immune cell function and proliferation (Fan et al., 2020). Studies suggest that MSCs can influence macrophage function in tissues to transform them from a pro-inflammatory (M1) to an anti-inflammatory (M2) phenotype (Cho et al., 2014). For example, in a mouse model of myocardial infarction, macrophages that infiltrated the cardiac tissue were primarily in the M1 phenotype. However, when MSCs were injected into the injured myocardium, macrophages adjacent to MSCs exhibited a strong preference for the M2 phenotype, leading to accelerated tissue repair and cardiac remodeling (Jin et al., 2019). Similarly, MSCs have been demonstrated to secrete immunomodulatory agents, such as indoleamine 2,3-dioxygenase (IDO), nitric oxide (NO), programmed death ligand 1 (PD-L1), Galectin-1, and transforming growth factor β1 (TGF-β1), which transform inflammatory effector T-cells into anti-inflammatory regulatory T cells. Notably, the phenotype of these T cells has been implicated in playing a crucial role in cancer.

5.9.2. Engineering MSCs for Targeting Brain Cancers

Mesenchymal stem cells (MSCs) have become an appealing option for cell-based therapies aimed at treating brain cancers, such as glioblastomas (GBMs), due to their ability to cross the blood-brain barrier (BBB), which naturally allows them to reach tumor cells, their capacity for paracrine functions, and their low immunogenicity (Mangraviti et al., 2016). As a result, researchers have extensively explored the use of MSCs as vehicles for delivering therapeutic agents in cancer treatments by genetically engineering them to release cytokines directly into tumors upon arrival, thus inhibiting tumor growth (Reagan & Kaplan, 2011). In these studies, MSCs have been equipped with tumor-targeted drug carriers containing cytokines and proteins, such as the anti-inflammatory cytokine interferon (IFN)-β (Altaner et al., 2014), TRAIL, and bone morphogenetic factor 4 (BMP4), a growth factor known to decrease the tumorigenicity of brain tumor-initiating cells (Li et al., 2014). Overall, these efforts have shown encouraging outcomes for improving the survival of patients with GBMs and other cancers (Zhang et al., 2018).

5.10. NEUTROPHILS AS DELIVERY SYSTEMS

Neutrophils, which are the most abundant type of white blood cells in the human circulatory system, play a crucial role in the immune response against pathogens and in the recruitment of cells to sites of injury (Mayadas et al., 2014). These cells migrate towards areas of inflammation by following chemical gradients of various substances, such as interleukin-8 (IL-8), complement component a (C5a), leukotriene B4 (LTB4), platelet-activating factor, and N-formyl peptides, which are produced by invading bacteria (Harvath, 1991). Neutrophils have several characteristics that make them suitable for use as delivery vehicles. For example, they are highly mobile, able to cross biological barriers like the blood-brain barrier where other cells or molecules cannot, and they are biocompatible. Additionally, they are attracted to invading pathogens and can deliver therapeutics directly to inflamed tissue (Chu et al., 2018). As a result, there has been a growing interest in using neutrophils to deliver small molecules, polymeric and metallic nanoparticles (Lin et al., 2018). However, the use of neutrophils as carriers for peptides, proteins, and nucleic acids has received relatively little attention, making this an area of research that is ripe for exploration.

5.10.1. Neutrophils as Drug Carriers

Delivering drugs to the brain is a common challenge with many drug delivery approaches (Fossati et al., 1999). Neutrophils, a type of white blood cell, have the ability to cross the blood-brain barrier (BBB) and enter tumor sites such as gliomas and glioblastomas. Researchers have exploited this property by loading neutrophils with chemotherapy drugs encapsulated in liposomes, allowing the cells to carry the drugs across the BBB and deliver them to the tumor site (Xue et al., 2017). In one study, neutrophils loaded with paclitaxel-filled liposomes were shown to slow the growth of recurrent tumors and improve survival in mice. Additionally, neutrophils can be used to track the delivery of drugs to tumor sites. For example, neutrophils loaded with magnetic mesoporous silica nanoparticles and doxorubicin can be tracked using magnetic resonance imaging in inflamed mouse models to verify proper delivery of the drug (Wu et al., 2018).

Several studies have utilized the cell surface of neutrophils to effectively "hitch-hike" to targeted sites in vivo (Zhang et al., 2017). Injury and disease cause the vascular endothelium to become inflamed and recruit neutrophils. This natural targeting of neutrophils to the injured endothelium has been harnessed to deliver nanoparticles across blood vessel barriers, which is otherwise a significant challenge. To accomplish this, drug-loaded nanoparticles were internalized by the neutrophils, which then carried the nanoparticles across blood vessel barriers and entered sites of inflammation or infected tissue to deliver anti-inflammatory or antibacterial therapeutics (Chu et al., 2015). Similarly, neutrophils have also been modified for targeted therapeutic delivery to tumors. Albumin nanoparticles loaded with pyropheophorbide-a, a drug used for tumor therapy, were bound to activated neutrophils for in vivo delivery to tumors. These Ppa-loaded nanoparticles attached to activated neutrophils and were delivered directly to tumor sites, improving outcomes by suppressing tumor growth and increasing survival rates (Chu et al., 2017). In another study, a poly(sialic acid) biopolymer that naturally binds to receptors expressed on neutrophils was conjugated to an octadecylamine compound, forming a PSA-ODA conjugate. This conjugate was then anchored onto the surface of liposomes filled with pixantrone, a chemotherapy drug. These PSA-ODA-filled liposomes were bound to neutrophil membranes to enhance neutrophil-targeting in in vivo tumor mouse models (Luo et al., 2018). This study showed considerably higher cellular uptake of the loaded neutrophils, leading to tumor growth inhibition.

A different approach that takes advantage of neutrophils' natural ability to home to infections and their capacity to recognize and engulf bacteria is to load neutrophils with deactivated nonpathogenic bacteria coupled to antimicrobial agents (Wendel et al., 2015). In this study, deactivated bacteria were incubated in a solution with chlorhexidine (CHX), an antimicrobial that attaches naturally to the bacterial cell wall. After attaching CHX to the bacterial cell wall, the bacteria were washed and incubated with neutrophils to be engulfed, thus loading the neutrophils with antibacterial drugs. These loaded neutrophils were then injected into a mouse model of liver abscess, which showed that the loaded neutrophils maintained their ability to home to sites of infection and released the antibacterial therapies upon reaching the site of infection.

5.11. BACTERIA AS DELIVERY SYSTEMS

Bacteria naturally inhabit various microbiomes within the human body, such as the nose, mouth, breast, and gut. These bacteria play a crucial role in maintaining immune and metabolic balance throughout a person's lifetime (Thursby & Juge, 2017). Since certain bacteria occupy specific niches in various body areas and can selectively target and thrive in tumor microenvironments, using them as a delivery system for drugs and other molecules is a promising option (Nejman et al., 2020). In the following section, we will explore how bacterial systems have been employed as therapeutic delivery vehicles.

5.11.1. Targeting Cancer

The tumor microenvironment presents significant obstacles to delivering therapeutics to the tumor site, including defective vascular architecture, elevated interstitial pressures, and increased permeation distances (Maeda et al., 2000). Recently, bacteria have been engineered to deliver drug-containing nanoliposomes (Felfoul et al., 2016), nanobody antagonists (Gurbatri et al., 2020), and functional nucleic acid molecules (Akin et al., 2007) to cells within tumor microenvironments. For instance, Magnetococcus marinus strain MC-1, a magnetotactic bacterium, was shown to deliver nanoliposomes loaded with the anti-tumor drug 7-ethyl-10-hydroxycamptothecin (SN38) to hypoxic areas of tumors. The bacterial cells were modified to attach the drug-loaded liposomes via chemical conjugation, and then injected near tumors in mice. An external magnetic field was used to guide the modified bacterial cells to the center of the tumor, where the magnetic field was turned off, allowing the cells to preferentially migrate deep into the tumor, towards areas of lower oxygen concentration. This demonstrates the ability of these bacterial cells to penetrate solid organ tumors and regions lacking vascularization, with up to 55% of the injected cells entering the hypoxic region of the tumor.

In a recent study, it was demonstrated that bacteria can be utilized to deliver functional nucleic acids into solid tumors in humans. The study found that nanoparticles attached to the surface of bacteria were successfully able to deliver reporters such as green fluorescence protein (GFP), firefly luciferase, and secreted alkaline phosphatase (SEAP) to mouse tumor models (Akin et al., 2007). Once the bacteria were phagocytosed by the tumor cells, they released the encapsulated DNA into the tumor cells, leading to successful gene expression. These findings suggest that bacteria could be a promising alternative for non-viral delivery of genes to tumor cells, potentially reprogramming them for cell death.

5.12. SYNTHETIC BIOLOGY TOWARDS IMPROVED DELIVERY DEVICES

The use of cells as drug delivery vehicles has greatly expanded the possibilities for treating various diseases, but sometimes native cells do not have the necessary properties to achieve the desired therapeutic outcomes (Xie & Fussenegger, 2018). To overcome this, synthetic biology can be used to engineer cells with the required phenotype by genetically modifying and reprogramming them with gene circuits. Synthetic biologists have been working on developing genetic tools to enable the reprogramming of cells with new functions using genetic circuits that allow for biosensing, oscillatory behavior, logic gates, and switches (Krawczyk et al., 2020). These tools have proven effective in many therapeutic applications. By combining advances in cell delivery and targeting with synthetic biology, autonomous in vivo delivery systems can be created that can sense and respond to diverse environments and deliver payloads in response to local microenvironment dynamics and the severity of disease or injury (Scheller and Fussenegger, 2019). For example, a recent study engineered *E. coli* with a genetic circuit that enabled them to regulate gene expression in response to population density within a tumor. Once the engineered bacteria reached a high cell number, they underwent lysis to release immunotherapeutic agents that promoted tumor regression (Chowdhury et al., 2019). Synthetic biology tools that can dynamically control gene expression in a spatiotemporal manner can be utilized to reprogram stem cells for the efficient production of large numbers of adult cells required for therapeutic purposes (Fitzgerald et al., 2017). The genetic circuits used during stem cell differentiation enable the precise control of gene expression, allowing for the creation of custom receptors on cell surfaces and the loading of cells with therapeutic proteins at specific stages of cell fate. These genetic circuits can also be used to engineer custom receptors as designer triggers for the activation and release of payloads at desired locations (Guye et al., 2016). Overall, genetic circuits can reprogram stem cells for custom loading of therapeutic proteins and engineering custom receptors, which can be activated and released at specific times and locations (Bush et al., 2019).

5.13. CELL ENGINEERING FOR CANCER IMMUNOTHERAPY

Cancer immunotherapy, which bolsters the immune system's response to tumor cells, has emerged as a new pillar in the fight against cancer (Wang et al., 2017). Among the various treatment strategies, the engineering of immune cells using chemical methods and/or synthetic nanomaterials without disrupting key cellular functions has gained considerable attention for enhancing cancer immunotherapy (Stephan & Irvine, 2011). For example, to overcome the rapid loss of T cell effector function in adoptive cell therapy (APC), researchers have used direct surface conjugation of T cells with nanoparticles containing cytokines or immunomodulators. This approach not only sustains T cell stimulation after transplantation but also avoids the side effects caused by high doses of adjuvants administered systemically (Zheng et al., 2013). These engineering strategies offer a new way to modulate signaling events at the T cell-APC interface. In the development of therapeutic cancer vaccines, the engineering of dendritic cells (DCs) through in vitro endocytosis of antigen-conjugated nanoparticles with bioimaging functionalities has been shown to enhance antigen-specific T cell responses and inhibit tumor growth. This approach also allows for tracking the fate of DCs during the processes (Xiang et al., 2015).

The use of checkpoint blockade immunotherapy, which works by interfering with inhibitory pathways in T cells, has resulted in durable clinical responses and long-term remissions in treating various types of cancer (Sharma & Allison, 2015). However, there are still limitations to improving its effectiveness, such as insufficient lymphocyte infiltration and off-target binding of intravenously administered checkpoint inhibitors (Figure 5.3). Some groups developed a checkpoint inhibitor delivery platform for postsurgical cancer immunotherapy by surface conjugation with aPDL1 (P-aPDL1) (Semple et al., 2011). Activated in vitro, aPDL1 is released in the form of platelet plasma membrane-derived microparticles (PMPs) (Figure 5.3B), accompanied by the release of several pro-inflammatory cytokines (Wang et al., 2017). After intravenous injection, P-aPDL1 is enriched at the surgical wounds around the residual tumors (Figure 5.3C), leveraging the injury homing nature of platelets in immune checkpoint delivery. The release of aPDL1-tagged PMPs and various inflammatory factors synergistically boost the activation of T cells for destroying tumor cells with high efficiency (Figure 5.3D). Furthermore, this strategy of engineering platelets with immune checkpoint inhibitors demonstrates the potential for treating tumor metastasis. The versatility of this engineering technique was proven by further inhibiting the recurrence of triple-negative breast cancer models. Overall, this strategy provides an effective and generalizable approach to targeted delivery of antibodies to enhance their immunotherapeutic potential in preventing postsurgical cancer recurrence while limiting side effects.

The surface of cells contains numerous reactive chemical groups, including thiols, amines, and carboxylates, which offer numerous sites for the covalent attachment of synthetic materials. Additionally, the interactions between nano-biological interfaces, such as the uptake of nanomaterials by cells, can be studied to give cells new functions. Notably, bioorthogonal chemistry, which has proven to be highly effective in selectively modifying biological species (Sletten et al., 2007), may provide new avenues for this field. These methods are both convenient and robust, allowing cells to be given new functions while maintaining their own biological capabilities, making them a powerful complement to traditional genetic engineering techniques for enhancing the functions of living cells.

Recently, a whole-tumor lysate was engineered containing melanin from B16F10 cells for cancer immunotherapy by incorporating it into near-infrared light (NIR)-irradiated microneedle patches along with granulocyte-macrophage colony-stimulating factor (GM-CSF) (Figure 5.4) (Ye et al., 2017). This sustained transdermal delivery of GM-CSF effectively recruited and activated dendritic cells (DCs), which then internalized the released lysate and presented the tumor antigens. Additionally, the NIR irradiation stimulated the melanin to generate a high local temperature, which facilitated the migration of DCs and other immune cells to the vaccination sites by increasing local microcirculatory blood/lymphatic perfusion and releasing danger signals, including reactive oxygen species and inflammatory cytokines. This process resulted in increased filtration of polarized T cells and the topical release of cytokines, ultimately initiating a B16F10-specific immune response. The feasibility of this vaccine design was demonstrated by the high survival rate of mice after tumor challenge and its antitumor effects on primary and distant tumors.

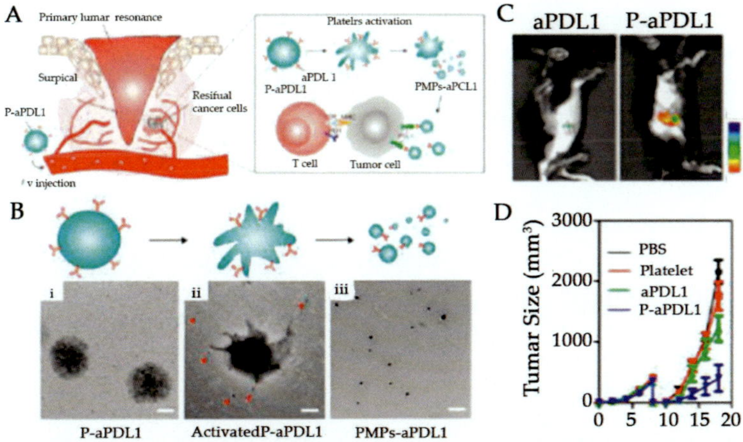

Figure 5.3. *(A) A schematic illustration depicting the delivery of an aPDL1-conjugated platelet to the surgical wound site is shown. (B) Transmission electron microscopy (TEM) images of aPDL1 (i) before and (ii) after activation and (iii) release of PMPs are also provided, with scale bars of 0.5 μm.*

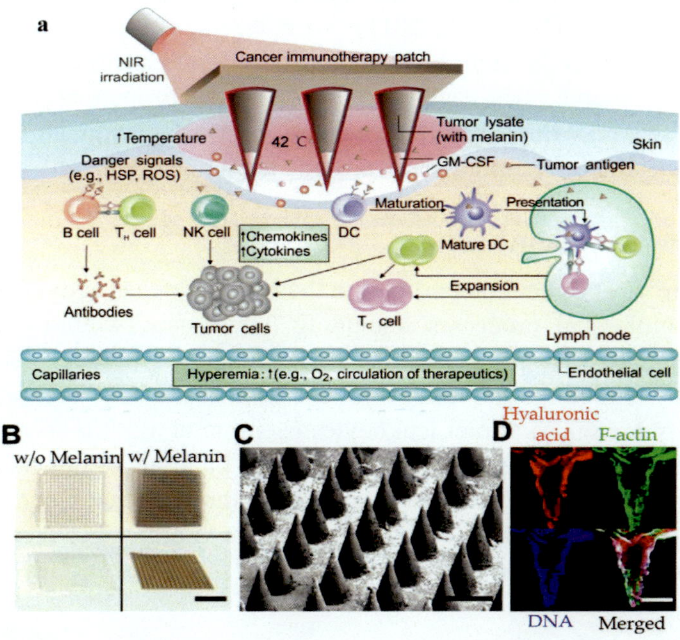

Figure 5.4. *(A) The following is a schematic representation of the use of microneedle patches to engineer tumor lysate for transdermal vaccination. T cell: T helper cell, Tc cell: cytotoxic T cell. (B) This photograph shows the microneedle patches without (w/o) and with (w/) melanin, with a scale bar of 4 mm. (C) The scanning electron microscopy image of the microneedle patch is provided. (D) The cross-*

sectional fluorescence images in this section show the presence of the lysate inside the microneedle patches. Actin filaments, DNA fragments, and the hyaluronic acid matrix are visualized using Alexa Fluor 488-phalloidin (green), Hoechst (blue), and rhodamine (red), respectively. The scale bar for this image is 200 μm.

Source: Chen et al., (2018).

In vivo fluorescence imaging of mice 2 hours after intravenous injection of free aPDL1-Cy5.5 (left) and P-aPDL1 Cy5.5-conjugated platelets (right) is demonstrated. The sizes of recurrent tumors following different postsurgical treatments are shown, with error bars representing the standard error of the mean. (Chen et al., 2018)

5.14. ENGINEERING NANOMATERIALS WITH BIO-MEMBRANES

The latest "top-down" engineering of cellular membranes to hide nanomaterials is providing insights that are helping to create biomimetic platforms for replicating natural cellular biological functions, such as targeting and evading the immune system (Hu et al., 2011). To date, various cell types, including leukocytes, cancer cells, stem cells, bacteria, and platelets, have been used as source cells for membranes, each with its own unique properties (Dehaini et al., 2017). This section will concentrate on the recent work in engineering synthetic nanocarriers using biomembranes to provide local therapeutic effects.

5.14.1. Platelet Membrane

As one of the most critical bioparticles in the body, platelets actively participate in various biological processes, including hemostasis, wound healing, and immune response (Semple et al., 2011). For instance, platelets can enhance immunity by directing themselves to the injury site and discharging pro-inflammatory factors to attract leukocytes (Semple et al., 2011). Additionally, recent studies have uncovered that platelets can facilitate tumor metastasis by accumulating on CTCs to conceal them from immune attack and migrate to new tissue for tumor growth, with biomolecular binding between P-selectin and the CD44 receptor playing a pivotal role in this process (Gay & Felding-Habermann, 2011).

The approach involved leveraging the interaction between platelets and cancer cells to develop a platelet-mimicking drug delivery system (Figure 5.5A) (Hu et al., 2015). The utilization of the "top-down" method to wrap platelet membranes around Dox-loaded nanogels, taking advantage of the abundant surface-associated proteins to replicate the specific interactions between platelets and cancer cells. Next, the platelet membranes with tumor necrosis factor-related apoptosis-inducing ligand (TRAIL) were decorated and created a platelet-membrane-coated nanovehicle (PM-NV) that could efficiently target the surface of cancer cells to enhance the therapeutic efficacy of TRAIL. The nanogel inside the PM-NV composite, containing acid-responsive cross-linkers, dissociated inside the acidic endosomes, releasing the loaded drugs. This sequential drug delivery fashion led to synergistic cytotoxicity (Figure 5.5B). Additionally, the preservation of the CD47 protein on the membrane minimized uptake by macrophage cells and prolonged the in vivo circulation time (Tsai et al., 2010). The superior anticancer potency of PM-NV in the MDA-MB-231 tumor model was demonstrated, attributing it to the selective cancer homing capability of PM-NV and synergistic anticancer effect based on the site-specific delivery of TRAIL and Dox (Figure 5.5C). Furthermore, the versatility of PM-NV was confirmed by a reduction in the number of circulating tumor cells after intravenous administration, as evidenced by decreased lung metastasis (Figure 5.5D).

Many studies have revealed a relay drug delivery strategy that amplifies tumor targeting signals by inducing local inflammation of tumor blood vessels (Hu et al., 2017). This approach was designed based on the inherent property of platelet homing to injured blood vessels and specific interactions between platelets and cancer cells. The strategy consists of two modules: the signaling transmission nanocarrier (NCA) and the execution nanocarrier (NCB). The NCA is made from an Arg-Gly-Asp peptide functionalized polymeric nanoparticle loaded with TNFα, an inflammation inducer, while the NCB is composed of platelet-membrane-coated nanoparticles encapsulated with paclitaxel to receive the broadcasting inflammation signal and accumulate at the tumor site. In vitro studies showed that NCA induced a strong inflammation response on human umbilical vein endothelial cells, as evidenced by upregulated cell adhesion molecules and increased release of cytokines. After in vivo administration, NCA selectively accumulated at the tumor site and triggered local inflammation of blood vessels. Furthermore, the sequential administration of NCA and NCB significantly increased drug availability and retention time at the tumor site compared to NCB alone, resulting in enhanced inhibition of tumor growth. Importantly, this relay delivery induced minimal systemic inflammation response and did not cause any pathological abnormalities in normal tissues.

In order to address the challenge of thrombus formation and reduce the mortality rate associated with the treatment of multiple myeloma, a nanoplatelet delivery system was developed (Hu et al., 2016). This system involves coating bortezomib-loaded nanoparticles with platelet membranes, which are then conjugated with tissue plasminogen activator (tPA) and alendronate (Ald) for clot lysing and bone targeting, respectively. The resulting nanoplatelets demonstrate increased drug accumulation at the myeloma site through a two-step process involving first binding to bone via the calcium chelating ability of Ald and then interacting with myeloma cells through cellular affinity, such as the interaction between P-selectin on the platelet membrane and CD44 receptors on tumor cells. This leads to enhanced therapy efficacy. Furthermore, the nanoplatelets are capable of preventing thrombus complications by dissolving the clot. Both in vitro and in vivo studies have confirmed the effectiveness of this approach.

Researchers are actively working on creating biomimetic nanoparticles that mimic the structural and functional characteristics of biological assemblies for therapeutic purposes (Yoo et al., 2011). The use of natural cells, which are inherently biocompatible and nonimmunogenic, in functionalizing synthetic nanoparticles is expected to help bridge the gap between natural cells and synthetic formulations. This approach may lead to the development of more effective drug delivery systems that are inspired by nature.

5.14.2. Subcellular Membranes

When cargo is internalized, the plasma membrane invaginates to form vesicles. These vesicles then merge with early endosomal systems, where the internalized substances are sorted (Scott et al., 2014). Most of the membrane components, including lipids and proteins, are recycled back to the plasma membrane via early and recycling endosomes (Scott et al., 2014). Here, the membrane-associated proteins and receptors are distributed randomly over the endosome membrane (Van Weering et al., 2014). Recently, it was discovered that multiple molecules on the plasma membrane mediate the aggregation of homotypic cells (Glinsky et al., 2003).

A new method for creating endosome membrane-encapsulated nanogels (EM-NGs) (Figure 5.6A) was proposed to deliver cancer drugs homotypically. Mesoporous silica-coated magnetic nanoparticles with cross-linkers and photoinitiators inside, and methacrylated hyaluronic acid on the surface, were used as model nanomaterials. UV irradiation caused the nanoparticles to cross-link and form nanogels, which appeared as round or oval shapes with multiple silica/magnetic nanoparticles clustered inside (Figure 5.6B), and the endosome membrane components were preserved on the EM-NG shell (Figure 5.6C). The significantly higher *in vitro* cytotoxicity induced by the DOX-loaded EM-NGs (DOX-EM-NGs) compared to DOX-loaded nanogels (DOX-NGs) and free DOX indicated the targeting effects of the EM-NGs. However, the negligible differences in cytotoxicity for human lung adenocarcinoma epithelial cells treated with DOX-EM-NGs and DOX-NGs underlined the homotypic targeting ability of the EM-NGs (Figure 5.6D). This method is more efficient than the traditional method of coating nanomaterials with plasma membranes using mechanical sonication or extrusion, as it relies on the internalization process, which can maintain the function of biomolecules (Hu et al., 2011). Although the detailed mechanism underlying the function of endosome-encapsulated nanogels still needs further study, this novel cell engineering method has the potential to develop alternative drug delivery systems, especially personalized therapeutics.

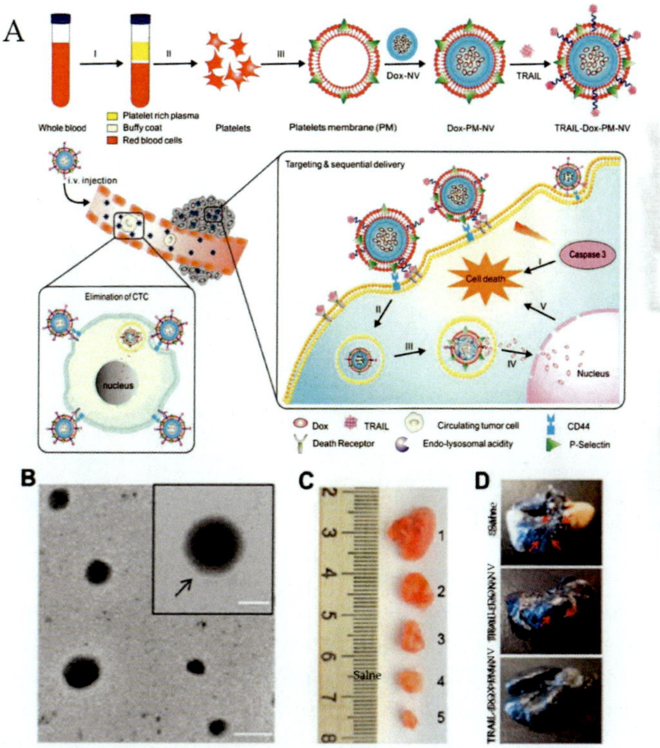

Figure 5.5. *(a) The design of the drug delivery system that mimics platelets. (b) The characterization of PM-NV using TEM, with the black arrow indicating the presence of the platelet membrane. Scale bars: 100 nm and (inset) 50 nm. (c) The in vivo antitumor efficacy of TRAIL-Dox-PM-NV. Tumors from mice treated with (1) saline, (2) TRAIL-Dox-NV, (3) TRAIL-PM-NV, (4) Dox-PM-NV, and (5) TRAIL-Dox-PM-NV are shown. (d) The in vivo antitumor metastasis efficacy, with red arrows indicating visible metastatic foci.*

Source: Chen et al., (2018).

Subcellular compartments beyond the conventional cytoplasm can be harnessed to create synthetic materials for diverse applications, as research on extracellular vesicles, which serve as crucial intercellular communication mediators, has demonstrated (El Andaloussi et al., 2013). The capacity of extracellular vesicles to regulate both normal and pathological processes, as well as their role in disease progression, offers valuable insights for developing and modifying these natural particles for therapeutic use (El Andaloussi et al., 2013). Furthermore, modifying cellular membranes with antigens or quantum dots has shown promise in creating micro/nanoparticles for theranostic applications (Chen et al., 2015). Many labs are actively pursuing research on engineering cell-derived nanovesicles for cancer immunotherapy.

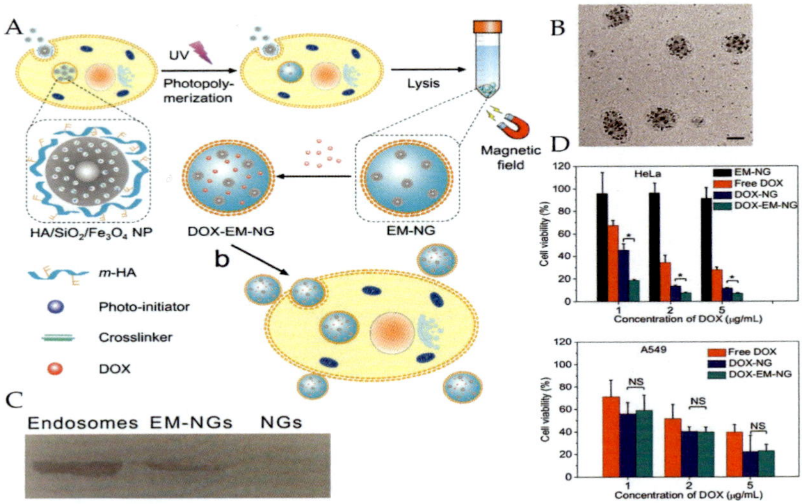

Figure 5.6. *(A) A schematic diagram depicts the procedure of engineering nanoparticles by incorporating endosome membranes for homotypic cell-targeted drug delivery. (B) A transmission electron microscopy (TEM) image shows the EM-NGs, with a scale bar of 200 nm. (C) Western blotting analysis of endosomes, EM-NGs, and bare NGs using early endosome antigen 1 as a marker. (D) In vitro cytotoxicity assays of DOX-EM-NGs against HeLa and A549 cells, with * indicating $p < 0.05$ and NS representing not significant.*

Source: Chen et al., (2018).

5.15. RED BLOOD CELL ENGINEERING FOR SMART INSULIN DELIVERY

Diabetes mellitus is a multifaceted and progressive condition marked by persistently elevated blood glucose levels, affecting a significant number of people globally. Despite insulin injections being the traditional treatment method, they are often accompanied by pain and difficulty in achieving strict glucose control (Xie et al., 2016). Therefore, considerable resources have been invested in developing alternative therapies such as cell-based treatments and insulin delivery using synthetic materials (Veiseh et al., 2015). Recent research has shown that glucose-responsive materials have significant potential for insulin delivery (Mo et al., 2014). However, creating a system that can swiftly respond to elevated blood glucose levels while closely mimicking the body's physiological needs for insulin remains a significant challenge (Bakh et al., 2017).

Many groups tackled this challenge by designing a new insulin delivery formulation that leverages the reversible interaction between glucose and the abundant glucose transporters (GLUTs) on the surface of red blood cell (RBC) membranes (Figure 5.7A) (Wang et al., 2017). They achieved this by engineering RBCs with a glucose derivative (glucosamine)-modified insulin (Glc-Insulin) that can specifically bind GLUT1 on human RBCs and GLUT4 on mice RBCs with high efficiency. The binding of Glc-Insulin to GLUTs was driven by competitive binding of free glucose, leading to quick insulin dissociation from the RBCs at 400 mg/dL glucose within 30 min, whereas slower release was observed under normal glycemic conditions (Figure 5.7 B, C). After intravenous injection of the Glc-Insulin-bound mice RBCs into diabetic mice, the injected cells remained stable for up to 72 hours, facilitating long-term insulin delivery. In vivo diabetes treatment showed that Glc-Insulin-bound mice RBCs could restore blood glucose levels for a much longer time than free Glc-Insulin or free insulin plus mice RBCs, due to the long-term stability and glucose-responsive property of the designed system (Figure 5.7 D). The dynamics of insulin release detected in intraperitoneal glucose tolerance tests further verified the glucose-dependent responsiveness. Furthermore, they also developed a bioinspired nanoformulation using RBC-membrane-coated nanoparticles for insulin delivery via the same mechanism, which can be conveniently integrated with microneedle patches for painless intradermal insulin delivery, with potential for clinical applications (Figure 5.7E).

The use of red blood cells (RBCs) in cell therapies has several advantages, as they have a long circulation time and are biocompatible, which helps to avoid some of the challenges associated with traditional cell therapies. However, it is important to note that RBCs are unable to replenish insulin, which means that new cell engineering methods or the development of alternative cell species may be necessary to create more sustainable treatment options for diabetes (Chen et al., 2018).

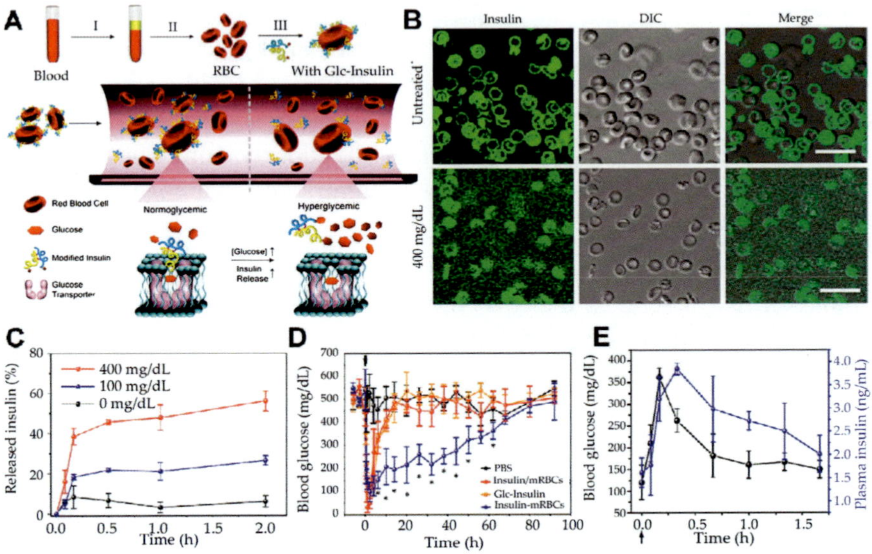

Figure 5.7. *(A) A schematic diagram illustrates the engineering of red blood cells (RBCs) with Glc-Insulin for diabetes treatment. (B) Confocal microscopy images show the release of fluorescence-labeled insulin from the cell surfaces under high blood sugar conditions. (C)The profiles of in vitro Glc-Insulin release from mRBCs at various glucose concentrations are shown. (D) The blood glucose levels (BGL) in diabetic mice after different treatments are shown in panel D, with statistical significance indicated by an asterisk and $p < 0.05$. (E) Plasma insulin levels and glucose levels after intraperitoneal glucose tolerance tests are shown in panel E.*

Source: Chen et al., (2018).

5.16. B-CELLS INTEGRATED WITH MICRONEEDLE PATCHES

Cellular therapy based on β-cells has long been considered the optimal treatment for diabetes due to their inherent ability to sense blood glucose levels and deliver insulin sustainably (Veiseh et al., 2015). To overcome the obstacles associated with conventional direct islet transplantation and immunoisolation techniques, some labs have developed a painless microneedle-patch strategy that externally engineers β-cells to control internal blood glucose levels (Figure 5.8A) (Ye et al., 2016). This design involves positioning alginate microgel-encapsulated β-cells and "glucose-signal amplifiers" (GSAs) on the top and within the matrix of the microneedle patches, respectively (Figure 5.8B). When exposed to high glucose levels, the GSAs trigger a tandem enzymatic process that generates a local high glucose concentration (approximately 800 mg/dL) (Figure 5.8C). This glucose then diffuses into the externally located β-cell microcapsules, stimulating the secretion and delivery of insulin back into the transdermal vascular networks (Figure 5.8D). In vivo diabetes studies have shown that this platform can quickly respond to high blood glucose levels and restore them to the normal range within two hours, without any hyperglycemic or hypoglycemic peaks observed in the following six hours (Figure 5.8E). This minimally invasive approach to engineering β-cells with microneedle patches provides an alternative to conventional cell transplantation for treating diabetes and highlights the potential of engineering other types of cells as living drug depots for treating diseases involving cell function deficiency. However, issues regarding the longevity and inevitable contamination of externally positioned cells should be taken into consideration.

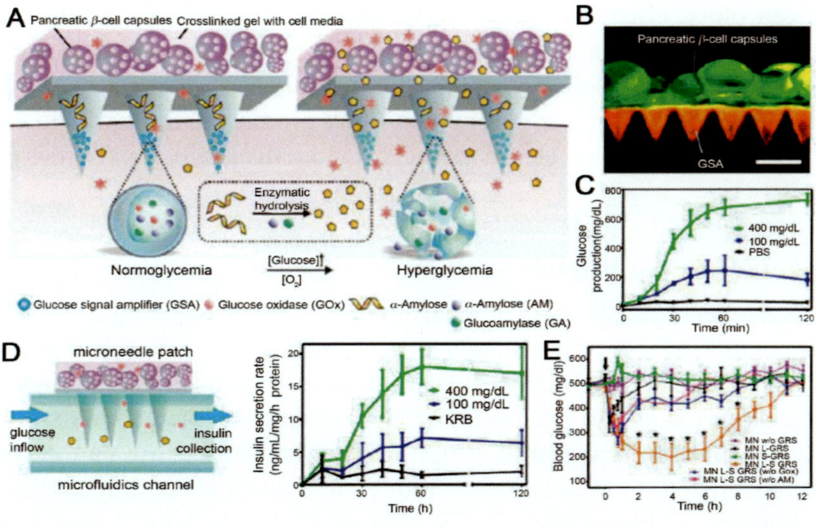

Figure 5.8. *(A) A schematic diagram demonstrates the integration of β-cells and GSAs with microneedle patches. (B) A fluorescence microscopy image shows the designed structure, with β-cells and GSAs labeled with calcium AM and rhodamine, respectively. The production of glucose by GSAs in various glucose solutions is shown in (C). The accumulated insulin secretion is simulated on the right, with the inflow of different glucose solutions through a microfluidics device shown on the left. (E) The in vivo therapeutic efficiency of the designed microneedle patch (MN L-S GRS) and control designs is compared in (E), with statistical significance indicated by asterisks and* $p < 0.05$.

Source: Chen et al., (2018).

5.17. ADVANTAGES OF LIVING CELL DELIVERY SYSTEM

The benefits of living cell delivery systems encompass various aspects.

5.17.1. Long Circulation, Serum Stability, and Immune Escape

Drugs encapsulated within cells will not diffuse through the plasma membrane and interact with external compounds, which reduces their ability to interact with homologous receptors and other substances. Additionally, the drug can avoid degradation in vivo (Glassman et al., 2020). By binding to the surface of living cells, prolonged circulation time and redistribution of drugs through non-specific adsorption of nanoparticles, antibodies, peptides, or other combinations can be achieved (Glassman et al., 2020). RBCs have been proven to be a safe and effective carrier for disease treatment (Rossi et al., 2016). Due to their prolonged blood circulation time in the vascular system, RBCs can circulate in the human body for about three months and distribute throughout the body. The RBCs-based drug delivery system significantly alters the pharmacokinetics of drugs by prolonging their circulation time, promoting the redistribution of blood from the plasma to cellular elements, and redistribution in tissues (Wang et al., 2018).

5.17.2. Large Quantities Can Be Obtained

Abundant natural platelets can be easily obtained without a complex production process, while MSCs preferentially inhabit perivascular niches in nearly all types of human tissue (Bhonde, 2021). These cells are easily separable from various tissues and can be proliferated in culture while maintaining their function (Almeida-Porada et al., 2020). Additionally, bacteria can be readily and genetically cultured in large quantities under suitable conditions (Sarkar et al., 2013).

5.17.3. Low Toxicity

The toxicity of cells as carriers is generally low. Neutrophils, for example, are minimally immunogenic, meaning they are well-characterized, stable chromosomally and functionally, and non-tumorigenic (Mooney et al., 2014). Additionally, they lack pro-angiogenic or immunomodulatory properties that promote tumor growth (Han et al., 2018). Endogenous cells undergo apoptosis when they become old or damaged, making them completely biodegradable (Yu et al., 2020). In many cases, using living cells for drug delivery can improve the bioavailability of drugs, reduce the dose required, and lower the risk of adverse reactions (Rossi et al., 2016).

5.17.4. Barrier Targeting and Crossing Ability

Red blood cells (RBCs) can be utilized as carriers for intravascular therapy and can also enter the extravascular cavity (Wang et al., 2018). Tumor cells can stimulate platelets to a certain extent, leading to platelet aggregation in the tumor (Ortiz-Otero et al., 2018). Platelets can release their contents in the tumor tissues due to the large gap between endothelial cells and the tumor vascular system, causing enhanced permeability and retention effect (EPR) and vascular leakage, making them a promising option for cancer treatment (Li et al., 2017). Mesenchymal stem cells (MSCs) are a promising method for developing gene transfer systems and drug loading strategies due to their inherent characteristics, including homing ability and tumor tropism (Wu et al., 2019). Immune cells are ideal carriers for drug delivery because they are naturally recruited to tumor sites during inflammation. Neutrophils are the first white blood cells to be recruited to inflammatory sites, and their permeability is the reason why they become carriers (Naumenko et al., 2019). During inflammation and cancer, endothelial cells are activated, and the expression of adhesion molecules is upregulated. Neutrophils adhere to activated endothelial cells and then crawl along blood vessels. Activated neutrophils move across the vascular barrier to reach the infected area (Han et al., 2018). Neutrophil-mediated transport has also shown tumor type-dependence and effectiveness in neutrophil-rich tumors (Naumenko et al., 2020). Neutrophils can pass through the blood-brain barrier and, although they cannot cure cancers, they can inhibit tumor growth (Xue et al., 2017). Macrophages possess the ability to penetrate deep into the hypoxic regions of tumors, while monocytes are effective drug delivery vehicles due to their capacity to traverse even the most rigorous biological barriers. They transport particles profoundly into the tumor tissue, where even small molecules, such as oxygen, can be challenging to access. For example, cyclo(Arg-Gly-Asp-D-Tyr-Lys) liposomes are engineered to recognize αvβ1 integrins on the surface of macrophages or monocytes, inducing the internalization of modified carriers. These engineered liposomes are transported into the penumbra of the brain, enhancing the targeting of therapeutic drugs and the therapeutic effect on stroke (Hou et al., 2019).

5.17.5. Immunotherapy

The cell membrane receptors can activate the immune system in the body. Ji et al. used engineered red blood cells to remove antigens, such as TNF-alpha, from serum (Ji et al., 2020). Tumors can be recognized and eliminated by the innate immune system through phagocytosis or antibody-dependent cell-mediated cytotoxicity. However, sometimes tumors can evade the immune system, leading to chronic inflammation that is difficult to resolve (Singh et al., 2019). Immune cells can infiltrate pre-tumor tissues and contribute to the establishment of the tumor's inflammatory microenvironment. As a result, immune cells play a crucial role in cancer treatment and are also suitable for drug delivery (Singh et al., 2019). Macrophages are the primary immune cells in most solid tumors and change in response to the tumor microenvironment (Zhu et al., 2021). Tumor-associated macrophages (TAMs) are the most abundant immune population in the tumor microenvironment and can even reach deep hypoxic areas of the tumor (Singh et al., 2019) to participate in tumor growth and metastasis (Petty & Yang, 2017). Neutrophils, as the first line of defense against infection, respond to diverse cancer inflammatory factors (Giese et al., 2019). While exerting basic anti-inflammatory effects, they are activated at the tumor site to secrete cytokines and interleukins, enhancing tumor therapy. Moreover, the transformation of monocytes to macrophages mediates the targeting of therapeutic agents to inflammatory tissues. A case study proved that modified monocytes retain important cellular functions, differentiate into macrophages, and distribute to various disease tissues by transmigrating through the endothelial monolayer (Anselmo et al.,2015).

5.17.6. Increased Drug Loading

The concentration of drugs in lesions and the efficiency of viral transfection in MSCs are crucial factors in genetic engineering (Almeida-Porada et al., 2020). M1 macrophages exhibit a stronger phagocytic capability than other macrophage types, which enables them to effectively absorb drug-loaded nanoparticles for treating diseases (Pang et al., 2018).

5.18. CHALLENGES

Living cell-based drug delivery systems often experience drug leakage, which can be challenging to prevent. Cytoprotective mechanisms can cause drugs to be prematurely released through the Pgp system and neutralize their endocytosis, resulting in therapeutic drugs being released before they reach the tumor site (Pessina et al., 2011). This can lead to uncontrolled premature drug release, causing cell death or dysfunction, and drug leakage in RBCs may occur through transmembrane transport (Zhang et al., 2015). Additionally, free drugs that are not bound to target sites are easily expeled from the tumor, making it difficult for them to remain there (Zhang et al., 2015). Secondly, surface modification of living cells or drug loading inside cells can alter their properties. When drugs or nanoparticles are attached to the surface of living cells, the cell surface proteins and fluidity of the cell membrane may be affected, leading to changes in cell activity and increased immunogenicity (Villa et al., 2016). Internalization can also damage cell membrane integrity and plasticity, partially destroying the drug delivery system, reducing the biocompatibility of modified cells, and leading to drug exposure. For instance, macrophages require circulating living cells to ensure migration and homing ability, and then release a large number of drugs after reaching the tumor site (Si et al., 2016). However, most anticancer drugs have high cytotoxicity, and in some cases, loading these drugs can kill the vehicles (Cao et al., 2019). Furthermore, storage time can affect RBCs, resulting in immunomodulatory and different clinical outcomes (Garcia-Rao et al., 2017). The selection of carrier cells must be done with great care, as their role in vivo may not always align with expectations. Although mesenchymal stromal cells (MSCs) have numerous therapeutic applications in regenerative medicine and contribute to tissue repair processes (Hmadcha et al., 2020), some studies have shown that MSCs extracted from lysate or supernatant can inhibit the growth of various carcinoma cell lines, such as breast, ovarian, and osteosarcoma cells (Arjunan et al., 2020). Conversely, in vivo cellular interactions have demonstrated that primary human MSCs cultured with human SK-OV-3 ovarian cancer cells promote the development and occurrence of tumors (Melzer et al., 2018). Lastly, safety is a critical factor in determining the clinical applicability of living cell drug delivery systems. In some clinical cases, pathological vascular permeability and vascular injury can be caused by red blood cells (RBCs). Additionally, the attached RBCs can hinder the free interaction of drugs with certain targets, even within blood vessels (Rossi et al., 2016).

5.19. FUTURE PERSPECTIVES

Biological cells, including red blood cells (RBCs), platelets, neutrophils, mesenchymal stem cells (MSCs), and bacteria, often possess inherent targeting mechanisms, are biocompatible, have a longer circulation lifespan, and the ability to cross biological barriers, making them ideal candidates for drug delivery vehicles. When selecting a cell type for targeted delivery, it is essential to consider each cell's innate biological function, the intended application, and any potential challenges associated with each cell type (Bush et al., 2021).

Red blood cells and platelets are circulated throughout the body, making them ideal for delivering therapeutics systemically or to a targeted location that can be controlled through cell surface modifications and intracellular encapsulation. However, there is currently no single "best" method for producing anucleate cells or their associated cellular mimetics, which limits the use of these cells for drug delivery to small-scale therapies. Until good manufacturing practice (GMP) grade scale-up of anucleate cells can be achieved, it will be challenging to realize the enormous potential of this technology in the clinic. MSCs have natural paracrine and targeting functions that make them ideal for therapeutic strategies (Parekkadan & Milwid 2010). By modifying MSCs, they can be used as carriers to deliver various therapeutics to specific locations. MSCs have shown potential therapeutic responses in a variety of diseases, but they have not been shown to exhibit long-term engraftment. Additionally, while MSCs have cancer-inhibiting paracrine properties, it has been suggested that cancer cells can hijack MSCs to promote their own survival (Lee & Hong, 2017). A better understanding of how cancer cells utilize MSCs is needed before widespread clinical use. One possible suggestion is that MSCs can be used as delivery vehicles to target the desired site, release their payload before being cleared from the system, which would not require long-term engraftment. Neutrophils are the most abundant leukocyte in the body and one of the first responders to sites of injury, making them ideal for carrying therapeutics to areas of inflammation. However, because they are non-dividing cells, they have poor survival after isolation, which makes it challenging to manipulate them for targeted therapies (Kean et al., 2013). Bacteria are commonly used as drug delivery vehicles due to their ability to inhabit various microbiomes throughout the body and preferentially target tumors. However, it is important to exercise caution when designing bacteria-based drug delivery systems because native bacteria play a crucial role in maintaining the balance of their microbiomes, and disruptions in this balance can lead to a range of diseases (Claesen & Fischbach, 2015). Additionally, lysis-driven bacterial drug delivery systems can result in the buildup and absorption of bacterial cell lysate byproducts into the bloodstream, which can cause toxic systemic effects. Moreover, the use of living cells for drug delivery raises concerns about immunogenicity and safety. The use of autologous cells, which are derived from the patient themselves, can minimize this risk because the immune system is less likely to mount an immunogenic response since it recognizes the cells as "self." A potential future approach to harvesting adult cells for drug delivery is the use of patient-specific induced pluripotent stem (iPS) cells, which can be generated from individual patients and expanded indefinitely to produce the desired cell types for targeted drug delivery.

5.20. CONCLUSION

The development and use of living cell-based drug delivery systems were comprehensively assessed. These cells exhibit high stability in serum, low toxicity, low immunogenicity, targeted delivery, and efficient drug loading. As a result, living cells show great potential as drug delivery systems. As the field continues to advance and combine multiple disciplines, we expect the creation of intelligent living cell-based drug delivery systems with multiple functions for use in the body, which could greatly benefit disease treatment (Wu et al., 2022).

REFERENCES

1. Agrahari, V., et al. (2017). Next generation drug delivery: Circulatory cells-mediated nanotherapeutic approaches. *Expert Opinion on Drug Delivery, 14*(3), 285–289.
2. Akin, D., et al. (2007). Bacteria-mediated delivery of nanoparticles and cargo into cells. *Nature Nanotechnology, 2*(7), 441–449.
3. Al-Essa, M. K., Melzer, S., & Tarnok, A. (2019). Two-color analysis of leukocytes labeled by modified RBCs and their fragments. *Cytometry. Part A, 95A* (3), 339–346.
4. Al-Essa, M. K., Melzer, S., Tarnok, A., Hadidi, K. A., & El-Khateeb, M. (2018). Fast RBC loading by fluorescent antibodies and nuclei staining dye and their potential bioanalytical applications. *Zeitschrift für Naturforschung. Section C, A Journal of Biosciences, 73*(3–4), 95–105.
5. Almeida-Porada, G., Atala, A., & Porada, C. D. (2020). Therapeutic mesenchymal stromal cells for immunotherapy and for gene and drug delivery. *Molecular Therapy. Methods & Clinical Development, 16*, 204–224.
6. Altaner, C., et al. (2014). Complete regression of glioblastoma by mesenchymal stem cells mediated prodrug gene therapy simulating clinical therapeutic scenario. *International Journal of Cancer, 134*(6), 1458–1465.
7. Anselmo, A. C., et al. (2015). Monocyte-mediated delivery of polymeric backpacks to inflamed tissues: A generalized strategy to deliver drugs to treat inflammation. *Journal of Controlled Release, 199*, 29–36.
8. Arjunan, S., et al. (2020). Inhibition of growth of Asian keloid cells with human umbilical cord Wharton's jelly stem cell-conditioned medium. *Stem Cell Research & Therapy, 11*(1), 78.
9. Bakh, N. A., et al. (2017). Glucose-responsive insulin by molecular and physical design. *Nature Chemistry, 9*, 937–943.
10. Baraniak, P. R., & McDevitt, T. C. (2010). Stem cell paracrine actions and tissue regeneration. *Regenerative Medicine, 5*(1), 121–143.
11. Barnett, B. P., et al. (2007). Magnetic resonance-guided, real-time targeted delivery and imaging of magneto capsules immunoprotecting pancreatic islet cells. *Nature Medicine, 13*, 986–991.
12. Berglund, A. K., et al. (2017). Immunoprivileged no more: Measuring the immunogenicity of allogeneic adult mesenchymal stem cells. *Stem Cell Research & Therapy, 8*(1), 288.
13. Bhonde, R. R. (2021). Mesenchymal stem cells in human health and diseases. *Current Science, 120*(5), 951–952.

14. Biagiotti, S., et al. (2011). Drug delivery by red blood cells. *IUBMB Life, 63*(8), 621–631.
15. Bianco, P., & Robey, P. G. (2001). Stem cells in tissue engineering. *Nature, 414*, 118–121.
16. Bush, L. M., et al. (2019). Synthetic biology: Paving the way with novel drug delivery. *The Biochemist, 41*(3), 24.
17. Bush, L. M., Healy, C. P., Javdan, S. B., Emmons, J. C., & Deans, T. L. (2021). Biological cells as therapeutic delivery vehicles. *Trends in Pharmacological Sciences, 42*(2), 106–118.
18. Cao, H. Q., et al. (2019). Bioengineered macrophages can responsively transform into nanovesicles to target lung metastasis. *Nano Letters, 19*(12), 9171–9172.
19. Chen, G., et al. (2015). Transformation of cell-derived microparticles into quantum-dot-labeled nanovectors for antitumor siRNA delivery. *Angewandte Chemie International Edition, 54*, 1036–1040.
20. Chen, Z., Hu, Q., & Gu, Z. (2018). Leveraging engineering of cells for drug delivery. *Accounts of Chemical Research, 51*(3), 668–677.
21. Cho, D. I., et al. (2014). Mesenchymal stem cells reciprocally regulate the M1/M2 balance in mouse bone marrow-derived macrophages. *Experimental & Molecular Medicine, 46(1)*, e70. https://doi.org/10.1038/emm.2013.135.
22. Chowdhury, S., et al. (2019). Programmable bacteria induce durable tumor regression and systemic antitumor immunity. *Nature Medicine, 25*(7), 1057–1063.
23. Chu, D., et al. (2015). Neutrophil-mediated delivery of therapeutic nanoparticles across blood vessel barrier for treatment of inflammation and infection. *ACS Nano, 9*(12), 11800–11811.
24. Chu, D., et al. (2017). Photosensitization priming of tumor microenvironments improves delivery of nanotherapeutics via neutrophil infiltration. *Advanced Materials, 29*(27), 1703084.
25. Chu, D., et al. (2018). Neutrophil-based drug delivery systems. *Advanced Materials, 30*(22), 1706245.
26. Claesen, J., & Fischbach, M. A. (2015). Synthetic microbes as drug delivery systems. *ACS Synthetic Biology, 4*(4), 358–364.
27. Coker, S. A., et al. (2018). A study of the pharmacokinetic properties and the in vivo kinetics of erythrocytes loaded with dexamethasone sodium phosphate in healthy volunteers. *Transfusion Medicine Reviews, 32*(2), 102–110.

28. Deans, T. L., & Elisseeff, J. H. (2009). Stem cells in musculoskeletal engineered tissue. *Current Opinion in Biotechnology, 20*(5), 537–544.

29. Dehaini, D., et al. (2017). Erythrocyte-platelet hybrid membrane coating for enhanced nanoparticle functionalization. *Advanced Materials, 29*(43), 1606209. https://doi.org/10.1002/adma.201606209.

30. Doshi, N., et al. (2012). Platelet mimetic particles for targeting thrombi in flowing blood. *Advanced Materials, 24*(28), 3864–3869. https://doi.org/10.1002/adma.201202063.

31. Du, Y., & Chen, B. (2019). Combination of drugs and carriers in drug delivery technology and its development. *Drug Design, Development and Therapy, 13*, 1401–1408. https://doi.org/10.2147/DDDT.S195624.

32. El Andaloussi, S., et al. (2013). Extracellular vesicles: Biology and emerging therapeutic opportunities. *Nature Reviews Drug Discovery, 12*, 347–357. https://doi.org/10.1038/nrd3978.

33. Escajadillo, T., et al. (2017). A red blood cell membrane-camouflaged nanoparticle counteracts streptolysin O-mediated virulence phenotypes of invasive Group A *Streptococcus*. *Frontiers in Pharmacology, 8*, 477. https://doi.org/10.3389/fphar.2017.00477.

34. Fan, X. L., et al. (2020). Mechanisms underlying the protective effects of mesenchymal stem cell-based therapy. *Cellular and Molecular Life Sciences, 77*(14), 2771–2794. https://doi.org/10.1007/s00018-020-03454-y.

35. Felfoul, O., et al. (2016). Magneto-aerotactic bacteria deliver drug-containing nanoliposomes to tumor hypoxic regions. *Nature Nanotechnology, 11*(11), 941–947. https://doi.org/10.1038/nnano.2016.137.

36. Fiorina, P., et al. (2008). The clinical impact of islet transplantation. *American Journal of Transplantation, 8*, 1990–1997. https://doi.org/10.1111/j.1600-6143.2008.02345.x.

37. Fitzgerald, M., et al. (2017). Adoption of the Q transcriptional system for regulating gene expression in stem cells. *ACS Synthetic Biology, 6*(11), 2014–2020. https://doi.org/10.1021/acssynbio.7b00143.

38. Fliervoet, L. A. L., & Mastrobattista, E. (2016). Drug delivery with living cells. *Advanced Drug Delivery Reviews, 106*(Pt A), 63–72. https://doi.org/10.1016/j.addr.2016.07.006.

39. Fossati, G., et al. (1999). Neutrophil infiltration into human gliomas. *Acta Neuropathologica., 98*(4), 349–354. https://doi.org/10.1007/s004010051132.

40. Gao, M., et al. (2017). Photosensitizer decorated red blood cells as an

ultrasensitive light responsive drug delivery system. *ACS Applied Materials & Interfaces, 9*(7), 5855–5863. https://doi.org/10.1021/acsami.6b16186.

41. Garcia-Roa, M., et al. (2017). Red blood cell storage time and transfusion: Current practice, concerns and future perspectives. *Blood Transfusion, 15*(3), 222–231. https://doi.org/10.2450/2017.0308-16.
42. Gay, L. J., & Felding-Habermann, B. (2011). Contribution of platelets to tumor metastasis. *Nature Reviews Cancer, 11*, 123–134. https://doi.org/10.1038/nrc3004.
43. Giese, M. A., Hind, L. E., & Huttenlocher, A. (2019). Neutrophil plasticity in the tumor microenvironment. *Blood, 133*(20), 2159–2167. https://doi.org/10.1182/blood.2019000698.
44. Glassman, P. M., et al. (2020). Vascular drug delivery using carrier red blood cells: Focus on RBC surface loading and pharmacokinetics. *Pharmaceutics, 12*(5), 440. https://doi.org/10.3390/pharmaceutics12050440.
45. Guo, J., et al. (2020). Biomimetic rebuilding of multifunctional red blood cells: Modular design using functional components. *ACS Nano, 14*(7), 7847–7859. https://doi.org/10.1021/acsnano.0c02743.
46. Gurbatri, C. R., et al. (2020). Engineered probiotics for local tumor delivery of checkpoint blockade nanobodies. *Science Translational Medicine, 12*(530), eaax0876. https://doi.org/10.1126/scitranslmed.aax0876.
47. Guye, P., et al. (2016). Genetically engineering self-organization of human pluripotent stem cells into a liver bud-like tissue using Gata6. *Nature Communications, 7*, 10243. https://doi.org/10.1038/ncomms10243.
48. Hammel, P., et al. (2020). Erythrocyte-encapsulated asparaginase (eryaspase) combined with chemotherapy in second-line treatment of advanced pancreatic cancer: An open-label, randomized Phase IIb trial. *European Journal of Cancer, 124*, 91–101. https://doi.org/10.1016/j.ejca.2019.11.005.
49. Han, Y., et al. (2018). Neutrophil-based delivery systems for nanotherapeutics. *Small, 14*(42), e1801674. https://doi.org/10.1002/smll.201801674.
50. Harvath, L. (1991). Neutrophil chemotactic factors. *EXS, 59*, 35–52.
51. Hmadcha, A., et al. (2020). Therapeutic potential of mesenchymal stem cells for cancer therapy. *Frontiers in Bioengineering and Biotechnology, 8*, 43. https://doi.org/10.3389/fbioe.2020.00043.
52. Hou, J., et a□ (2019). Accessing neuroinflammation sites: Monocyte/neutrophil-mediated drug delivery for cerebral ischemia. *Science Advances, 5*(7), eaau8301. https://doi.org/10.1126/sciadv.aau8301.

53. Hu, C. M., et al. (2013). Erythrocyte membrane-camouflaged polymeric nanoparticles as a biomimetic delivery platform. *Proceedings of the National Academy of Sciences of the United States of America, 108*(27), 10980–10985. https://doi.org/10.1073/pnas.1106634108.
54. Hu, C.-M. J., et al. (2011). 'Marker-of-self' functionalization of nanoscale particles through a top-down cellular membrane coating approach. *Nanoscale, 5*(7), 2664–2668. https://doi.org/10.1039/c3nr34295k.
55. Hu, C.-M. J., Fang, R. H., Wang, K.-C., Luk, B. T., Thamphiwatana, S., Dehaini, D., & Zhang, L. (2015). Nanoparticle biointerfacing by platelet membrane cloaking. *Nature, 526*(7571), 118–121. https://doi.org/10.1038/nature15373.
56. Hu, Q., Qian, C., Sun, W., Wang, J., Chen, Z., Bomba, H. N., & Gu, Z. (2016). Engineered nanoplatelets for enhanced treatment of multiple myeloma and thrombus. *Advanced Materials, 28*(45), 9573–9580. https://doi.org/10.1002/adma.201603803.
57. Hu, Q., Sun, W., Qian, C., Bomba, H. N., Xin, H., & Gu, Z. (2017). Relay Drug Delivery for Amplifying Targeting Signal and Enhancing Anticancer Efficacy. *Advanced Materials, 29*(1), 1605803. https://doi.org/10.1002/adma.201605803.
58. Hu, Q., Sun, W., Qian, C., Wang, C., Bomba, H. N., & Gu, Z. (2015). Anticancer Platelet-Mimicking Nanovehicles. *Advanced Materials, 27*(46), 7043–7050. https://doi.org/10.1002/adma.201503323.
59. Ji, W. H., Smith, P. N., Koepsel, R. R., Andersen, J. D., Baker, S. L., Zhang, L. B., & Russell, A. J. (2020). Erythrocytes as carriers of immunoglobulin-based therapeutics. *Acta Biomaterialia, 101*, 422–435. https://doi.org/10.1016/j.actbio.2019.10.035.
60. Jin, L., et al. (2019). Mesenchymal stem cells promote type 2 macrophage polarization to ameliorate the myocardial injury caused by diabetic cardiomyopathy. *Journal of Translational Medicine, 17*(1), 251. https://doi.org/10.1186/s12967-019-1993-6.
61. Kean, T. J., et al. (2013). MSCs: Delivery routes and engraftment, cell-targeting strategies, and immune modulation. *Stem Cells International, 2013*, Article 732742. https://doi.org/10.1155/2013/732742.
62. Klyachko, N. L., Polak, R., Haney, M. J., Zhao, Y. L., Neto, R. J. G., Hill, M. C., & Batrakova, E. V. (2017). Macrophages with cellular backpacks for targeted drug delivery to the brain. *Biomaterials, 140*, 79–87. https://doi.org/10.1016/j.biomaterials.2017.06.016.
63. Krawczyk, K., et al. (2020). Rewiring of endogenous signaling

pathways to genomic targets for therapeutic cell reprogramming. *Nature Communications, 11*(1), 608. https://doi.org/10.1038/s41467-020-14429-4.

64. Layek, B., Sadhukha, T., & Prabha, S. (2016). Glycoengineered mesenchymal stem cells as an enabling platform for two-step targeting of solid tumors. *Biomaterials, 88*, 97–109. https://doi.org/10.1016/j.biomaterials.2016.02.029.

65. Layek, B., Shetty, M., Nethi, S. K., Sehgal, D., Starr, T. K., & Prabha, S. (2020). Mesenchymal stem cells as guideposts for nanoparticle-mediated targeted drug delivery in ovarian cancer. *Cancers (Basel), 12*(4), 965. https://doi.org/10.3390/cancers12040965.

66. Lee, H. Y., & Hong, I. S. (2017). Double-edged sword of mesenchymal stem cells: Cancer-promoting versus therapeutic potential. *Cancer Science, 108*(10), 1939–1946. https://doi.org/10.1111/cas.13364.

67. Li, J., et al. (2016). Genetic engineering of platelets to neutralize circulating tumor cells. *Journal of Controlled Release, 228*, 38–47. https://doi.org/10.1016/j.jconrel.2016.03.007.

68. Li, Q., et al. (2014). Mesenchymal stem cells from human fat engineered to secrete BMP4 are nononcogenic, suppress brain cancer, and prolong survival. *Clinical Cancer Research, 20*(9), 2375–2387. https://doi.org/10.1158/1078-0432.CCR-13-2789.

69. Li, S. P., Zhang, Y. L., Wang, J., Zhao, Y., Ji, T. J., Zhao, X., & Nie, G. J. (2017). Nanoparticle-mediated local depletion of tumor-associated platelets disrupts vascular barriers and augments drug accumulation in tumors (vol 1, pg. 680, 2017). *Nature Biomedical Engineering, 1*(8), 680. https://doi.org/10.1038/s41551-017-0068-0.

70. Lin, M. H., et al. (2018). The interplay between nanoparticles and neutrophils. *Journal of Biomedical Nanotechnology, 14*(1), 66–85. https://doi.org/10.1166/jbn.2018.2492.

71. Lo, C. Y., Weil, B. R., Palka, B. A., Momeni, A., Canty, J. M., & Neelamegham, S. (2016). Cell surface glycoengineering improves selectin-mediated adhesion of mesenchymal stem cells (MSCs) and cardiosphere-derived cells (CDCs): Pilot validation in porcine ischemia-reperfusion model. *Biomaterials, 74*, 19–30. https://doi.org/10.1016/j.biomaterials.2015.09.040.

72. Luk, B. T., & Zhang, L. (2015). Cell membrane-camouflaged nanoparticles for drug delivery. *Journal of Controlled Release, 220*(Pt B), 600–607. https://doi.org/10.1016/j.jconrel.2015.09.018.

73. Luo, X., et al. (2018). Neutrophil-mediated delivery of pixantrone-loaded liposomes decorated with poly (sialic acid)-octadecylamine conjugate for

lung cancer treatment. *Drug Delivery, 25*(1), 1200–1212. https://doi.org/10.1080/10717544.2018.1487111.

74. Lyde, R. B., et al. (2019). Infused factor VIII-expressing platelets or megakaryocytes as a novel therapeutic strategy for hemophilia A. *Blood Advances, 3*(9), 1368–1378. https://doi.org/10.1182/bloodadvances.2018025854.

75. Maeda, H., et al. (2000). Tumor vascular permeability and the EPR effect in macromolecular therapeutics: A review. *Journal of Controlled Release, 65*(1–2), 271–284. https://doi.org/10.1016/S0168-3659(99)00248-5.

76. Mammadova-Bach, E., et al. (2020). Platelet glycoprotein VI promotes metastasis through interaction with cancer cell-derived galectin-3. *Blood, 135*(14), 1146–1160. https://doi.org/10.1182/blood.2019002275.

77. Mangraviti, A., et al. (2016). Non-virally engineered human adipose mesenchymal stem cells produce BMP4, target brain tumors, and extend survival. *Biomaterials, 100*, 53–66. https://doi.org/10.1016/j.biomaterials.2016.05.010.

78. Mayadas, T. N., et al. (2014). The multifaceted functions of neutrophils. *Annual Review of Pathology, 9*, 181–218. https://doi.org/10.1146/annurev-pathol-020712-164023.

79. Melzer, C., von der Ohe, J., & Hass, R. (2018). MSC stimulate ovarian tumor growth during intercellular communication but reduce tumorigenicity after fusion with ovarian cancer cells. *Cell Communication and Signaling, 16*(1), 67. https://doi.org/10.1186/s12964-018-0284-7.

80. Merkel, T. J., et al. (2011). Using mechanobiological mimicry of red blood cells to extend circulation times of hydrogel microparticles. *Proceedings of the National Academy of Sciences of the United States of America, 108*(2), 586–591. https://doi.org/10.1073/pnas.1010013108.

81. Miller, M. A., et al. (2015). Tumor-associated macrophages act as a slow-release reservoir of nano-therapeutic Pt (IV) pro-drug. *Nature Communications, 6*, Article 8692. https://doi.org/10.1038/ncomms9692.

82. Mo, R., Jiang, T., Di, J., Tai, W., & Gu, Z. (2014). Emerging micro- and nanotechnology based synthetic approaches for insulin delivery. *Chemical Society Reviews, 43*, 3595–3629. https://doi.org/10.1039/C3CS60373D.

83. Mooney, R., et al. (2014). Conjugation of pH-responsive nanoparticles to neural stem cells improves intratumoral therapy. *Journal of Controlled Release, 191*, 82–89. https://doi.org/10.1016/j.jconrel.2014.06.031.

84. Mooney, R., Weng, Y. M., Tirughana-Sambandan, R., Valenzuela, V., Aramburo, S., Garcia, E., & Aboody, K. S. (2014). Neural stem

cells improve intracranial nanoparticle retention and tumor-selective distribution. *Future Oncology, 10*(3), 401–415. https://doi.org/10.2217/fon.13.224.

85. Nandi, S., & Brown, A. C. (2016). Platelet-mimetic strategies for modulating the wound environment and inflammatory responses. *Experimental Biology and Medicine, 241*(10), 1138–1148. https://doi.org/10.1177/1535370216647230.
86. Naumenko, V. A., Vlasova, K. Y., Garanina, A. S., Melnikov, P. A., Potashnikova, D. M., Vishnevskiy, D. A., & Majouga, A. G. (2019). Extravasating neutrophils open vascular barrier and improve liposomes delivery to tumors. *ACS Nano, 13*(11), 12599–12612. https://doi.org/10.1021/acsnano.9b05876.
87. Naumenko, V., Nikitin, A., Garanina, A., Melnikov, P., Vodopyanov, S., Kapitanova, K., & Majouga, A. (2020). Neutrophil-mediated transport is crucial for delivery of short-circulating magnetic nanoparticles to tumors. *Acta Biomaterialia, 104*, 176–187. https://doi.org/10.1016/j.actbio.2020.01.027.
88. Nejman, D., et al. (2020). The human tumor microbiome is composed of tumor type-specific intracellular bacteria. *Science, 368*(6494), 973–980. https://doi.org/10.1126/science.aay9189.
89. Ortiz-Otero, N., Mohamed, Z., & King, M. R. (2018). Platelet-based drug delivery for cancer applications. In C. Dong, N. Zahir, & K. Konstantopoulos (Eds.), *Biomechanics in Oncology 2018* (pp. 235–251). Springer.
90. Pang, L., Zhu, Y., Qin, J., Zhao, W. J., & Wang, J. X. (2018). Primary M1 macrophages as multifunctional carrier combined with PLGA nanoparticle delivering anti-cancer drug for efficient glioma therapy. *Drug Delivery, 25*(1), 1922–1931. https://doi.org/10.1080/10717544.2018.1498190.
91. Parekkadan, B., & Milwid, J. M. (2010). Mesenchymal stem cells as therapeutics. *Annual Review of Biomedical Engineering, 12*, 87–117. https://doi.org/10.1146/annurev-bioeng-071910-124743.
92. Park, K. (2019). The beginning of the end of the nanomedicine hype. *Journal of Controlled Release, 305*, 221–222. https://doi.org/10.1016/j.jconrel.2019.05.037.
93. Pessina, A., Bonomi, A., Cocce, V., Invernici, G., Navone, S., Cavicchini, L., & Alessandri, G. (2011). Mesenchymal stromal cells primed with paclitaxel provide a new approach for cancer therapy. *PLoS ONE, 6*(12), e28321. https://doi.org/10.1371/journal.pone.0028321.
94. Petty, A. J., & Yang, Y. (2017). Tumor-associated macrophages:

Implications in cancer immunotherapy. *Immunotherapy, 9*(3), 289–302. https://doi.org/10.2217/imt-2016-0132.

95. Pishesha, N., Bilate, A. M., Wibowo, M. C., Huang, N. J., Li, Z. Y., Deshycka, R., & Ploegh, H. L. (2017). Engineered erythrocytes covalently linked to antigenic peptides can protect against autoimmune disease. *Proceedings of the National Academy of Sciences of the United States of America, 114*(12), 3157–3162. https://doi.org/10.1073/pnas.1701746114.
96. Polak, R., Lim, R. M., Beppu, M. M., Pitombo, R. N. M., Cohen, R. E., & Rubner, M. F. (2015). Liposome-loaded cell backpacks. *Advanced Healthcare Materials, 4*(18), 2832–2841. https://doi.org/10.1002/adhm.201500607.
97. Reagan, M. R., & Kaplan, D. L. (2011). Concise review: Mesenchymal stem cell tumor homing: detection methods in disease model systems. *Stem Cells, 29*(6), 920–927. https://doi.org/10.1002/stem.643.
98. Ren, X., et al. (2016). Red blood cell membrane camouflaged magnetic nanoclusters for imaging-guided photothermal therapy. *Biomaterials, 92*, 13–24. https://doi.org/10.1016/j.biomaterials.2016.04.023.
99. Rossi, L., Pierige, F., Antonelli, A., Bigini, N., Gabucci, C., Peiretti, E., & Magnani, M. (2016). Engineering erythrocytes for the modulation of drugs' and contrasting agents' pharmacokinetics and biodistribution. *Advanced Drug Delivery Reviews, 106*, 73–87. https://doi.org/10.1016/j.addr.2016.01.004.
100. Santiago, T., & da Silva, J. A. (2014). Safety of low- to medium-dose glucocorticoid treatment in rheumatoid arthritis: Myths and reality over the years. *Annals of the New York Academy of Sciences, 1318*, 41–49. https://doi.org/10.1111/nyas.12440.
101. Sarkar, S., Alam, M. A., Shaw, J., & Dasgupta, A. K. (2013). Drug delivery using platelet cancer cell interaction. *Pharmaceutical Research, 30*(11), 2785–2794. https://doi.org/10.1007/s11095-013-1115-y.
102. Scheller, L., & Fussenegger, M. (2019). From synthetic biology to human therapy: Engineered mammalian cells. *Current Opinion in Biotechnology, 58*, 108–116. https://doi.org/10.1016/j.copbio.2019.02.006.
103. Scott, C. C., Vacca, F., & Gruenberg, J. (2014). Endosome maturation, transport and functions. *Seminars in Cell & Developmental Biology, 31*, 2–10. https://doi.org/10.1016/j.semcdb.2014.05.013.
104. Semple, J. W., Italiano, J. E., & Freedman, J. (2011). Platelets and the immune continuum. *Nature Reviews Immunology, 11*, 264–274. https://doi.org/10.1038/nri2956.

105. Sharma, P., & Allison, J. P. (2015). The future of immune checkpoint therapy. *Science, 348*, 56–61. https://doi.org/10.1126/science.aaa8172.

106. Shi, J., et al. (2014). Engineered red blood cells as carriers for systemic delivery of a wide array of functional probes. *Proceedings of the National Academy of Sciences of the United States of America, 111*(28), 10131–10136. https://doi.org/10.1073/pnas.1409861111.

107. Shi, Y., et al. (2011). Mechanisms and management of doxorubicin cardiotoxicity. *Herz, 36*(4), 296–305. https://doi.org/10.1007/s00059-011-3481-6.

108. Shields, C. W., Evans, M. A., Wang, L. L. W., Baugh, N., Iyer, S., Wu, D., & Mitragotri, S. (2020). Cellular backpacks for macrophage immunotherapy. *Science Advances, 6*(18), eaaz6579. https://doi.org/10.1126/sciadv.aaz6579.

109. Shum, T., & Heslop, H. E. (2018). A backpack revs up T-cell activity. *Nature Biotechnology, 36*(8), 702–703. https://doi.org/10.1038/nbt.4249.

110. Si, J. X., Shao, S. Q., Shen, Y. Q., & Wang, K. (2016). Macrophages as active nanocarriers for targeted early and adjuvant cancer chemotherapy. *Small, 12*(37), 5108–5119. https://doi.org/10.1002/smll.201601538.

111. Singh, N., Baby, D., Rajguru, J. P., Patil, P. B., Thakkannavar, S. S., & Pujari, V. B. (2019). Inflammation and Cancer. *Annals of African Medicine, 18*(3), 121–126. https://doi.org/10.4103/aam.aam_59_18.

112. Sletten, E. M., & Bertozzi, C. R. (2009). Bioorthogonal chemistry: Fishing for selectivity in a sea of functionality. *Angewandte Chemie International Edition, 48*, 6974–6998. https://doi.org/10.1002/anie.200900942.

113. Stephan, M. T., & Irvine, D. J. (2011). Enhancing cell therapies from the outside in: Cell surface engineering using synthetic nanomaterials. *Nano Today, 6*, 309–325. https://doi.org/10.1016/j.nantod.2011.06.004.

114. Stephan, M. T., Moon, J. J., Um, S. H., Bershteyn, A., & Irvine, D. J. (2010). Therapeutic cell engineering with surface-conjugated synthetic nanoparticles. *Nature Medicine, 16*, 1035–1041. https://doi.org/10.1038/nm.2208.

115. Sun, D., et al. (2019). Advances in refunctionalization of erythrocyte-based nanomedicine for enhancing cancer-targeted drug delivery. *Theranostics, 9*(23), 6885–6900. https://doi.org/10.7150/thno.38585.

116. Tang, L., Zheng, Y. R., Melo, M. B., Mabardi, L., Castano, A. P., Xie, Y. Q., & Irvine, D. J. (2018). Enhancing T cell therapy through TCR-signaling-responsive nanoparticle drug delivery. *Nature Biotechnology, 36*(8), 707–716. https://doi.org/10.1038/nbt.4198.

117. Thursby, E., & Juge, N. (2017). Introduction to the human gut microbiota. *Biochemical Journal, 474*(11), 1823–1836. https://doi.org/10.1042/BCJ20160510.

118. Tsai, R. K., Rodriguez, P. L., & Discher, D. E. (2010). Self-inhibition of phagocytosis: The affinity of 'marker of self' CD47 for SIRPα dictates potency of inhibition but only at low expression levels. *Blood Cells, Molecules, & Diseases, 45*, 67−74. https://doi.org/10.1016/j.bcmd.2010.02.010.

119. van Weering, J. R. T., & Cullen, P. J. (2014). Membrane-associated cargo recycling by tubule-based endosomal sorting. *Seminar in Cell & Developmental Biology, 31*, 40−47. https://doi.org/10.1016/j.semcdb.2014.04.021.

120. Vegas, A. J., Veiseh, O., Gurtler, M., Millman, J. R., Pagliuca, F. W., Bader, A. R., & Anderson, D. G. (2016). Long-term glycemic control using polymer-encapsulated human stem cell-derived beta cells in immune-competent mice. *Nature Medicine, 22*, 306−311. https://doi.org/10.1038/nm.4030.

121. Veiseh, O., Tang, B. C., Whitehead, K. A., Anderson, D. G., & Langer, R. (2015). Managing diabetes with nanomedicine: Challenges and opportunities. *Nature Reviews Drug Discovery, 14*, 45−57. https://doi.org/10.1038/nrd4477.

122. Villa, C. H., Anselmo, A. C., Mitragotri, S., & Muzykantov, V. (2016). Red blood cells: Supercarriers for drugs, biologicals, and nanoparticles and inspiration for advanced delivery systems. *Advanced Drug Delivery Reviews, 106*, 88–103. https://doi.org/10.1016/j.addr.2016.04.023.

123. Villa, C. H., Seghatchian, J., & Muzykantov, V. (2016). Drug delivery by erythrocytes: "Primum non nocere." *Transfusion and Apheresis Science, 55*(3), 275–280. https://doi.org/10.1016/j.transci.2016.09.019.

124. Wang, C., Sun, W., Ye, Y., Hu, Q., Bomba, H. N., & Gu, Z. (2017). In situ activation of platelets with checkpoint inhibitors for post-surgical cancer immunotherapy. *Nature Biomedical Engineering, 1*, 0011. https://doi.org/10.1038/s41551-016-0011.

125. Wang, C., Ye, Y., Hu, Q., Bellotti, A., & Gu, Z. (2017). Tailoring biomaterials for cancer immunotherapy: Emerging trends and future outlook. *Advanced Materials, 29*, 1606036. https://doi.org/10.1002/adma.201606036.

126. Wang, C., Ye, Y., Sun, W., Yu, J., Wang, J., Lawrence, D. S., & Gu, Z. (2017). Red blood cells for glucose-responsive insulin delivery. *Advanced Materials, 29*, 1606617. https://doi.org/10.1002/adma.201606617.

127. Wang, Y. J., Hoinka, J., Liang, Y., Adamus, T., Swiderski, P., & Przytycka, T. M. (2018). Apt-aBlocks: Designing RNA complexes and accelerating RNA-based drug delivery systems. *Nucleic Acids Research, 46*(16), 8133–8142. https://doi.org/10.1093/nar/gky646.

128. Wendel, S. O., et al. (2015). Cell based drug delivery: *Micrococcus luteus* loaded neutrophils as chlorhexidine delivery vehicles in a mouse model of liver abscesses in cattle. *PLoS One, 10*(5), e0128144. https://doi.org/10.1371/journal.pone.0128144.

129. Wilhelm, S., et al. (2016). Analysis of nanoparticle delivery to tumors. *Nature Reviews Materials, 1*, 16014. https://doi.org/10.1038/natrevmats.2016.14.

130. Wu, H. H., Zhou, Y., Tabata, Y., & Gao, J. Q. (2019). Mesenchymal stem cell-based drug delivery strategy: From cells to biomimetic. *Journal of Controlled Release, 294*, 102–113. https://doi.org/10.1016/j.jconrel.2018.12.024.

131. Wu, M., et al. (2018). MR imaging tracking of inflammation-activatable engineered neutrophils for targeted therapy of surgically treated glioma. *Nature Communications, 9*(1), 4777. https://doi.org/10.1038/s41467-018-07208-0.

132. Wu, Y., Liu, Y., Wang, T., Jiang, Q., Xu, F., & Liu, Z. (2022). Living cell for drug delivery. *Engineered Regeneration, 3*(2), 131–148. https://doi.org/10.1016/j.engreg.2022.03.003.

133. Xiang, J., Xu, L., Gong, H., Zhu, W., Wang, C., Xu, J., & Liu, Z. (2015). Antigen-loaded up conversion nanoparticles for dendritic cell stimulation, tracking, and vaccination in dendritic cell-based immunotherapy. *ACS Nano, 9*, 6401−6411. https://doi.org/10.1021/acsnano.5b01580.

134. Xie, M., & Fussenegger, M. (2018). Designing cell function: Assembly of synthetic gene circuits for cell biology applications. *Nature Reviews Molecular Cell Biology, 19*(8), 507–525. https://doi.org/10.1038/s41580-018-0007-z.

135. Xie, M., Ye, H., Wang, H., Charpin-El Hamri, G., Lormeau, C., Saxena, P., & Fussenegger, M. (2016). β-cell-mimetic designer cells provide closed-loop glycemic control. *Science, 354*, 1296−1301. https://doi.org/10.1126/science.aah6180.

136. Xu, P., et al. (2017). Doxorubicin-loaded platelets as a smart drug delivery system: An improved therapy for lymphoma. *Scientific Reports, 7*, 42632. https://doi.org/10.1038/srep42632.

137. Xue, J. W., Zhao, Z. K., Zhang, L., Xue, L. J., Shen, S. Y., Wen, Y. J., & Zhang, C. (2017). Neutrophil-mediated anticancer drug delivery

for suppression of postoperative malignant glioma recurrence. *Nature Nanotechnology, 12*(7), 692–700. https://doi.org/10.1038/nnano.2017.37.

138. Yao, S., Li, X. Q., Liu, J. X., Sun, Y. Q., Wang, Z. H., & Jiang, Y. Y. (2017). Maximized nanodrug-loaded mesenchymal stem cells by a dual drug-loaded mode for the systemic treatment of metastatic lung cancer. *Drug Delivery, 24*(1), 1372–1383. https://doi.org/10.1080/10717544.2017.1365444.
139. Ye, Y., Wang, C., Zhang, X., Hu, Q., Zhang, Y., Liu, Q., & Gu, Z. (2017). A melanin-mediated cancer immunotherapy patch. *Science Immunology, 2*, eaan5692. https://doi.org/10.1126/sciimmunol.aan5692.
140. Ye, Y., Yu, J., Wang, C., Nguyen, N.-Y., Walker, G. M., Buse, J. B., & Gu, Z. (2016). Microneedles integrated with pancreatic cells and synthetic g□ucose-signa□amp□ifiers for smart insulin delivery. *Advanced Materials, 28*, 3115−3121. https://doi.org/10.1002/adma.201505554.
141. Yoo, J.-W., Irvine, D. J., Discher, D. E., & Mitragotri, S. (2011). Bioinspired, bioengineered and biomimetic drug delivery carriers. *Nature Reviews Drug Discovery, 10*, 521−535. https://doi.org/10.1038/nrd3499.
142. Yu, H. L., Yang, Z. H., Li, F., Xu, L. S., & Sun, Y. (2020). Cell-mediated targeting drugs delivery systems. *Drug Delivery, 27*(1), 1425–1437. https://doi.org/10.1080/10717544.2020.1803383.
143. Zhang, C., et al. (2017). Direct macromolecular drug delivery to cerebral ischemia area using neutrophil-mediated nanoparticles. *Theranostics, 7*(13), 3260–3275. https://doi.org/10.7150/thno.20098.
144. Zhang, Q., et al. (2018). Current status and potential challenges of mesenchymal stem cell-based therapy for malignant gliomas. *Stem Cell Research & Therapy, 9*(1), 228. https://doi.org/10.1186/s13287-018-0962-0.
145. Zhang, X. F., Yao, S., Liu, C., & Jiang, Y. Y. (2015). Tumor tropic delivery of doxorubicin-polymer conjugates using mesenchymal stem cells for glioma therapy. *Biomaterials, 39*, 269–281. https://doi.org/10.1016/j.biomaterials.2015.01.014.
146. Zhang, X., et al. (2018). Engineering PD-1-presenting platelets for cancer immunotherapy. *Nano Letters, 18*(9), 5716–5725. https://doi.org/10.1021/acs.nanolett.8b01831.
147. Zheng, Y., Stephan, M. T., Gai, S. A., Abraham, W., Shearer, A., & Irvine, D. J. (2013). In vivo targeting of adoptively transferred T-cells with antibody- and cytokine-conjugated liposomes. *Journal of Controlled Release, 172*, 426−435. https://doi.org/10.1016/j.jconre□2013.09.001.

148. Zhou, Y., et al. (2019). The immunomodulatory functions of mesenchymal stromal/stem cells mediated via paracrine activity. *Journal of Clinical Medicine, 8*(7), 1025. https://doi.org/10.3390/jcm8071025.

149. Zhu, S. Y., Luo, Z. Y., Li, X. X., Han, X., Shi, S. L., & Zhan, T. (2021). Tumor-associated macrophages: Role in tumorigenesis and immunotherapy implications. *Journal of Cancer, 12*(1), 54–64. https://doi.org/10.7150/jca.51332.

Chapter **6**

ENGINEERING MICROBIAL CELLS

Shiv Sanjeevi Sripathi

6.1. INTRODUCTION

Microbial cell factories provide a cost-effective and eco-friendly method for producing valuable chemicals, such as biofuels, fine chemicals, and pharmaceuticals, from renewable feedstocks (Lee & Kim, 2015). Utilizing the intrinsic cellular synthetic capacity of a microorganism is a powerful approach for achieving high-value chemical production, which is influenced by factors such as catalytic efficiency, strain performance, and cell-cell communication. To enhance the synthetic capacity of microorganisms, various techniques have been employed, including genome-scale engineering (Yu et al., 2018), promoter engineering (Zhu et al., 2020), modular pathway engineering (Xu et al., 2013), transporter engineering (Doong et al., 2018), and other approaches (Chen et al., 2018). Although pathway engineering is highly effective, improving the synthetic capacity remains a significant challenge due to the lack of comprehensive understanding of cellular functions at various levels, including the organelle, cell, and consortium levels (Figure 6.1).

Several studies have aimed to illuminate cellular functions at three different levels, such as using yeast mitochondria to confine Ehrlich pathway enzymes and increase the concentration of the crucial α-ketoisovalerate intermediate (Avalos et al., 2013), regulating cell shape to work around the cell size constraint and expand the cell volume for inclusion body accumulation (Jiang & Chen, 2016), and designing the cooperative arrangement of the consortium to utilize alkali-extracted corn cobs for butanol production (Shong et al., 2012). In this study, we mainly focus on engineering methods for improving catalytic efficiency, strain performance, and cell-cell communication at the whole-cell level, collectively referred to as microbial cell engineering. This approach can be integrated with synthetic biology strategies and the exploitation of microbial physiological mechanisms through metabolic pathway engineering. Pathway engineering primarily involves modifying, rewiring, and constructing target chemical synthetic processes in metabolic pathways, whereas microbial cell engineering mainly focuses on enhancing the overall cellular synthetic capacity (as shown in Figure 6.1).

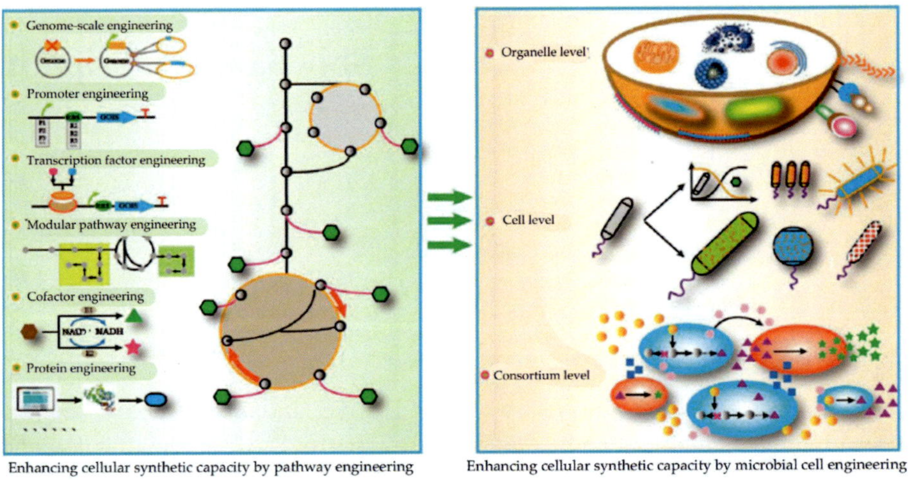

Figure 6.1. *Conventional microbial engineering involves manipulating multiple strains genetically to produce a targeted chemical through a static process that relies on adjusting pathway efficiency. In contrast, microbial cell engineering aims to enhance the synthetic capacity at the organelle, cell, and consortium levels, and adapt to different conditions based on various chemical and processing modules. This approach was introduced to overcome the limitations of conventional microbial engineering.*

Source: Ding et al., (2020).

The global health crisis triggered by the surge in clinical cases of bacterial infections with multiple antibiotic resistance has been exacerbated by the overuse, inappropriate prescribing, and lack of innovation in antibiotics.

Pathogenic bacteria such as *Enterococcus faecium* (*E. faecium*), *Staphylococcus aureus* (*S. aureus*), *Klebsiella pneumoniae* (*K. pneumoniae*), *Acinetobacter baumannii* (*A. baumannii*), *Pseudomonas aeruginosa* (*P. aeruginosa*), and *Escherichia coli* (*E. coli)* (ESKAPE) have become resistant to traditional antimicrobial treatments and pose a significant threat to public health (World Health Organization, 2017). To combat this crisis, there is an urgent need for new therapeutic strategies, and synthetic biology offers a promising approach to developing novel antimicrobial therapies.

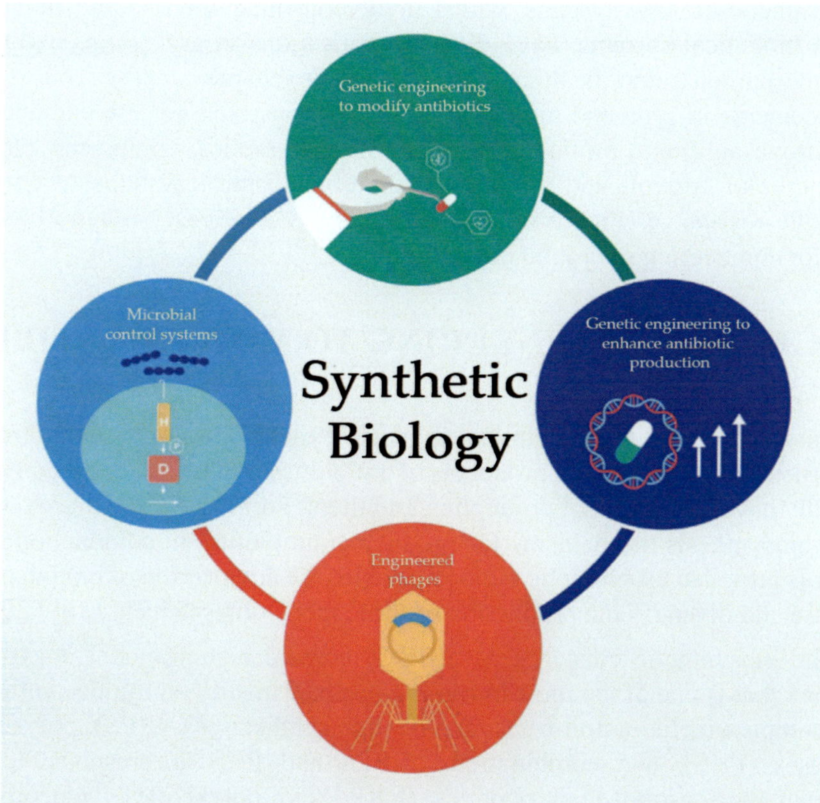

Figure 6.2. *The influence of synthetic biology and its tools and techniques, which have made it possible to modify genetically antibiotics, increase antibiotic production, engineer phages, and design microbial control systems, has been demonstrated by recent studies.*

Source: León-Buitimea et al., (2022).

Synthetic biology is built on a robust theoretical foundation that offers methods for envisioning, designing, and constructing synthetic biological systems (Perrino et al., 2021). Recent progress in synthetic biology has made it possible to create synthetic control systems within microbial cells. By employing protein engineering, DNA synthesis, and in silico design, researchers can construct metabolic pathways and biological circuits that regulate cellular behavior (McCarty and Ledesma-Amaro, 2019). As a result, synthetic biology advancements have enabled the development of communication networks between organisms, allowing exogenous molecules to control specific population behaviors, trigger intracellular signaling, and establish interdependent relationships among microorganisms (McCarty and Ledesma-Amaro, 2019).

Synthetic biology systems, which involve artificial circuits and functions within biological systems, have diverse applications in agriculture, industry, the environment, and healthcare studies. This chapter focuses on recent advancements in synthetic biology, such as its impact on genetic engineering to improve antibiotic production and modify antibiotics, engineered phages for microbial control, and alternative microbial control systems to combat antibiotic resistance (Figure 6.2). Additionally, it explores synthetic biology tools for engineering microbial communities.

6.2. GENETIC ENGINEERING TO MODIFY ANTIBIOTICS

Organisms have the inherent ability to generate metabolites with antimicrobial properties. However, natural producers of antibiotics, referred to as "wild type," typically have low levels of production and titers. Fortunately, the discovery of novel biosynthesis mechanisms for producing antibiotics, in conjunction with the tools provided by synthetic biology, enables the engineering of organisms to increase and diversify the production of these metabolites (Zhang et al., 2020).

The biosynthetic cluster genes (BCG) contain a multimodal enzymatic complex that is one of the most intriguing mechanisms for antibiotic synthesis. This complex includes non-ribosomal peptide synthetases (NRPS), polyketide synthases (PKS), and combinations. NRPS and PKS are responsible for constructing non-ribosomal peptides (NRP) and polyketides, respectively, and endowing them with active chemical properties (Hwang et al., 2020). The challenge of low antibiotic production and limited diversity can be addressed by utilizing synthetic biology tools and engineering these enzyme complexes.

In order to address the issue of low production and diversity of antimicrobial compounds, the NRP group of antibiotics, particularly the glycopeptides, has been studied (Yim et al., 2014). Some researchers have focused on diversifying these antibiotics, such as Yim et al., who used a combination of 13 scaffold-modifying enzymes from 7 GPA BCGs to introduce them into *Streptomyces coelicolor (S. coelicolor)*. As a result, nine new compounds were reported, and interestingly, eight of them showed antimicrobial activity against vancomycin-resistant *Enterococcus fecalis* (*E. fecalis*) with a MIC ranging from 0.5 to 4 µg/mL (Yim et al., 2016). One promising member of the glycopeptide family is corbomycin, which has outstanding clinical performance but poor natural production with *Streptomyces* sp. (Culp et al., 2020). To overcome this limitation, Xu et al. used a glycopeptide antibiotic heterologous expression system (GPAHex) to enhance the expression of genes for corbomycin synthesis in *S. coelicolor*, resulting in a 19-fold increase in titers using this platform (Xu et al., 2020).

It is also possible to discover that combinations between NRPS and PKS modules can result in more classes of antibiotics, such as lipopeptides. Daptomycin is a clinically significant lipopeptide antibiotic primarily used against methicillin-resistant *Staphylococcus aureus* (MRSA), which is naturally produced in *Streptomyces roseosporus (S. roseosporus).* However, similar to other antibiotic natural producers, the "wild-type" *S. roseosporus* has low daptomycin titers (Ye et al., 2014). Recently, Ji et al. used a "top-down" synthetic biology approach to increase total lipopeptide production by up to 2,300%, with up to 40% being daptomycin (Ji et al., 2022).

The production of antibiotics via the combination of NRPS and PKS in certain microorganisms is not only restricted to these enzyme complexes. Additional families of enzymes such as fatty acid synthases (FAS) may also be present. One example of this is the new family of zeamines, a natural antibiotic discovered in *Dickeya zeae,* which was found to result from the interaction between NRPS, PKS, and FAS (Wu et al., 2010). These zeamines are known for their potent microbicidal activity against both Gram-positive and Gram-negative bacteria (Liao et al., 2014). However, the production of zeamines in *Dickeya zeae* is limited. To address this, researchers such as Masschelein et al. have studied other natural producers of zeamines, such as *Serratia plymuthica* (*S. plymuthica*) RVH1. By using in-frame deletion of biosynthetic genes, they have uncovered some of the mechanisms involved in synthesizing these antibiotics, contributing significantly to synthetic biology for future combinatorial biosynthesis and bioengineering to produce new antibacterial compounds (Masschelein et al., 2015).

In addition to using enzyme complexes, metabolic engineering offers another approach to increase and diversify the production of antibiotics. Metabolic engineering aims to understand and redesign the cellular metabolism networks to enhance productivity (Palazzotto et al., 2019). This approach typically involves redirecting metabolic fluxes and regulating enzymes and BGCs that cause bottlenecks. The scientific community has made advancements in antibiotic production by studying the relationships between intermediate metabolites and antibiotic biosynthesis. One example is the regulation of S-Adenosylmethionine (SAM), which is involved in methylation processes and antibiotic production in various species (Wang et al., 2007). Researchers such as Cai et al. have increased bacitracin production yields in Bacillus licheniformis by up to 28.97% through the use of synthetic biology techniques like heterologous expression, gene deletion, and overexpression, by analyzing the SAM synthetic pathway (Cai et al., 2020).

To date, the research conducted has demonstrated that uncovering the various processes that contribute to the production of antibiotics, in conjunction with the resources offered by synthetic biology, have facilitated the modification of organisms by redesigning their genetic makeup to enhance their ability to generate antibiotics. This effort contributes to the battle against antimicrobial resistance. While there are still some significant obstacles to be overcome in order to routinely employ synthetic biology techniques to develop new antibiotics, the future appears promising for making new discoveries in this field.

6.3. GENETIC ENGINEERING TO ENHANCE ANTIBIOTIC PRODUCTION

Various microorganisms produce compounds that can be utilized as antimicrobial agents. The genetic material responsible for the production of these metabolites is diverse, and accessing this genetic diversity may lead to the discovery of new and more effective ways to combat drug-resistant microorganisms. Techniques such as gene mutation and the ability to control cellular functions at the genetic level can enhance antimicrobial biosynthesis. For instance, exposing *Streptomyces graminofaciens* to UV light resulted in changes in antibiotic production, as observed by Arafat et al. (2021). Researchers have also studied the cellular function of various genes to produce higher amounts and better antibiotics. For example, the gene cluster of the nucleoside antibiotic A201A in *Marinactinospora thermotolerans* was studied by Zhu et al. (2012), leading to improved production. Similarly, genetic modification was used to disrupt amphDIII and amphL in *Streptomyces nodosus* by Byrne et al. (2003) to improve the biosynthesis of amphotericin analogs. Makitrynskyy developed a method to manipulate the AdpA regions of *Streptomyces ghangensis* (Makitrynskyy et al., 2021), providing researchers with insight into a metabolic pathway to improve specific antimicrobial biosynthesis. By understanding metabolic pathways and applying gene engineering techniques, researchers can enhance antimicrobial production through targeted modifications to regulate specific genes. This demonstrates that genetic modifications can control biosynthesis pathways to improve antimicrobial production and its activities (Li et al., 2021). Genetic modifications can be targeted to enhance antimicrobial production, improve the availability of precursor molecules, and boost biosynthesis (Moosmann et al., 2020). These modifications can increase the levels of antimicrobial precursors, such as increasing the expression of the gene cluster for carbapenem in *E. coli,* which leads to a 60-fold increase in antibiotic production (Shomar et al., 2018). Additionally, modifying metabolic pathways, such as redirecting carbon flux from the pentose phosphate pathway (PPP) to glycolysis in *Streptomyces lividans* by modifying the zwf1 and zwf2 genes, can increase the production of glycolysis intermediates needed for antibiotic synthesis.

A more efficient bio-factory could be developed by introducing genes from or in a different host organism that utilizes a different synthesis pathway (Sakai et al., 2012). Chen et al. (Chen et al., 2009) investigated the expression of a gene cluster from *Streptomyces cacaoi* (*S. cacaoi*) in *S. lividans* TK24. However, polyoxin production by *S. lividans* TK24 was only of the polyoxin H derivative due to the absence of genes for hydroxylation/carboxylation, resulting in the production of polyoxin A/F derivatives. Eustáquio et al. (Eustáquio et al., 2004) expressed the gene cluster responsible for novobiocin synthesis in S. coelicolor. Two *S. coelicolor* mutants were generated by modifying the novO gene, resulting in an 8′-unsubstituted novobiocin by inactivating novO, and adding a chlorine atom at C-8′ by expressing the clohal gene from the clorobiocin gene cluster. Wang et al. (Wang et al., 2021) expressed cytochrome P450 (CYP) genes from the mushroom Ganoderma lucidum in *Saccharomyces cerevisiae* (*S. cerevisiae*) to produce a derivative of 3,28-dihydroxy-lanosta-8,24-dien-26-oic acid. By increasing the copy number of two resistance plasmids, the production of this new antibiotic was increased 8.2-fold compared to the control strain. Eppelmann et al. (2001) studied the expression of non-ribosomal peptide synthetases in *Bacillus subtilis* by introducing the bacitracin biosynthesis gene cluster of *B. licheniformis*. *B. subtilis* showed similar self-resistance to bacitracin due to gene replacement and increased bacitracin A production due to the high-level expression of the bacitracin synthetase and a higher growth rate compared to *B. licheniformis*. Similarly, Wu et al. (2015) expressed the ram29 gene from the ramoplanin producer *Actinoplanes sp.* in *Streptomyces fungicidius* to produce monomannosylated enduracin derivatives.

The primary objective of creating an engineered microorganism is to increase the production of a specific antimicrobial. However, when genetic engineering focuses on manipulating foreign genes in an organism, various factors can affect its biosynthesis. Modifying genes in a specific biosynthesis or metabolic pathway may enhance production, but it could also decrease antimicrobial activity. Even within the same species, differences between strains can allow for the production of derivatives of the intended antimicrobial, which could alter its effectiveness. Therefore, understanding how a microorganism's genetic material enables it to survive in its microenvironment can provide us with the tools to manipulate it and increase the likelihood of developing new and improved antibiotics.

6.4. ENGINEERED PHAGES TO FIGHT SUPERBUGS

Phages, which are viruses that specifically target bacteria, are the most prolific organisms on Earth, with up to 2.5×10^8 phages per milliliter of water (Bergh et al., 1989). Since phages require bacteria to replicate and thrive, they naturally function as bacterial controllers. Although they were first discovered by Frederick Twort and Felix d'Herelle at the dawn of the 20th century (Sharma et al., 2017), antibiotics quickly overshadowed phages in treating microbial infections. However, due to the surge in multi-antibiotic-resistant bacterial pathogens, phages are now seen as a promising alternative for treating such resistant bacteria.

In recent times, scientists have been exploring the use of single phages, the discovery of new phages, and the formulation of phage cocktails as potential treatments for multi-resistant bacterial infections. Research has shown that natural phages can be effective in combating superbugs, as demonstrated by a study in which patients infected with a drug-resistant A. baumannii strain were successfully treated with phage cocktails, leading to full recovery (Schooley et al., 2017). However, synthetic biology techniques can be used to enhance or broaden the capabilities of phages, such as designing and engineering them to create variants with unique properties (Edgar et al., 2012). One approach involves designing and constructing phages with a broad host range (Lin et al., 2012). Since phages are highly specific to particular bacteria strains, the use of cocktails is necessary to combat infections and reduce the chances of bacterial resistance to any phage. Researchers have developed a high-throughput method for engineering host range-determining regions (HRDRs) in T3 phage through site-directed mutagenesis. This approach, inspired by antibody-specific engineering, aims to minimize disruptions to the tail structure and is known as "phage bodies" (Yehl et al., 2019).

Employing the monophage method has the potential to decrease the abundance of phages in cocktails, thereby simplifying preparation and purification procedures. Furthermore, this approach presents fewer complications typically associated with phage cocktails, such as phage purification and compatibility. Consequently, it becomes more accessible for use in treatment. Additionally, the use of cocktails may drive bacteria to develop broader-spectrum mechanisms of phage resistance, such as capsules that obstruct phage binding (Schooley et al., 2017).

Moreover, Ando et al. (2015) transferred tail fiber genes to make genetically modified *E. coli* target *Klebsiella* and vice versa. The authors also showed that synthetic phage cocktails with a similar scaffold but different tails can selectively eliminate bacteria from multi-species communities. Bacterial biofilms are notoriously difficult to eliminate due to the physical and physicochemical properties of the biofilm matrix and exopolysaccharides, as well as the heterogeneity of the bacterial cells within the biofilm. These properties make biofilms highly resistant to chemicals, including antibiotics, immunological approaches, and phages. However, genetically engineered phages can be used to overcome this challenge. Chen et al. (2021) designed an engineered T4 phage called T4 Rn11 that exhibits antibiofilm activity against *Streptococcus mutans*. This phage was successful in reducing biofilm biomass and forming microcolonies. In a similar study, Pei and Lamas-Samanamud (2014) constructed a T7 phage that produces a lactonase enzyme with broad-range activity for quenching quorum sensing. The modified phage effectively degrades acyl-homoserine lactones (AHLs) from many bacteria and inhibits biofilm formation by *P. aeruginosa* and *E. coli.* Reducing biofilm is crucial in treating infections caused by microorganisms in cystic fibrosis disease (Dedrick et al., 2019) or wound infections, as well as in preventing biofilm formation on medical devices such as catheters (Fu et al., 2010). These devices are a significant contributor to morbidity and mortality among patients.

As previously discussed, integrating phages' inherent abilities with synthetic biology techniques can enhance their effectiveness in combating multi-resistant bacteria. By broadening their range of infection and facilitating bacterial communication or expressing enzymes (as demonstrated by Bikard et al., 2014), these phages can better eliminate such resistant bacteria.

6.5. MICROBIAL CONTROL SYSTEMS

In recent years, advancements in synthetic biology have enabled the development of microbial control systems through the engineering of entire living cells to function as biosensors. These cells are capable of detecting and responding to internal and external signals, including quorum sensing (QS) signals secreted by pathogens (Benítez-Chao et al., 2021). The QS-based phenotypical behaviors exhibited by bacteria include the formation of sporulation, the production of virulence factors related to invasion, the display of bioluminescence, and the regulation of population size (Benítez-Chao et al., 2021). Synthetic genetic circuits have been developed and utilized in various studies to analyze and guide the design of biological systems based on Quorum Sensing (QS). Researchers have genetica□y modified *E. coli* to detect wild-type P. aeruginosa (PAO1) through its QS molecule, resulting in the successful inhibition of PAO1 growth through the secretion of a pathogen-specific engineered chimeric bacteriocin (Gupta et al., 2013). In another study, *Lactococcus lactis (L. lactis)* was designed to detect *E. fecalis*, producing and secreting peptides that inhibit enterococcal growth and decrease the viability of enterococci in the surrounding area of *L. lactis*. This engineered system was effective against multidrug-resistant *E. faecium* strains (Borrero et al., 2015). Holowko et al. (2016) created a synthetic genetic sensing system using nonpathogenic *E. coli* as the host based on CRISPRi technology, demonstrating high sensitivity to the presence of Vibrio cholerae (V. cholerae) supernatant with tight control of output GFP protein expression (Holowko et al., 2016). Lubkowicz et al. (2018) developed a probiotic lactic acid bacteria (*Lactobacillus reuteri*) engineered to detect autoinducer peptide-I (AIP-I), a QS molecule produced by *Staphylococcus* sp. during pathogenesis. Their results showed that the engineered biosensor could detect AIP-I levels under various challenging conditions in the *S. aureus* (Lubkowicz et al., 2018). Recently, Wu et al. (2021) developed a novel whole-cell biosensor to detect bacterial pathogens (*P. aeruginosa* and *Burkholderia pseudomallei*) by responding to relevant QS signal molecules, demonstrating that engineered whole-cell biosensors provide rapid and cost-effective detection of waterborne pathogens. Quorum sensing is a communication process through which bacteria coordinate their cellular behavior with their own species or other species, as previously mentioned. In the past decade, synthetic biology has utilized the properties of quorum sensing to develop genetic circuits for population control. The goal of scientists is to create biological systems that exhibit predictable behavior.

6.6. IMPROVING CATALYTIC EFFICIENCY AT ORGANELLE LEVEL

The efficiency of catalytic reactions within organelles is influenced by both external and internal processes, as reported by Jiang et al. (2018). Therefore, enhancing the catalytic efficiency of microorganisms can be achieved by targeting both extracellular and intracellular organelles, as illustrated in Figure 6.3.

6.6.1. Boosting Catalytic Efficiency by Harnessing Extracellular Organelles

The efficiency of catalytic processes can be enhanced by relocating extracellular organelles from the cell surface to the cell interior, which involves reconstructing the catalytic system of surface organelles, harnessing the synthesis mechanism of the cell wall, and developing a control system for membrane microdomains (as shown in Figure 6.3A). By immobilizing enzymes through surface organelles, substrate utilization can be significantly increased. These organelles are typically classified into two types: native and non-native. In native organelles, the biofilm can serve as a platform for self-assembly of extracellular enzymes to improve catalytic efficiency. For instance, the rate of trehalose formation was identified as the limiting factor in the conversion of starch to maltose using free β-amylase (Nguyen et al., 2014). To overcome this limitation, the biofilm protein (CsgA and β-amylase) was fused with Spy-catcher, creating a capsule-like particle in *Escherichia coli* to immobilize β-amylase and starch. The total amount of trehalose produced in the capsule-like particles reached up to 59.0 g/L, which also increased the trehalose formation rate by 2.64-fold (Jiang et al., 2018). Another example is the biofilm of *Bacillus subtilis*, which was limited in its capacity for protein secretion due to its weaker secretion system compared to *E. coli.* To address this, the TasA protein was fused with Spy-catcher in *B. subtilis*, allowing for efficient secretion of the MHETase. The toxic compound mono (2-hydroxyethyl) terephthalic acid was converted into the less toxic terephthalic acid, increasing to 7 mM (Huang et al., 2019). In contrast, non-native surface organelles have used surface display molecules to enhance catalytic efficiency through various approaches, such as a biohybrid system (Wei et al., 2018), peptide library screening (Becker et al., 2008), and whole-cell biocatalyst (Wu et al., 2006). These organelles enable exogenous peptides and proteins to be displayed on the cell surface through fusion with anchoring motifs. For example, efficient solar-to-chemical conversion was achieved through native surface adsorption in a nanoparticle precipitation system. A surface-displaying biohybrid system facilitated the controllable biosynthesis and adsorption of biocompatible cadmium sulfide semiconductors via a biomimetic silica encapsulation strategy. The content of cadmium sulfide on the cell surface reached 8.09 ± 0.70 mg/108 cells, increasing hydrogen production (~0.52 mmol/108 cells) through self-photosynthesis in engineered *E. coli* (Wei et al., 2018).

Several approaches have been reported to enhance extracellular secretion of proteins and chemicals by altering the physiological characteristics of the cell wall, including cell wall synthesis-based secretion (Yang et al., 2018), cell wall integrity-directed fermentation (Zhang et al., 2018), and cell wall-based glycozylphosphatidylinositol anchoring (Inokuma et al., 2020). Among these strategies, engineering cell wall synthesis-based protein secretion has shown the greatest improvement, achieving better results by altering the cell's shape (Yang et al., 2018). For instance, *E. coli* does not efficiently secrete amylase into the extracellular space due to the stable peptidoglycan structure in its cell wall. By deleting *dacA* and *dacB* genes in *E. coli* BL21, the rigid structure of the peptidoglycan in the cell wall was disrupted, which enabled efficient secretion of the protein.

The mutant strains BL21 *ΔdacA* and BL21 *ΔdacB* exhibited enhanced production of soluble peptidoglycan, with concentration increases of 4.1 and 1.4 times, respectively, compared to the wild-type, which increased recombinant amylase activity by 2.5 and 3.1 times, respectively (Yang et al., 2018). As the reaction rate determines the efficiency of chemical production, other researchers have attempted to improve the reaction rate by synthesizing multi-enzyme complex systems, including the development of endogenous membrane spatial scaffolds (Lee et al., 2018), TAT-thylakoid membrane scaffolds (Ramos, 2017), and lipid droplet membrane scaffolds of micro-scaffolds (Myhrvold et al., 2016). Among these strategies, anchoring metabolic pathway enzymes to endogenous membrane spatial scaffolds significantly improved reaction efficiency (Lv et al., 2020). For instance, the bottleneck of N-acetylglucosamine (GlcNAc) production is the relatively low catalytic reaction rate (Wheeldon et al., 2016). Therefore, the GlcNAc pathway enzyme was anchored onto endogenous spatial scaffold-functional membrane microdomains by constructing a multi-enzyme complex system in *B. subtilis*, resulting in a 1.8-fold increase in the GlcNAc titer. The maximized reaction rate was also verified using a metabolic model (Lv et al., 2020).

6.6.2. Increasing Catalytic Efficiency by Engineering Intracellular Organelles

The efficiency of catalytic reactions can be improved by designing microcompartments, both natural (endogenous) and artificial organelles. This can be applied to various microorganisms, not just yeast or bacteria (as shown in Figure 6.3B). In yeast, microcompartments can enhance the efficiency of enzymes involved in metabolic pathways and protein secretion by utilizing the sequestration organelles and protein synthesis organelles respectively. To increase the efficiency of enzymes in the organelle sequestration pathway, the metabolic pathway can be compartmentalized within the mitochondria (Avalos et al., 2013), lipid droplets (Wagner et al., 2018), or peroxisomes (Liu et al., 2020).

The most successful strategy reported so far for enhancing metabolic pathways has been compartmentalizing them within mitochondria (Avalos et al., 2013). In this study, overexpression of enzymes in the metabolic pathway reduced isobutanol production by decreasing the transmembrane transport of intermediates. Specifically, the researchers engineered Saccharomyces cerevisiae to compartmentalize the Ehrlich pathway enzymes in the mitochondria to increase the availability of the key α-ketoisovalerate intermediate. This resulted in an isobutanol titer of 635 mg/L, a 260% increase compared to the control (Avalos et al., 2013). Another study found that fatty acid synthetic pathways for microbial production were often hindered by competing pathways in the cytosol. Therefore, the researchers targeted fatty acyl-CoA-derived fatty alcohol synthetic pathways to peroxisomes to eliminate side-pathway competition. This strategy increased the production of fatty-acid-derived fatty alcohols, alkanes, and olefins by 700% compared to the control (Zhou et al., 2016). Protein synthesis organelles are another notable yeast compartment target. By synthesizing heterologous proteins in vesicle trafficking and extending secretion proteins in the Golgi apparatus and endoplasmic reticulum, the secretion capacity of the protein can be enhanced (Hou et al., 2012; Kim et al., 2019; Besada-Lombana & Da Silva, 2019). Among these strategies, extending secretion proteins has been the most widely applied because the endoplasmic reticulum and Golgi apparatus are relatively stable intracellular structures (Huang et al., 2018). For example, the production of recombinant proteins in yeast is limited to the processes associated with intracellular protein retention. To overcome this limitation, the researchers engineered endosome-to-Golgi transport to transport and secrete available Kex2p for amylase maturation in *S. cerevisiae* K17. This resulted in a 12% increase in amylase secretion efficiency, improved the amylase titer and productivity to 2.5 g/L and 0.5 mg/g dry cell weight per hour (Huang et al., 2018).

To reduce metabolic crosstalk in a pathway, researchers have employed various strategies such as using bacterial microcompartments as physical barriers, engineering the pathway in the periplasmic compartment, controlling the function of metabolosomes, and regulating the assembly of carboxysomes. Among these approaches, modifying the periplasm may be the most effective as it enhances the efficiency of targeted metabolite production, preventing metabolite exchange while avoiding undesirable interactions. For example, in *E. coli,* metabolic crosstalk significantly affects the utilization of precursor (pyruvate) for malate production. To overcome this, researchers used periplasmic engineering to direct the reductive tricarboxylic acid (rTCA) pathway towards the periplasm for full utilization of pyruvate, resulting in increased malate titer up to 193 mM and 0.5 mol/mol (Guo et al., 2018). In *Cyanobacteria,* carboxysomes can be employed as microcompartments to facilitate CO_2 fixation. However, central carbon metabolism and nitrogen metabolism are affected by CO_2 fixation, leading to the creation of a conditional auxotroph by repressing carboxysome synthesis through CRISPR interference of the key subunit's ccm operon and nitrogen assimilation pathway. This resulted in *Synechococcus* sp. cells producing 2-fold more lactate than control cells (Gordon et al., 2016).

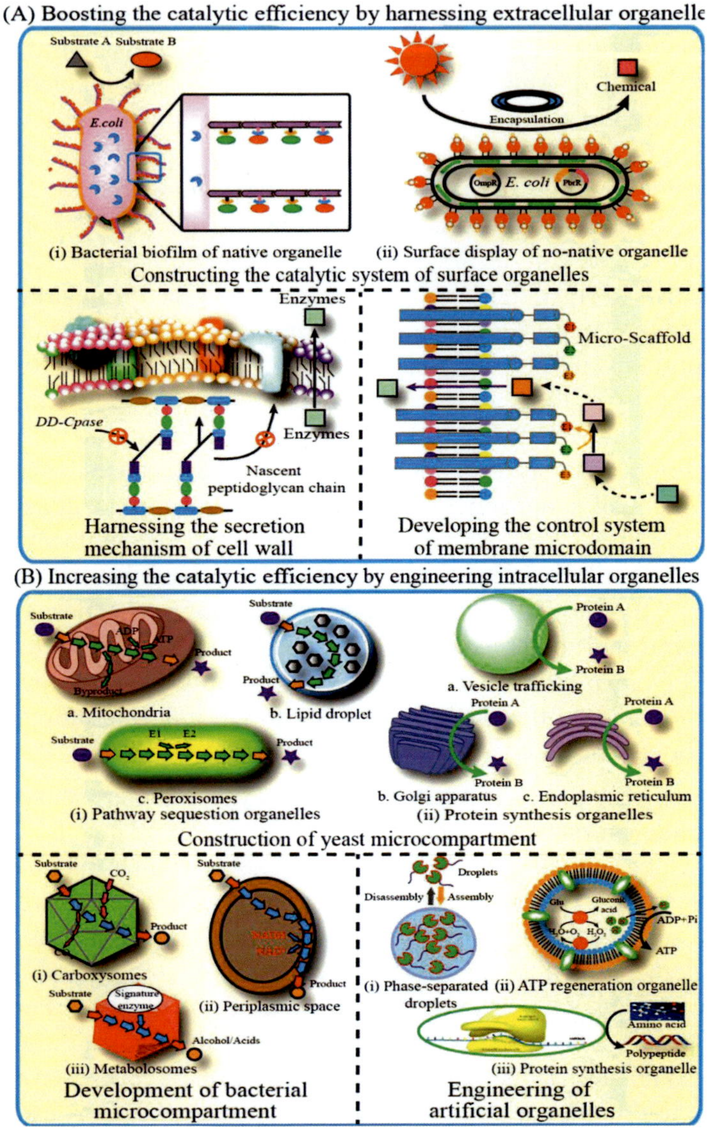

Figure 6.3. *Strategies for enhancing catalytic efficiency at the organelle level include: (A) Boosting the catalytic efficiency of microorganisms through the use of extracellular organelles, constructing catalytic systems on surface organelles, harnessing the synthesis mechanism of the cell wall, and developing control systems for membrane microdomains. These approaches can improve chemical production efficiency. (B) Increasing catalytic efficiency can also be achieved through the construction of yeast microcompartments (including pathway sequestration organelles such as mitochondria, lipid droplets, and peroxisomes; protein synthesis organelles such as vesicle trafficking, golgi apparatus, and endoplasmic reticulum), bacte-*

rial microcompartments (such as periplasmic, metabolosomes, and carboxysomes), and artificial organelles (such as phase-separated droplets, encapsulin nanocompartments, and organelles for ATP regeneration).

Source: Ding et al., (2020).

Recent research has focused on engineering artificial organelles to endow cells with compartmentalization functions that enhance catalytic efficiency. One approach has involved assembling organelles in phase-separated droplets, which can enhance metabolic flux by co-localizing sequential enzymes (Zhao et al., 2019). However, this approach can result in the accumulation of toxic intermediates. To address this, optogenetic control can be used to control the assembly and disassembly of pathway enzyme droplets, reducing the concentrations of intermediates and efficiently enhancing deoxyviolacein formation by 6-fold (Zhao et al., 2019).

6.7. INCREASING STRAIN PERFORMANCE AT CELL LEVEL

Improving the performance of a strain at the cellular level involves increasing its titer, yield, and productivity of chemical production, which are all influenced by the cellular physiological state (Guo et al., 2020). To enhance strain performance, it is possible to improve the cell growth capacity and adjust the cell morphology (as illustrated in Figure 6.4).

6.7.1. Improving the Cell Growth Capacity

Increasing cell growth capacity can be achieved by decoupling cell growth from chemical synthesis, enhancing cell division to increase growth density, and employing adaptive laboratory evolution strategies to boost growth rate. Decoupling cell growth from chemical synthesis can also improve the efficiency of microbial cell factories by optimizing carbon flux. Various methods are available for decoupling cell growth, such as protein degradation, optogenetics regulation, temperature switch, and others. Among these methods, optogenetics regulation is a promising approach to manipulate cell metabolism in a reversible and convenient manner. One study aimed to overcome the limitations of isobutanol and 2-methyl-1-butanol production by targeting coupled cell growth and chemical synthesis. By using a newly developed optogenetic circuit to shift cells from a light-induced growth phase to a dark-induced production phase through periodic light pulses, the researchers controlled pdh with blue light to decouple cell growth. As a result, the mitochondrial isobutanol pathway produced 8.49 g/L of isobutanol and 2.38 g/L of 2-methyl-1-butanol through optogenetic regulation (Zhao et al., 2018). To enhance the density of cell growth, it is possible to target cell division by regulating the cell division cycle. This can be achieved by manipulating the C and D periods, which are key stages in the bacterial cell division cycle. The C period involves the initiation to the termination of DNA replication, while the D period is the bacterial division process. Stimulating dNTP synthesis and divisome assembly through the use of isopropyl-β-D-thiogalactoside or anhydrous tetracycline induction during the C and D periods can increase the division frequency of *E. coli* cells and ultimately lead to increased protein production. Overproduction of recombinant proteins often results in the growth cessation of *E. coli,* so one group overexpressed *ftsZ* and *ftsA*, which are related to cell division in the D period, to enhance cell growth density and protein production by suppressing cell filamentation. This resulted in a 57.14% increase in growth density and a 30% increase in specific growth rate, ultimately increasing the human leptin titer by 2.27 folds. Another strategy is to use fast-growing strains, as the productivity of industrial fermentation processes is limited by biomass accumulation. Vibrio natriegens, the fastest-growing non-pathogenic bacterium, is a promising candidate for shortening the doubling time. By using this strain, the doubling time was decreased to 9.4 min and biomass accumulated to 20 g/L, resulting in a 0.56 g/L/min productivity of alanine (Hoffart et al., 2017).

The growth rate of cells mainly depends on their endogenous growth performance, which can be enhanced through the use of adaptive laboratory evolution approaches such as mutation breeding-based adaptive laboratory evolution and environment stress-based adaptive laboratory evolution (Yin et al., 2020; Sandberg et al., 2019). Mutation breeding-based adaptive laboratory evolution is an effective technique that not only significantly increases the growth rate of a strain but also targets specific rate-limiting system components. This approach was used to efficiently screen a strain with a faster growth rate in glucose minimal medium through a multiplex automatable genome engineering strategy, leading to the removal of over 300 amber stop codons in a genomically recoded *E. coli* strain. As a result, the strain was evolved to generate strains with a 40% increase in growth rates (Wannier et al., 2018).

6.7.2. Adjusting the Cell Morphology

The effects of endogenous morphology manipulations and exogenous fermentation conditions on cell morphology can vary, with endogenous manipulations having a more significant impact (Wang et al., 2019). To manipulate endogenous morphology, researchers can engineer divisome assembly to create longer cells, control elongasome formation for wider cells, utilize microbial lifespan to alter resource distribution within cells, manage cell lysis to disrupt the cell's closed state, and regulate community morphology (as shown in Figure 6.4B).

A strategy to increase cell volume and overcome the constraint of cell length for inclusion body accumulation is to control cell morphology by inhibiting divisome assembly (*FtsZ*) or expressing cell division inhibitors (SulA, MinCD) (Jiang & Chen, 2016). Currently, cell division inhibitors are more commonly used for this purpose. For instance, the production of inclusion bodies in rod-shaped bacteria is largely affected by the limited cell length. To overcome this limitation, one research group overexpressed the cell division inhibitor MinCD in *Halomonas* TD08, which resulted in an increase in cell length by at least 1.4-fold compared to its original length. This increase in length led to an improvement in PHB accumulation from 69 wt% to 82 wt% (Tan et al., 2014).

Engineering elongasome assembly for extending the width of *E. coli* cells can be achieved through various strategies, including augmenting transpeptidase, cytoskeletal proteins, penicillin-binding proteins, transglycozylase, and transmembrane proteins, as cell width affects cell volume (Jiang & Chen, 2016). Disrupting cytoskeletal proteins is a fruitful strategy as it significantly disturbs the helical filamentous structures that serve as a scaffold for peptidoglycan synthesis. For instance, CRISPRi technology was employed to optimize the expression of mreB and ftsZ using different single-guide RNAs targeting various locations in the *E. coli* chromosome, which increased the accumulation of intracellular PHB, reaching 40–80% bacterial cell dry weight (Elhadi et al., 2016). Another example of increasing cell width is observed in Pseudomonas mendocina NK-01, where the transport pathway for alginate oligosaccharides secretion is limited by the cell space. Overexpression of mreB in the pBBR vector and its location in the cell membrane improved the cell membrane secretion capacity, resulting in a 1.49-fold increase in cell width and an increased titer of alginate oligosaccharides to 0.961 g/L (Fan et al., 2019).

The aim is to distribute intracellular inclusion bodies efficiently. To achieve this, the lifespan of microorganisms was manipulated to result in asymmetric division, taking into account both chronological and replicative lifespan. The replicative lifespan can be used to regulate the intracellular distribution of inclusion bodies, thereby enhancing chemical production. In a previous study (Guo et al., 2020), the intracellular distribution of inclusion bodies was not considered when constructing the poly(lactate-co-3-hydroxybutyrate) pathway in *E. coli.* Therefore, a two-output recombinase-based state machine was employed to shorten the replicative lifespan by separating the mother and daughter cells. The mother cell accumulated poly(lactate-co-3-hydroxybutyrate) while the daughter cell was used for growth, resulting in a 52 wt% poly(lactate-co-3-hydroxybutyrate) content in a 5-L fermenter.

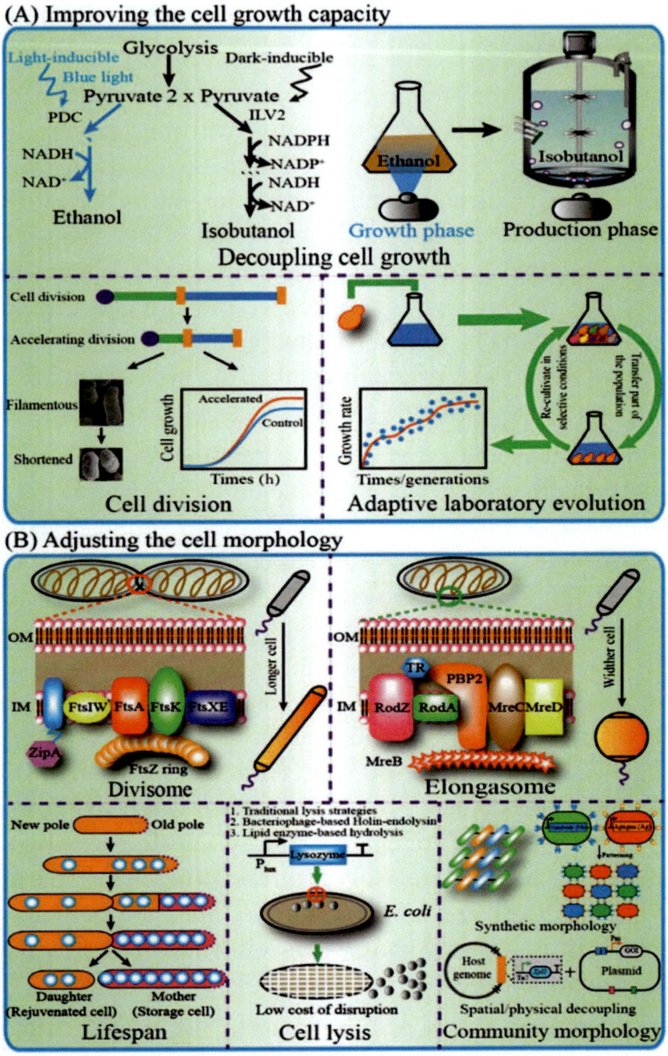

Figure 6.4. *Improving strain performance at the cellular level involves several strategies. (A) Enhancing cell growth capacity can be achieved by decoupling cell growth from chemical synthesis, engineering cell division, and adaptive laboratory evolution. These methods can efficiently boost strain performance. (B) Engineering adjustable cell morphology is another approach. This can be achieved by constructing divisome assembly, elongasome formation, adjusting lifespan, cell lysis, and community morphology to change cell shapes. FtsZ, a bacterial tubulin homolog, and ZipA, a membrane-anchored protein that binds to FtsZ, are important components in this process. Additionally, SulA, a cell division inhibitor, MinCD, an FtsZ inhibitor, PBP2, a murein synthase, RodA, a transglycozylase, RodZ, a transmembrane protein, MreB, MreC, and MreD cytoskeletal proteins, FtsK, a fila-*

ment temperature-sensitive mutant K, FtsI, a monofunctional transpeptidase, FtsA, a filamentous temperature-sensitive A, OM, outer membrane, and IM, inner membrane are also involved.

Source: Ding et al., (2020).

Proposed to simplify downstream processing and reduce engineering costs, three cell lysis methods have been introduced: traditional lysis strategies (Wang et al., 2019), bacteriophage-based holin-endolysin strategies (Gao et al., 2013), and lipid enzyme-based hydrolysis systems (Short et al., 2012). The most effective microbial cell lysis was achieved by introducing into the cells a lysozyme from *E. coli* phage λ, known as endolysin, and the pre-forming protein holin (Gao et al., 2013). Inclusion bodies cannot be effectively released into the extracellular space once they accumulate in the closed cell; therefore, endolysin and holin were inserted to disrupt the peptidoglycan in the cell wall and thus the cell structure. Lastly, accumulated intracellular PHB was effectively released into the extracellular space to reduce the costs associated with cell breakage (Zhang et al., 2009).

The performance of the non-modal strain *Pseudomonas putida* was enhanced by regulating its cell community morphology through synthetic morphology manipulations and spatial decoupling, as reported by Glass and Riedel-Kruse (2018) and Volke et al. (2020), respectively. To achieve this, natural and synthetic catalytic biofilms were utilized in bioprocesses, as demonstrated by Volke and Nikel (2018). For example, the biodegradation of haloalkanes was limited by low dehalogenase activity, which was addressed by regulating c-di-GMP through the yedQ or yhjH genes to optimize the whole-cell catalyst system. The levels of dehalogenase activity in the biofilm fraction ranged from 0.75–1.85 units mg/protein, showing a 2.1- and 4.3-fold higher activity ratio compared to glucose and succinate cultures, respectively. In spatial decoupling, the genomic and plasmid parts are physically separated to express genes, as described by Benedetti et al. (2016). For instance, the expression of the target gene was reduced to close-to-zero levels of leaky expression by constructing an expression system that integrated *xylS* into the chromosome of *P. putida* and replaced the Pm promoter in a plasmid vector for physical decoupling. This resulted in a 170-fold increase in GFP fluorescence.

6.8. REINFORCING CELL-CELL COMMUNICATION AT THE CONSORTIUM LEVEL

Cell-cell communication plays a crucial role in regulating the behavior of microbial consortia by modulating the exchange of metabolites and signaling molecules between cells. This process is influenced by the composition of the consortium and can be improved by designing multi-strain interactions and creating systems that focus on population interaction (Jiang, 2020) (Figure 6.5).

6.8.1. Engineering The Pattern of Multi-Strain Interactions

Engineering altruistic interactions, cooperative interactions, and commensal interactions can rewire the pattern of multiple-strain interactions (Shong et al., 2012; Xin et al., 2019; Zhou et al., 2015) (Figure 5A). Specifically, two designs have been proposed for constructing a cell consortium to achieve altruistic interactions, including lysis-based altruistic interactions (Honjo et al., 2019) and self-limiting-based altruistic interactions (Din et al., 2016). Among these, lysis-based altruistic interactions are the most commonly used pathway to enhance microbial production. For instance, cellobiose utilization is a limiting factor in isopropanol production, and an enzyme-producing strain was self-lyzed to produce glucosidase through a quorum-sensing promoter, which subsequently degraded cellobiose to glucose for isopropanol production. This approach reduced the time required for cellobiose saccharification by fourfold, resulting in an isopropanol titer of 16–19 mM, a 2.7–3.28-fold increase compared to the control (Honjo et al., 2019).

In an effort to strengthen cell-cell communication and promote cooperative interactions, researchers have explored the use of consolidated bioprocessing-based cooperative strategies (Wen et al., 2014) and resource partitioning-based approaches (Xin et al., 2019). Among these, consolidated bioprocessing has been more widely applied in microbial consortium fermentation, as it involves different microorganisms working together to degrade lignocellulose and produce biofuels. For example, butanol is a difficult solvent to produce on a large scale from alkali-extracted corn cobs without adding butyrate. By engineering *Clostridium thermocellum* ATCC 27405 and *Clostridium beijerinckii* NCIMB 8052 for sequential and cooperative co-culture in a one-pot reaction with optimized culture conditions, the consolidated bioprocessing system was able to decompose 88.9 g/L of alkali-extracted corn cobs and produce 3.96 g/L of acetone, 10.9 g/L of butanol, and 5.04 g/L of ethanol (Wen et al., 2014).

Engineered commensalism interactions have been employed to efficiently distribute different metabolic targeted pathways among various strains by dividing multi-metabolic pathways and establishing multi-strain syntrophic interactions (Shong et al., 2012). To date, this approach has proven to be the most successful method for engineering commensalism interactions, as it allows for the selection of the most suitable organism for each partial transformation, thereby simplifying the reuse of individual partial pathways (Zhou et al., 2015). For instance, the synthesis of acetylated diol paclitaxel involves long and complex metabolic pathways, some of which require specialized environments or compartments for optimal function. To overcome this challenge, the synthetic pathway for the monoacetylated dioxygenated taxane precursor (oxygenated taxadiene) was divided between *S. cerevisiae* and *E. coli,* with *E. coli* producing taxadiene and *S. cerevisiae* utilizing it to obtain oxygenated taxadiene. Finally, a stable co-culture was achieved in the same bioreactor by designing a commensal relationship to produce 33mg/L of monoacetylated dioxygenated taxane (Zhou et al., 2015).

6.8.2. Developing a Population Interaction Systems

To achieve population-level interactions, it is possible to engineer the quorum-sensing system with population quality control and effective consortium structure, as well as design a rational metabolic model. A quorum-sensing genetic circuit can be used to engineer pathway-independent dynamic regulation, cell growth-response regulation, and quorum quenching. Among these options, pathway-independent dynamic regulation is a good strategy for achieving precise tunability of an engineered pathway with minimal upfront effort in any growth environment. Since metabolic engineering of microorganisms to produce desired products on an industrial scale can lead to unbalanced metabolic flux in *E. coli,* a pathway-independent genetic mode was established to dynamically regulate the expression of the *pfk-1* gene, which was identified as the optimal point to redirect glycolytic flux by the quorum-sensing system. This approach resulted in a 5.5-fold increase in myo-inositol titers and glucaric acid titers from unmeasurable to >0.8 g/L (Gupta et al., 2017). *B. subtilis* typically employs quorum sensing systems that predominantly involve down-regulation of gene expression rather than up-regulation. To address this imbalance, a bifunctional and modular system was developed, involving the Phr60-Rap60-Spo0A pathway, which was based on two native promoters (PabrB & PspoiiA) to optimize cell growth and chemical production. This innovative approach resulted in a significant improvement in MK-7 production, increasing it 40-fold from 9 mg/L to 360 mg/L in shake flasks and 200 mg/L in a 15-L bioreactor (Cui et al., 2019).

Engineering population quality control can help enrich high-performing subpopulations and reduce low-performing ones, which has been successfully applied in various bioprocesses, such as the production of free fatty acids and tyrosines (Lv et al., 2019; Zhang et al., 2012; Kim et al., 2018). One effective strategy is to engineer the free fatty acid-responsive transcriptional repressor *FadR* to regulate the expression of the tetracycline efflux pump. By contro□ing the expression of a growth-dependent gene (*leuABCD*) using a free fatty acid signal-responsive promoter (PAR), a population with high free fatty acid production can be selected and a population with low free fatty acid production can be eliminated. This approach resulted in a 3-fold increase in both free fatty acid titer and yield compared to the control (Xiao et al., 2016).

To address the issue of signal molecule diffusion limitation in coordinating gene expression in spatially extended consortia, various strategies such as task division, cyclic consortium control, and consortium component control can be employed. Among these, task division has been the preferred approach for optimizing consortium structure in bio-manufacturing.

For instance, a study by Tsai et al. (2010) demonstrated that a single culture with expression of both parts resulted in a lower yield of ethanol compared to proportional consortium cultivation. Subsequently, three different yeast strains were developed to secrete three distinct proteins with docking tags, enabling their assembly onto an extracellular scaffold at a final reported ratio of 7:2:4:2 scaffold:endoglucanase:exoglucanase:glucosidase.

This optimized ratio led to 87% theoretical ethanol production, which was three times higher than that obtained using a similar consortium with a control strain (Garima, 2011). In another study, the specific microorganisms responsible for acetoin formation in vinegar remained unclear. However, the diacetyl/acetoin metabolic pathway of vinegar microbiota was elucidated through a combination of metagenomic sequencing and clone library analysis. Furthermore, the acetoin content in vinegar Pei that was augmented with *A. pasteurianus* G3–2L plus *L. buchneri* F2–5 (15.4 ± 2.6 mg/g Pei) was significantly higher compared to the control group (6.8 ± 0.1 mg/g Pei) (Lu et al., 2016). In addition to introducing a metabolic model, another potential strategy to improve production efficiency through rational cell-cell communication is to select the most appropriate consortium members.

This can be achieved using various metabolic models such as genome-scale metabolic models (Xu et al., 2019), elementary mode analysis (Taffs et al., 2009), and flux balance analysis (Niels, 2010). Genome-scale metabolic models were developed to simulate metabolic features and predict optimal solutions, as previously mentioned (Xu et al., 2019).

These models were used in conjunction with sequencing technology to identify the direct degrader of atrazine within Arthrobacter and optimize 16 combinations of endogenous consortia for improved growth and degradation rates. Through simulations, the biomass was increased by approximately 110% and the time required for atrazine degradation was shortened by 16.7% (Xu et al., 2019).

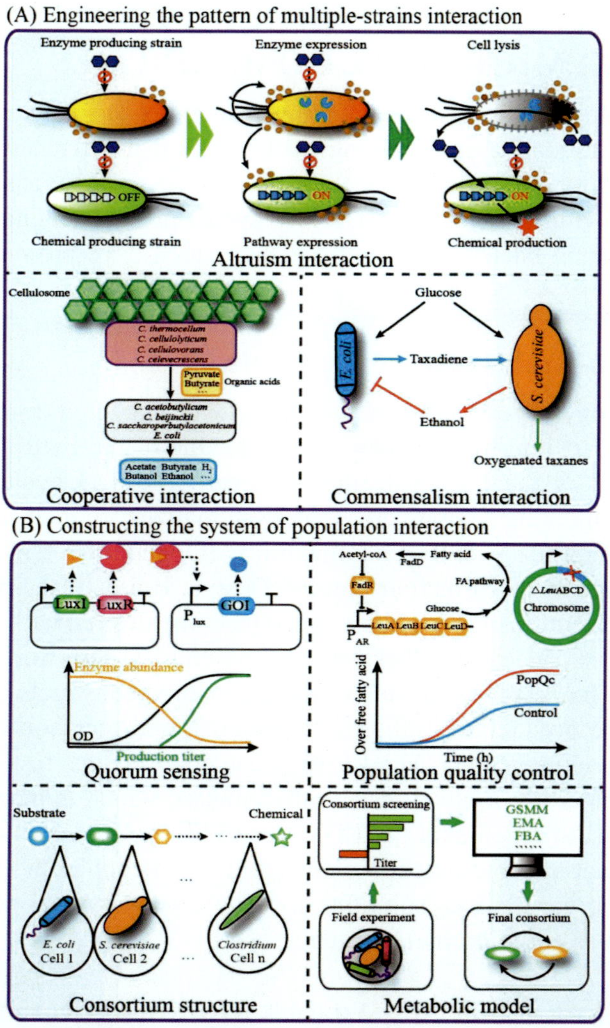

Figure 6.5. *The strategies to enhance cell-cell communication at the consortium level involve (A) engineering the pattern of interactions between multiple strains through the development of altruistic, cooperative, and commensal interactions at the consortium level, and (B) constructing a system of population interaction by designing control methods for quorum sensing, establishing population quality control, managing the effective consortium structure, and creating a rational metabolic model. Key terms: GOI (gene of interest), GSMM (genome-scale metabolic models), EMA (elementary mode analysis), FBA (flux balance analysis), PopQc (population quality control), LuxI and LuxR (quorum sensing components), Plux (quorum sensing promoter).*

Source: Ding et al., (2020).

6.9. OTHER SYNTHETIC BIOLOGY TOOLS TO ENGINEER MICROBIAL COMMUNITIES

Synthetic biology employs engineering principles to create predictable, controllable, and standardized biological systems that offer new insights and enhance cellular capabilities. This field utilizes advanced technologies, such as the Design-Build-Test-Learn (DBTL) cycle, protein engineering against multidrug-resistant bacteria, and logic gates, which have promising potential applications.

6.9.1. Design-Build-Test-Learn Cycle

Synthetic biology is a discipline that aims to design or redesign biological systems to impart improved or new qualities. One of the SynBio engineering principles used is the Design-Build-Test-Learn (DBTL) cycle, which is a metabolic engineering framework that systematizes cellular activities and enhances their efficacy and generalizability (Opgenorth et al., 2019). The DBTL cycle is composed of four phases: Design, Build, Test, and Learn. The Design phase identifies the problem and selects the desired pathway and host. The Build phase selects, synthesizes, and assembles parts for incorporation into the host. The Test phase validates the engineered constructs and strains for target molecule production, transcripts, proteins, and metabolites, generating a significant amount of data. The Learn phase analyzes the Test data and uses the learnings to create a novel testable hypothesis, which is incorporated into the next cycle (Carbonell et al., 2018). The DBTL cycle enables researchers to rapidly construct new biological systems, and when coupled with engineering technologies, it can deliver solutions in drug discovery and address global problems such as antimicrobial resistance.

6.9.2. Protein Engineering Against Multidrug-Resistant Bacteria

Proteins, such as endolysins encoded by bacterial viruses, can disrupt the bacterial cell wall or other proteins that regulate genes related to antibiotic resistance, leading to cell death. They can also alter microbial behavior and interrupt signals that confer multi-drug resistance. By using synthetic biology techniques, proteins can be engineered to make them more stable or efficient. Additionally, synthetic proteins like artilysins, which are outer membrane-penetrating endolysins, can be created and manufactured. However, one drawback is that endolysins are not effective against Gram-negative bacteria due to their impermeable lipopolysaccharide layer surrounding the cell wall. Nonetheless, artilysins are highly bactericidal against Gram-negative pathogens, including *P. aeruginosa* and *A. baumannii,* as demonstrated by Briers et al. (2014). An instance of this is dCas9, which functions as a "dead" catalytic protein. It operates in a manner similar to that of CRISPR-Cas9 technology, except that dCas9 can attach to the DNA without severing the strands. This feature enables dCas9 to be utilized for suppressing gene transcription. In their 2017 study, Wang and Nicholaou designed two CRISPR-dCas9 systems to specifically target the mecA promoter in MRSA, with the aim of inhibiting the gene's transcription (Wang & Nicholaou, 2017). Although these alternatives have not yet been employed in clinical settings, they exhibit great potential for use individually or in conjunction with antibiotics or other techniques.

6.9.3. Logic Gates

A noteworthy area of research in synthetic biology involves employing microorganisms as control elements, enabling them to transmit and receive signals via logic gates. These gates, commonly found in electronic devices, facilitate the transformation of input signals into alternative output signals or the consolidation of two input signals into a singular output signal (Wang et al., 2011). In synthetic biology, living cells function as control elements, allowing them to be programmed to exhibit precise behaviors in response to internal or external signals (Morris et al., 2010). By harnessing gene regulatory networks, which cells utilize to communicate and respond to environmental cues, researchers can devise logic circuits that connect cellular sensors and actuators. Genetic logic gates depend on the direct regulation of transcription, activating a promoter. One illustration is the application of these genetic components to program cells to recognize quorum-sensing signals. Benítez-Chao et al. (2021) designed a whole-cell biosensor to detect and eradicate MRSA by employing genetic logic gates. They devised a genetic circuit in which *E. coli* recognizes quorum-sensing molecules from MRSA, subsequently triggering the expression of a bacteriocin that eliminates MRSA. Genetic circuits can be designed to disrupt metabolic pathways in multi-drug resistant bacteria, or to create whole-cell biosensors that respond to molecules produced by pathogenic organisms and produce antimicrobial compounds. While whole-cell biosensors are still in the research phase, they show promise for controlling pathogenic bacteria in the future.

6.10. MOLECULAR TOOLS FOR THE INTRODUCTION OF BIOSYNTHETIC PATHWAYS

Advances in genetic engineering tools have significantly enhanced the introduction and optimization of pathways in host strains. Plasmids continue to play a crucial role in this process due to their ease of manipulation and introduction into microbial host cells (Henke et al., 2021). These non-native pathways can be rapidly constructed and tested in microbial cell factories using plasmids. Researchers are actively studying tools and strategies for fine-tuning gene expression levels with plasmids. The development of synthetic promoters and ribosome binding site (RBS) sequences is on the rise to fine-tune the expression of biosynthetic genes at the plasmid level. While previous promoter libraries were based on random sequences, new synthetic promoter libraries have been constructed for yeast using model-guided design strategies and for *E. coli* using a deep generative network. Additionally, synthetic RBS sequences with diverse translation initiation levels can be designed using computational tools such as RBS calculator, UTR designer, and RBS designer (Reeve et al., 2014).

The use of recombinase systems, such as Lambda Red or RecET from *E. coli,* has been widely employed to delete or introduce target genes in various microbial hosts for industrial applications, due to the instability and maintenance problems associated with plasmid-based expression. However, these conventional methods are laborious and time-consuming, leading to much effort being exerted to improve recombination-based methods or develop alternative efficient chromosomal engineering methods to expedite strain development (Wannier et al., 2021). Recently, CRISPR-Cas systems, an adaptive immune system of microorganisms, have gained much attention as genome engineering tools due to their ability to introduce double-stranded breaks at a precise DNA sequence in the genome, enabling simple and rapid engineering compared to conventional methods. The Type II and class 2 CRISPR/Cas9 system derived from *Streptococcus pyogenes* has been most widely employed for genome engineering applications, in combination with existing molecular tools to engineer microorganisms (Strecker et al., 2019). More recently, Tn7-like transposons associated with CRISPR-Cas systems have been developed as powerful tools for inserting DNA into genomes. The CRISPR-associated transposase (CAST) system derived from cyanobacteria *Scytonema hofmanni* and *Anabaena cylindrical* and the INTEGRATE system derived from *Vibrio cholerae* have enabled the insertion of large gene clusters into targeted sites in the chromosome (Vo et al., 2021).

Numerous efforts have been made to develop genetic manipulation tools for non-model organisms in engineering, such as mobile-CRISPRi (Klompe et al., 2019), CRAGE (Wang et al., 2019), and XPORT (Brophy et al., 2018). These tools have been created to improve our understanding of diverse bacterial species, characterize essential pathways, and generate desired phenotypes (Peters et al., 2019). For instance, mobile-CRISPRi employs large guide-RNA libraries to rapidly screen various bacteria for essential pathways and desired phenotypes (Peters et al., 2019). CRAGE uses transposon and conjugation systems to introduce biosynthetic gene clusters into diverse bacteria (Wang et al., 2019), while XPORT is an engineered donor strain designed to transfer miniaturized integrative and conjugative elements to undomesticated organisms (Brophy et al., 2018). As molecular tools for engineering non-model organisms become more widely available, these organisms may be metabolically engineered to produce industrial quantities of certain chemicals, potentially becoming the production strains of the future.

6.11. ENZYME ENGINEERING

The development of enzyme engineering has opened up new possibilities for modifying enzyme characteristics such as activity, specificity, and stability to suit our needs. The selection and design of enzymes are important steps in constructing metabolic pathways, which impact the efficiency of metabolic reactions and ultimately affect the overall metabolic flow of the pathway. When heterologous enzymes are introduced into microbial hosts, issues such as low activity, low stability, and poor solubility may arise (Li et al., 2020). However, enzymes can be rationally designed to overcome these obstacles. For instance, fusion tags that have been previously studied can be attached to target enzymes to address stability or solubility problems (Ki and Pack, 2020). In addition, eukaryotic enzymes, particularly those from plants, often contain transit peptide sequences that are responsible for transporting proteins to specific organelles, such as chloroplasts. When expressed in microbial hosts, these enzymes are typically non-functional. Removing these transit peptide sequences is a widely used technique for successfully expressing the enzymes in active forms in microbes (Park et al., 2020). Rational enzyme engineering can also involve the engineering of TesA, a thioesterase from *E. coli* (Grisewood et al., 2017), which was developed through structure-guided mutagenesis to broaden its range of substrate selectivity. The engineered TesA improved the production of medium-chain-length fatty acids while maintaining high enzymatic activity when introduced into *E. coli* (Grisewood et al., 2017). Similarly, computational methods such as homology modeling and protein docking simulation have been used to create mutant enzymes with improved activities. Introducing these mutant versions of enzymes, such as DnrF and GtCGT, resulted in increased production of carminic acid in *E. coli* (Yang et al., 2021).

Although the previously mentioned techniques are valuable, they cannot be universally applied to optimize all enzymes. Instead, directed evolution can be used to enhance enzyme performance. This process involves the creation of a library of mutated enzymes and screening them through multiple rounds of random mutagenesis. From the screened mutated enzymes, those with desired characteristics, such as increased activity, can be selected under various selection pressures (Ko et al., 2020). A recent example of directed evolution is the engineering of carboligase for synthetic one-carbon metabolism (Nattermann et al., 2021). In this case, oxalyl-CoA decarboxylase was converted into glycolyl-CoA synthase through a high-throughput screening of saturation mutagenesis libraries (Nattermann et al., 2021).

Rational design and directed evolution are methods used to modify naturally occurring proteins. In contrast, de novo proteins can be created through computational design to perform reactions that do not exist or have not yet been discovered in nature. With the increasing knowledge of proteins and the availability of X-ray crystallography and NMR databases, customized proteins can now be produced on demand (Huang et al., 2016). For example, formolase, a computationally designed protein, can carry out carboligation reactions that directly fix one-carbon units into three-carbon units, leading to the creation of a more efficient carbon fixation pathway than naturally occurring one-carbon assimilation pathways (Siegel et al., 2015). Although there have not been many reported cases of de novo protein design in the field of metabolic engineering, it has the potential to open up a new avenue for the construction of novel biosynthetic pathways (Kuhlman et al., 2019). When the bottleneck of a metabolic reaction is due to substrate toxicity or low substrate availability instead of low enzyme activity, multiple enzymes can be simultaneously engineered to promote substrate channeling, allowing for improved metabolic conversion. Substrate channeling involves physically recruiting proteins in close proximity through covalent fusion of two or more proteins using linkers or post-translational scaffold formation (Kang et al., 2019). Enhancing the proximity of multiple catalytic domains can create an artificial metabolic flux node that improves reaction efficiency by minimizing metabolite escape via diffusion and results in higher levels of target chemical production (Yang et al., 2020). This approach also rescues cells from toxic effects by rapidly converting a harmful product into a substrate for the next enzyme reaction (Huttanus & Feng, 2017). Additionally, proteins can be directed to specific membranes or organelles, particularly in eukaryotic cells, to achieve substrate channeling effects (Dusseaux et al., 2020). By localizing enzymes and substrates within compartmentalized organelles, eukaryotic strains can streamline reaction cascades and protect intermediates from competing pathways (Dusseaux et al., 2020). For instance, the squalene synthesis pathway was compartmentalized within peroxisomes of S. cerevisiae, leading to significant enhancement of squalene production (Dusseaux et al., 2020). Compartmentalization can also be used to confine toxic enzymes, such as norcoclaurine synthase, to peroxisomes, reducing toxicity and improving alkaloid production (Dusseaux et al., 2020). As our understanding of enzyme assembly mechanisms and substrate channeling deepens, this strategy will likely remain a promising approach for directing metabolic flux towards desired chemical production.

6.12. GENOME-SCALE METABOLIC MODELS (GEMS)

The design of microbial cell factories that achieve success requires optimizing the entire biological network, rather than just focusing on central metabolism and biosynthetic pathways. Genome-scale metabolic models (GEMs), which are mathematical representations of metabolic networks derived from gene-protein-reaction associations in organisms, have been employed to design metabolic engineering strategies at the systems level (Fang et al., 2020). Additionally, the development of omics integration algorithms has enabled GEMs to serve as platforms for integrating various omics layers to comprehend the metabolism of host cells, rather than being used separately (Chen & Nielson, 2021). In a recent study, proteomic data was integrated into iYO844, a GEM of *Bacillus subtilis*, to account for the proteome allocation in the metabolic network (Massaiu et al., 2019). The integrated GEM predicted and experimentally validated gene knockout targets, odhAB and sucCD, for enhanced production of poly-γ-glutamic acid, resulting in a 2.10-fold (6.5 g/L) and 2.3-fold (7.2 g/L) increase in titer, respectively.

The incorporation of omics data into GEMs can be enhanced by employing machine learning algorithms with bio-big data. For instance, machine learning algorithms were used to predict optimal combinations of target genes' promoters (such as *PCK1, TAL1, TKL1, CDC19*, and *PFK1*) in yeast, resulting in improved tryptophan production. The application of high-throughput technologies has expanded the types of data that can be used with GEMs, leading to increased use of GEMs in systems metabolic engineering. In the future, the integration of multiomics and cultivation data with machine learning approaches will enable the automated design of superior microbial cell factories (Cho et al., 2022).

6.13. RANDOM MUTAGENESIS OF CHASSIS STRAIN

Classical strain engineering has traditionally utilized random mutagenesis to identify strains that produce higher levels of the desired chemical. This method continues to be employed for identifying strains with improved production performance. Random mutagenesis presents several benefits, as the resulting strains are classified as genetically unmodified. Consequently, they are subject to less legal regulation when utilized for producing chemicals for human consumption. Moreover, products labeled as "genetically modified organism (GMO)-free" are more readily accepted by the public (Ottenheim et al., 2018).

Mutations can be induced in microorganisms using chemical mutagens, such as ethyl methanesulfonate (EMS), N-methyl-N′-nitro-N-nitrosoguanidine (NTG), and N-ethyl-N-nitrosourea (ENU), or through physical mutagens like UV light or plasma. Among the physical mutagens, atmospheric and room temperature plasma (ARTP) is a more recent technology that generates breaks in plasmid DNA and oligonucleotides with variations in plasma dosage. ARTP is a popular mutation tool for microbial breeding due to its low plasma temperature and cost-effectiveness, especially for developing mutant strains with complex genetic regulations that cannot be improved through other means (Ottenheim et al., 2018). Another technology that has received attention is genome shuffling, which combines the genomes of multiple parental strains using recursive protoplast fusion to improve the phenotypes of industrially used microbial strains.

6.14. TOLERANCE AND TRANSPORTER ENGINEERING

To achieve higher production, increasing the tolerance of host strains to toxic target chemicals is an important strategy. Several approaches have been successfully demonstrated to improve tolerance and enhance production performance. These strategies include enhancing the transport of toxic chemicals to the extracellular space, blocking biocatalytic conversions toward toxic metabolites, and membrane engineering (Besada-Lomnana et al., 2017). Additionally, process engineering approaches such as two-phase (aqueous/organic) fermentation and in situ product recovery techniques (Xue et al., 2016) can help overcome toxicity and improve overall production performance.

Adaptive Laboratory Evolution (ALE) is an efficient method that enables rapid strain evolution to improve tolerance to target chemical toxicity. Furthermore, the strains obtained from ALE can be reverse-engineered to uncover the cellular mechanism of tolerance for subsequent further engineering. For instance, an engineered *E. coli* deficient in L-serine degradation pathway was subjected to ALE with gradually increasing L-serine concentration, and the strains that were evolved to grow without inhibition at 50 g/L of L-serine were isolated. These isolated strains were analyzed by whole-genome sequencing, which led to the proposal of the mechanism of L-serine inhibition (Mundhada et al., 2017). ALE can also be combined with transcriptome analysis for a more in-depth understanding of tolerance mechanisms at the genetic level. In recent times, ALE has been coupled with strategies such as GREACE (Genome Replication Engineering Assisted Continuous Evolution), OrthoRep, eMutaT7 for increasing genotypic diversity, eVOLVER, and MMC (Microliter-Scale Microbial Microdroplet Culture System) for multiplexed automated culture and subsequent selection. By combining these tools with ALE, it can be a more powerful method to develop and characterize strains tolerant to toxic chemicals.

The process of transporter engineering can be utilized to enhance the influx of substrates and the efflux of the desired product. Exporter engineering plays a crucial role not only in overcoming the toxicity of target chemicals, as previously discussed, but also in improving the production of these chemicals by decreasing intracellular product concentration, allowing for continuous synthesis. This strategy is particularly advantageous in reducing purification costs, as the target chemicals are produced extracellularly, simplifying the purification process. For instance, the overexpression of ccmABC genes encoding the heme exporter in *E. coli* has facilitated the production of a significant portion of secreted free heme in the medium. Additionally, deleting intermediate metabolite exporters has been shown to prevent the loss of precursor metabolites to the extracellular space or organelles, thereby reinforcing metabolic flux toward target chemical production (Wang et al., 2021).

Importing engineering components is just as critical as managing exporters. Disrupting native importers that consume secreted products can increase production yield (Kim et al., 2018). On the other hand, re-importing leaked intermediate metabolites can be advantageous for increasing the output of target chemicals. The process of actively importing substrates has been extensively studied to enable the use of various carbon sources and mixtures as substrates. For instance, the introduction of different sugar importers in biofuel production allowed for the uptake of various sugars. Readers are encouraged to refer to a comprehensive review for more information on enhancing sugar uptake (Choi et al., 2019).

A lot of work has been done to find and study new transporters that can specifically carry certain chemicals. To discover potential transporters, researchers can look for sequences in literature and databases that are similar to known transporters for similar substrates. Additionally, transcriptomic analysis can be done on cells exposed to high levels of the target chemical to identify overexpressed genes with membrane protein characteristics, which can help identify potential native transporters of the host strain (Foo et al., 2014). Once candidate transporters have been shortlisted, they can be characterized using knockout assays, stress-based selection, and biosensor-guided screening (Bali et al., 2018). Furthermore, directed evolution and ALE are recently used methods to increase the efficiency of transporters. Although transporter engineering was previously underused, it is becoming an important strategy for improving the production of various chemicals.

6.15. LIPID AND MEMBRANE MORPHOLOGY ENGINEERING

The production of hydrophobic chemicals in microbial strains can sometimes be limited by a lack of accumulation space within the cell. This is because hydrophobic chemicals tend to accumulate within cell membranes due to their lipophilic nature. This can interfere with the normal functioning of membrane proteins and further deter the production of target hydrophobic compounds. To address this problem, an approach has been developed to increase the intracellular accumulation of triacylglycerols (TAGs), which can serve to solubilize hydrophobic compounds. This strategy has been successful in overproducing lycopene in *S. cerevisiae* by allowing its accumulation within lipid bodies. In addition, TAG accumulation can also be advantageous in terms of storing energy, as it can be degraded to metabolic intermediates such as acetyl-CoA, which can provide high carbon flux towards polyketides production in *Y. lipolytica* strains. Such oleaginous microorganisms are useful for accumulating hydrophobic compounds and driving precursors and energy from accumulated lipids, as demonstrated by the overproduction of lipophilic β-carotene (Ma et al., 2019). Reshaping cell morphology can be a useful approach to influence the production of target chemicals, like polyhydroxyalkanoates (PHAs), a group of biodegradable polymers (Zheng et al., 2020). By altering cell shapes, we can boost cell volume or membrane area, potentially leading to increased accumulation of target compounds. For example, a high accumulation of PHA granules in cells can trigger filamentation, slowing down cell growth. Overexpressing the *ftsZ* gene, which plays a role in cell division, can prevent filamentation and enhance overall poly-3-hydroxybutyrate (PHB) production. Similarly, boosting the expression of *mreB* and *ftsZ* can elevate PHB levels in *Halomonas campaniensis*. It's important to mention that the buildup of hydrophobic compounds in the cell membrane can hinder cell growth and product formation (Yang et al., 2021). Recently, natural rainbow colorants like astaxanthin, β-carotene, zeaxanthin, proviolacein, prodeoxyviolacein, violacein, and deoxyviolacein, corresponding to red, orange, yellow, green, blue, navy, and purple colors, have been successfully produced in *E. coli* (Yang et al., 2021). By engineering cells to generate intracellular membrane vesicles or outer membrane vesicles, we can significantly enhance the production of these hydrophobic rainbow colorants.

6.16. CONCLUSION

Microbial cell engineering is a promising approach for improving the synthetic capacity of cells. This method not only eliminates undesirable traits such as metabolic crosstalk, limitations in natural cell shape, and pathway competition, but also enhances the cellular synthetic capacity for chemical production. Metabolic engineering, which involves optimizing genetic and regulatory processes within cells to increase production of a specific substance, is a key component of microbial cell engineering. This approach can lead to higher levels of chemical production, overcome limitations such as undesirable energy use and resource shortages, and has numerous potential applications in various fields. While the aim of metabolic engineering is to regulate target pathways or strain robustness for increased product titer and productivity, microbial cell engineering focuses on enhancing the cellular synthetic capacity. Ultimately, microbial cell engineering is an important aspect of broader metabolic engineering, which can benefit from metabolic engineering to improve cellular synthetic capacity through a design-build-test-learn cycle (Chen et al., 2018).

Microbial cell engineering presents numerous benefits for improving catalytic efficiency, strain performance, and cell-cell communication. However, certain key questions must be considered, and their potential solutions can be utilized to further enhance the cellular synthetic capacity. The following are some of the questions that need to be addressed: (1) What design principles can be used to broaden the limited range of bacterial species? Multi-engineered species such as *K. vulgare* and *B. megaterium* for 2-keto-L-gulonic acid production can be leveraged to expand the range and numbers of bacterial species to construct complex communities for bioproduction (McCarty & Ledesma-Amaro, 2019). Additionally, synthetic small molecules such as nisin and lcnA or synthetic biology tools such as CRISPR/Cas tools, DNA assembly, and computer-aided design for non-modal strains can be utilized to regulate gene expression in different species to obtain defined social interactions (Langan et al., 2019). Furthermore, advanced bioinformatics and modeling such as synthetic-bioinformatic natural product for consortium interaction should be built to predict multiple cellular behaviors for engineering consortium interactions (Miano et al., 2020). Based on these solutions, synthetic consortia with desirable features can be designed to engineer multicellular behavior or learn from natural communities such as systems involving leguminous plants and *Rhizobium* (Liao et al., 2019). How can we engineer novel and artificial organelles for high-valued chemical production? One approach is to expand membraneless microcompartments and synthetic organelles for application in microbial systems such as engineering spider silk and spherical droplets for compartmentalizing pathway enzymes. Another approach is to use organelles in eukaryotic microorganisms such as yeast and fungus to build synthetic organelles in prokaryotic bacteria to perform physiological functions (Wei et al., 2020).

Furthermore, the application of bionic principles can facilitate the engineering of microcompartments in heterotrophic microorganisms for CO_2 fixation. This can be achieved by employing plant stomatal switches and artificial chloroplasts to capture CO_2 (Miller et al., 2020). Additionally, new synthetic biology tools can be utilized to enhance microbial cell engineering in microorganism systems. These tools, which have been developed in the fields of biomedicine, mammalian cells, and plants, can be employed to construct microbial systems that exhibit high temporal and spatial resolution. For instance, electric, ultrasound, pressure, and cold switching can be used to achieve spatial and temporal gene expression (Lakshmanan et al., 2020). Moreover, machine learning and protein structure theory can be employed to design and predict novel protein switches. This can be accomplished through computer simulations and analysis for protein function design (Ng et al., 2019). To construct genetic circuits that dynamically regulate gene expression in complex communities, native or heterologous signal transduction pathways should be introduced (Mushnikov et al., 2019). By addressing these issues, microbial cell engineering can be used to create increasingly sophisticated and predictable systems that enhance the cellular synthetic capacity and significantly increase chemical production by industrial microorganisms.

REFERENCES

1. Ando, H., Lemire, S., Pires, D. P., & Lu, T. K. (2015). Engineering modular viral scaffolds for targeted bacterial population editing. *Cell Systems, 1*(3), 187–196. https://doi.org/10.1016/j.cels.2015.09.001.
2. Arafat, H. H., Abu-Tahon, M. A., & Isaac, G. S. (2021). Production of antibiotic carbomycin from *Streptomyces graminofaciens* with high lipid content mutation. *Archives of Microbiology, 203*, 901–911. https://doi.org/10.1007/s00203-020-02134-y.
3. Avalos, J. L., Fink, G. R., & Stephanopoulos, G. (2013). Compartmentalization of metabolic pathways in yeast mitochondria improves the production of branched-chain alcohols. *Nature Biotechnology, 31*(4), 335–341. https://doi.org/10.1038/nbt.2509.
4. Bali, A. P., Genee, H. J., & Sommer, M. O. A. (2018). Directed evolution of membrane transport using synthetic selections. *ACS Synthetic Biology, 7*(3), 789−793. https://doi.org/10.1021/acssynbio.7b00417.
5. Becker, S., Hobenreich, H., Vogel, A., Knorr, J., Wilhelm, S., Rosenau, F., Jaeger, K. E., Reetz, M. T., & Kolmar, H. (2008). Single-cell high-throughput screening to identify enantioselective hydrolytic enzymes. *Angewandte Chemie International Edition, 47*(27), 5085–5098. https://doi.org/10.1002/anie.200800023.
6. Benedetti, I., de Lorenzo, V., & Nikel, P. I. (2016). Genetic programming of cata□ytic Pseudomonas putida biofi□ms for boosting biodegradation of haloalkanes. *Metabolic Engineering, 33*, 109–118. https://doi.org/10.1016/j.ymben.2015.10.010.
7. Benítez-Chao, D. F., Balderas-Cisneros, F. J., León-Buitimea, A., & Morones-Ramírez, J. R. (2021). Design and in silico analysis of a whole-cell biosensor able to kill methicillin-resistant *Staphylococcus aureus. Biotechnology and Applied Biochemistry. 69*(4), 1373–1382. https://doi.org/10.1002/bab.2210.
8. Bergh, Ø., BØrsheim, K. Y., Bratbak, G., & Heldal, M. (1989). High abundance of viruses found in aquatic environments. *Nature, 340*(6233), 467–468. https://doi.org/10.1038/340467a0.
9. Besada-Lombana, P. B., & Da Silva, N. A. (2019). Engineering the early secretory pathway for increased protein secretion in *Saccharomyces cerevisiae. Metabolic Engineering, 55*, 142–151. https://doi.org/10.1016/j.ymben.2019.06.005.
10. Besada-Lombana, P. B., Fernandez-Moya, R., Fenster, J., & Da Silva, N. A. (2017). Engineering *Saccharomyces cerevisiae* fatty acid composition for

increased tolerance to octanoic acid. *Biotechnology and Bioengineering, 114*(7), 1531−1538. https://doi.org/10.1002/bit.26276.

11. Bikard, D., Euler, C. W., Jiang, W., Nussenzweig, P. M., Goldberg, G. W., Duportet, X., et al. (2014). Exploiting CRISPR-Cas nucleases to produce sequence-specific antimicrobials. *Nature Biotechnology, 32*(11), 1146–1150. https://doi.org/10.1038/nbt.3043.
12. Borrero, J., Chen, Y., Dunny, G. M., & Kaznessis, Y. N. (2015). Modified lactic acid bacteria detect and inhibit multiresistant enterococci. *ACS Synthetic Biology, 4*(3), 299–306. https://doi.org/10.1021/sb500090b.
13. Briers, Y., Walmagh, M., van Puyenbroeck, V., Cornelissen, A., Cenens, W., Aertsen, A., et al. (2014). Engineered endolysin-based "Artilysins" to combat multidrug-resistant gram-negative pathogens. *mBio, 5*(4), e01379–14. https://doi.org/10.1128/mBio.01379-14.
14. Brophy, J. A. N., Triassi, A. J., Adams, B. L., Renberg, R. L., Stratis-Cullum, D. N., Grossman, A. D., & Voigt, C. A. (2018). Engineered integrative and conjugative elements for efficient and inducible DNA transfer to undomesticated bacteria. *Nature Microbiology, 3*(9), 1043−1053. https://doi.org/10.1038/s41564-018-0220-3.
15. Butler, M. J., Bruheim, P., Jovetic, S., Marinelli, F., Postma, P. W., & Bibb, M. J. (2002). Engineering of primary carbon metabolism for improved antibiotic production in *Streptomyces lividans*. *Applied and Environmental Microbiology, 68*(10), 4731–4739. https://doi.org/10.1128/AEM.68.10.4731-4739.2002.
16. Byrne, B., Carmody, M., Gibson, E., Rawlings, B., & Caffrey, P. (2003). Biosynthesis of deoxyamphotericins and deoxyamphoteronolides by engineered strains of *Streptomyces nodosus*. *Chemistry & Biology, 10*(12), 1215–1224. https://doi.org/10.1016/j.chembiol.2003.11.010.
17. Cai, D., Zhang, B., Zhu, J., Xu, H., Liu, P., Wang, Z., et al. (2020). Enhanced bacitracin production by systematically engineering S-adenosylmethionine supply modules in *Bacillus licheniformis*. *Frontiers in Bioengineering and Biotechnology, 8*, 305. https://doi.org/10.3389/fbioe.2020.00305.
18. Carbonell, P., Jervis, A. J., Robinson, C. J., Yan, C., Dunstan, M., Swainston, N., et al. (2018). An automated design-build-test-learn pipeline for enhanced microbial production of fine chemicals. *Communications Biology, 1*, 66. https://doi.org/10.1038/s42003-018-0076-9.
19. Chen, J., Chen, Z., Yuan, K., Huang, Z., & Mao, M. (2021). Recombinant bacteriophage T4 Rnl1 impacts *Streptococcus mutans* biofilm formation. *Journal of Oral Microbiology, 13*(1), 1–8. https://doi.org/10.1080/20002297.2021.1893080.

20. Chen, X. L., Gao, C., Guo, L., Hu, G. P., Luo, Q. Q., Liu, J., Nielsen, J., Chen, J., & Liu, L. M. (2018). DCEO biotechnology: Tools to design, construct, evaluate, and optimize the metabolic pathway for biosynthesis of chemicals. *Chemical Reviews, 118*(1), 4–72. https://doi.org/10.1021/acs.chemrev.7b00061.
21. Chen, Y., & Nielsen, J. (2021). Mathematical modeling of proteome constraints within metabolism. *Current Opinion in Systems Biology, 25*, 50−56. https://doi.org/10.1016/j.coisb.2021.05.005.
22. Cho, J. S., Kim, G. B., Eun, H., Moon, C. W., & Lee, S. Y. (2022). Designing microbial cell factories for the production of chemicals. *JACS Au, 2*(8), 1781–1799. https://doi.org/10.1021/jacsau.2c00182.
23. Choi, J. W., Jeon, E. J., & Jeong, K. J. (2019). Recent advances in engineering *Corynebacterium glutamicum* for utilization of hemicellulosic biomass. *Current Opinion in Biotechnology, 57*, 17−24. https://doi.org/10.1016/j.copbio.2018.11.012.
24. Cui, S., Lv, X., Wu, Y., Li, J., Du, G., Ledesma-Amaro, R., & Liu, L. (2019). Engineering a bifunctional Phr60-Rap60-Spo0A quorum-sensing mo□ecu□ar switch for dynamic fine-tuning of menaquinone-7 synthesis in *Bacillus subtilis*. *ACS Synthetic Biology, 8*(8), 1826–1837. https://doi.org/10.1021/acssynbio.9b00140.
25. Culp, E. J., Waglechner, N., Wang, W., Fiebig-Comyn, A. A., Hsu, Y. P., Koteva, K., et al. (2020). Evolution-guided discovery of antibiotics that inhibit peptidoglycan remodeling. *Nature, 578*(7796), 582–587. https://doi.org/10.1038/s41586-020-1990-9.
26. Dedrick, R. M., Guerrero-Bustamante, C. A., Garlena, R. A., Russell, D. A., Ford, K., Harris, K., et al. (2019). Engineered bacteriophages for treatment of a patient with a disseminated drug-resistant *Mycobacterium abscessus*. *Nature Medicine, 25*(5), 730–733. https://doi.org/10.1038/s41591-019-0437-z.
27. Din, M. O., Danino, T., Prindle, A., Skalak, M., Selimkhanov, J., Allen, K., Julio, E., Atolia, E., Tsimring, L. S., Bhatia, S. N., & Hasty, J. (2016). Synchronized cycles of bacterial lysis for in vivo delivery. *Nature, 536*(7614), 81–85. https://doi.org/10.1038/nature19090.
28. Ding, Q., Diao, W., Gao, C., Chen, X., & Liu, L. (2020). Microbial cell engineering to improve cellular synthetic capacity. *Biotechnology Advances, 45*, 107649. https://doi.org/10.1016/j.biotechadv.2020.107649.
29. Dusseaux, S., Wajn, W. T., Liu, Y., & Ignea, C., & Kampranis, S. C. (2020). Transforming yeast peroxisomes into microfactories for the

efficient production of high-value isoprenoids. *Proceedings of the National Academy of Sciences of the United States of America, 117*(50), 31789–31799. https://doi.org/10.1073/pnas.2008775117.

30. Edgar, R., Friedman, N., Molshanski-Mor, S., & Qimron, U. (2012). Reversing bacterial resistance to antibiotics by phage-mediated delivery of dominant sensitive genes. *Applied and Environmental Microbiology, 78*(3), 744–751. https://doi.org/10.1128/AEM.05741-11.
31. Elhadi, D., Lv, L., Jiang, X. R., Wu, H., & Chen, G. Q. (2016). CRISPRi engineering *E. coli* for morphology diversification. *Metabolic Engineering, 38*, 358–369. https://doi.org/10.1016/j.ymben.2016.09.007.
32. Eppelmann, K., Doekel, S., & Marahiel, M. A. (2001). Engineered biosynthesis of the peptide antibiotic bacitracin in the surrogate host *Bacillus subtilis*. *Journal of Biological Chemistry, 276*(37), 34824–34831. https://doi.org/10.1074/jbc.M105249200.
33. Eustáquio, A. S., Gust, B., Li, S. M., Pelzer, S., Wohlleben, W., Chater, K. F., et al. (2004). Production of 8′-halogenated and 8′-unsubstituted novobiocin derivatives in genetically engineered *Streptomyces coelicolor* strains. *Chemistry & Biology, 11*(11), 1561–1572. https://doi.org/10.1016/j.chembiol.2004.08.023.
34. Fan, X., Gong, T., Wu, Y., Zhao, F., & Yang, C. (2019). Enhanced synthesis of alginate oligosaccharides in *Pseudomonas mendocina* NK-01 by overexpressing MreB. *Microbial Cell Factories, 18*, 115. https://doi.org/10.1186/s12934-019-1160-4.
35. Fang, Y., Wang, J., Ma, W., Yang, J., Zhang, H., Zhao, L., Chen, S., Zhang, S., Hu, X., Li, Y., & Wang, X. (2020). Rebalancing microbial carbon distribution for L-threonine maximization using a thermal switch system. *Metabolic Engineering, 61*, 1–13. https://doi.org/10.1016/j.ymben.2020.04.002.
36. Faust, K. (2019). Microbial consortium design benefits from metabolic modeling. *Trends in Biotechnology, 37*(2), 123–125. https://doi.org/10.1016/j.tibtech.2018.09.009.
37. Foo, J. L., Jensen, H. M., Dahl, R. H., George, K., Keasling, J. D., Lee, T. S., Leong, S., Mukhopadhyay, A., Papoutsakis, E. T., & Newman, D. K. (2014). Improving microbial biogasoline production in *Escherichia coli* using tolerance engineering. *mBio, 5*(6), e01932. https://doi.org/10.1128/mBio.01932-14.
38. Fu, W., Forster, T., Mayer, O., Curtin, J. J., Lehman, S. M., Donlan, R. M., et al. (2010). Bacteriophage cocktail for the prevention of biofilm formation by *Pseudomonas aeruginosa* on catheters in an in vitro model

system. *Antimicrobial Agents and Chemotherapy, 54*(1), 397–404. https://doi.org/10.1128/AAC.00669-09.

39. Gao, Y., Feng, X., Xian, M., & Wang, Q., Zhao, G. (2013). Inducible cell lysis systems in microbial production of bio-based chemicals. *Applied Microbiology and Biotechnology, 97*(16), 7121–7129. https://doi.org/10.1007/s00253-013-5034-8.
40. Garima, G. Y., Tsai, S. L., Bhawna, M., Nanc, A. D., & Chen, W. (2011). Simultaneous cell growth and ethanol production from cellulose by an engineered yeast consortium displaying a functional mini-cellulosome. *Microbial Cell Factories, 10*(1), 89. https://doi.org/10.1186/1475-2859-10-89.
41. Glass, D. S., & Riedel-Kruse, I. H. (2018). A synthetic bacterial cell-cell adhesion toolbox for programming multicellular morphologies and patterns. *Cell, 174*(3), 649–658.e16. https://doi.org/10.1016/j.cell.2018.06.041.
42. Gordon, G. C., Korosh, T. C., Cameron, J. C., Markley, A. L., Begemann, M. B., & Pfleger, B. F. (2016). CRISPR interference as a titratable, trans-acting regulatory tool for metabolic engineering in the cyanobacterium *Synechococcus* sp. strain PCC 7002. *Metabolic Engineering, 38*, 170–179. https://doi.org/10.1016/j.ymben.2016.07.003.
43. Grisewood, M. J., Hernandez-Lozada, N. J., Thoden, J. B., Gifford, N. P., Mendez-Perez, D., Schoenberger, H. A., Allan, M. F., Floy, M. E., Lai, R.-Y., Holden, H. M., Pfleger, B. F., & Maranas, C. D. (2017). Computational redesign of acyl-ACP thioesterase with improved selectivity toward medium-chain-length fatty acids. *ACS Catalysis, 7*(6), 3837–3849. https://doi.org/10.1021/acscatal.7b00855.
44. Guo, L., Diao, W., Gao, C., Hu, G., Ding, Q., Ye, C., Chen, X., Liu, J., & Liu, L. (2020). Engineering *Escherichia coli* lifespan for enhancing chemical production. *Nature Catalysis, 3*(3), 307–318. https://doi.org/10.1038/s41929-020-0434-8.
45. Guo, L., Zhang, F., Zhang, C., Hu, G., Gao, C., Chen, X., & Liu, L. (2018). Enhancement of malate production through engineering of the periplasmic rTCA pathway in *Escherichia coli. Biotechnology and Bioengineering, 115*(6), 1571–1580. https://doi.org/10.1002/bit.26684.
46. Gupta, A., Reizman, I. M., Reisch, C. R., & Prather, K. L. (2017). Dynamic regulation of metabolic flux in engineered bacteria using a pathway-independent quorum-sensing circuit. *Nature Biotechnology, 35*(3), 273–279. https://doi.org/10.1038/nbt.3809.
47. Gupta, S., Bram, E. E., & Weiss, R. (2013). Genetically programmable

pathogen sense and destroy. *ACS Synthetic Biology, 2*(12), 715–723. https://doi.org/10.1021/sb4000516.

48. Henke, N. A., Krahn, I., & Wendisch, V. F. (2021). Improved plasmid-based inducible and constitutive gene expression in *Corynebacterium glutamicum*. *Microorganisms, 9*(1), 204. https://doi.org/10.3390/microorganisms9010204.

49. Hoffart, E., Grenz, S., Lange, J., Nitschel, R., Muller, F., Schwentner, A., Feith, A., Lenfers-Lucker, M., Takors, R., & Blombach, B. (2017). High substrate uptake rates empower Vibrio natriegens as production host for industrial biotechnology. *Applied and Environmental Microbiology, 83*(22), e01614–16. https://doi.org/10.1128/AEM.01614-17.

50. Holowko, M. B., Wang, H., Jayaraman, P., & Poh, C. L. (2016). Biosensing *Vibrio cholerae* with genetically engineered *Escherichia coli*. *ACS Synthetic Biology, 5*(12), 1275–1283. https://doi.org/10.1021/acssynbio.6b00079.

51. Honjo, H., Iwasaki, K., Soma, Y., Tsuruno, K., Hamada, H., & Hanai, T. (2019). Synthetic microbial consortium with specific roles designated by genetic circuits for cooperative chemical production. *Metabolic Engineering, 55*, 268–275. https://doi.org/10.1016/j.ymben.2019.07.007.

52. Hou, J., Tyo, K., Liu, Z., Petranovic, D., & Nielsen, J. (2012). Engineering of vesicle trafficking improves heterologous protein secretion in *Saccharomyces cerevisiae*. *Metabolic Engineering, 14*(2), 120–127. https://doi.org/10.1016/j.ymben.2012.01.007.

53. Huang, J., Liu, S., Zhang, C., Wang, X., Pu, J., Ba, F., Xue, S., Ye, H., Zhao, T., Li, K., Wang, Y., Zhang, J., Wang, L., Fan, C., Lu, T. K., & Zhong, C. (2019). Programmable and printable *Bacillus subtilis* biofilms as engineered living materials. *Nature Chemical Biology, 15*(1), 34–41. https://doi.org/10.1038/s41589-018-0169-2.

54. Huang, M., Wang, G., Qin, J., Petranovic, D., & Nielsen, J. (2018). Engineering the protein secretory pathway of *Saccharomyces cerevisiae* enables improved protein production. *Proceedings of the National Academy of Sciences of the United States of America, 115*(47), E11025–E11032. https://doi.org/10.1073/pnas.1809921115.

55. Huang, P.-S., Boyken, S. E., & Baker, D. (2016). The coming of age of de novo protein design. *Nature, 537*(7620), 320–327. https://doi.org/10.1038/nature19946.

56. Huttanus, H. M., & Feng, X. (2017). Compartmentalized metabolic engineering for biochemical and biofuel production. *Biotechnology Journal, 12*(6), 1700052. https://doi.org/10.1002/biot.201700052.

57. Hwang, S., Lee, N., Cho, S., Palsson, B., & Cho, B. K. (2020). Repurposing modular polyketide synthases and non-ribosomal peptide synthetases for novel chemical biosynthesis. *Frontiers in Molecular Biosciences, 7*, 87. https://doi.org/10.3389/fmolb.2020.00087.

58. Inokuma, K., Kurono, H., Haan, R. D., Heber van Zyl, W., Hasunuma, T., & Kondo, A. (2020). Novel strategy for anchorage position control of GPI-attached proteins in the yeast cell wall using different GPI-anchoring domains. *Metabolic Engineering, 57*, 110–117. https://doi.org/10.1016/j.ymben.2019.12.001.

59. Ji, C.-H., Kim, H., Je, H.-W., Kwon, H., Lee, D., & Kang, H.-S. (2022). Top-down synthetic biology approach for titer improvement of clinically important antibiotic daptomycin in *Streptomyces roseosporus*. *Metabolic Engineering, 69*, 40–49. https://doi.org/10.1016/j.ymben.2021.12.001.

60. Jiang, L., Song, X., Li, Y., Xu, Q., Pu, J., Huang, H., & Zhong, C. (2018). Programming integrative extracellular and intracellular biocatalysis for rapid, robust, and recyclable synthesis of trehalose. *ACS Catalysis, 8*(3), 1837–1842. https://doi.org/10.1021/acscatal.7b03234.

61. Jiang, X. R., & Chen, G. Q. (2016). Morphology engineering of bacteria for bio-production. *Biotechnology Advances, 34*(4), 435–440. https://doi.org/10.1016/j.biotechadv.2016.02.007.

62. Jiang, Y. J., Dong, W. L., Xin, F. X., & Jiang, M. (2020). Designing synthetic microbial consortia for biofuel production. *Trends in Biotechnology, 38*(8), 828–831. https://doi.org/10.1016/j.tibtech.2020.02.002.

63. Kang, W., Ma, T., Liu, M., Qu, J., Liu, Z., Zhang, H., Shi, B., Fu, S., Ma, J., Lai, L. T. F., He, S., Qu, J., Wing-Ngor Au, S., Ho Kang, B., Yu Lau, W. C., Deng, Z., Xia, J., & Liu, T. (2019). Modular enzyme assembly for enhanced cascade biocatalysis and metabolic flux. *Nature Communications, 10*(1), 4248. https://doi.org/10.1038/s41467-019-12211-2.

64. Ki, M.-R., & Pack, S. P. (2020). Fusion tags to enhance heterologous protein expression. *Applied Microbiology and Biotechnology, 104*(6), 2411–2425. https://doi.org/10.1007/s00253-020-10395-5.

65. Kim, B., Binkley, R., Kim, H. U., & Lee, S. Y. (2018). Metabolic engineering of *Escherichia coli* for the enhanced production of L-tyrosine. *Biotechnology and Bioengineering, 115*(10), 2554–2564. https://doi.org/10.1002/bit.26791.

66. Kim, J.-E., Jang, I.-S., Son, S.-H., Ko, Y.-J., Cho, B.-K., Chang Kim, S., & Lee, J. Y. (2019). Tailoring the *Saccharomyces cerevisiae* endoplasmic reticulum for functional assembly of terpene synthesis pathway. *Metabolic Engineering, 56*, 50–59. https://doi.org/10.1016/j.ymben.2019.08.009.

67. Kim, S. G., Noh, M. H., Lim, H. G., Jang, S., Jang, S., Koffas, M. A. G., & Jung, G. Y. (2018). Molecular parts and genetic circuits for metabolic engineering of microorganisms. *FEMS Microbiology Letters, 365*(17), 15–27. https://doi.org/10.1093/femsle/fny175.

68. Klompe, S. E., Vo, P. L. H., Halpin-Healy, T. S., & Sternberg, S. H. (2019). Transposon-encoded CRISPR-Cas systems direct RNA-guided DNA integration. *Nature, 571*(7764), 219–225. https://doi.org/10.1038/s41586-019-1369-8.

69. Ko, Y. S., Kim, J. W., Lee, J. A., Han, T., Kim, G. B., Park, J. E., & Lee, S. Y. (2020). Tools and strategies of systems metabolic engineering for the development of microbial cell factories for chemical production. *Chemical Society Reviews, 49*(14), 4615–4636. https://doi.org/10.1039/d0cs00087b.

70. Kuhlman, B., & Bradley, P. (2019). Advances in protein structure prediction and design. *Nature Reviews Molecular Cell Biology, 20*(11), 681–697. https://doi.org/10.1038/s41580-019-0163-x.

71. Lakshmanan, A., Jin, Z., Nety, S. P., Sawyer, D. P., & Shapiro, M. G. (2020). Acoustic biosensors for ultrasound imaging of enzyme activity. *Nature Chemical Biology, 16*, 988–996. https://doi.org/10.1038/s41589-020-0570-1.

72. Langan, R. A., Boyken, S. E., Ng, A. H., Samson, J. A., Dods, G., Westbrook, A. M., Nguyen, T. H., Lajoie, M. J., Chen, Z., Berger, S., Mulligan, V. K., Dueber, J. E., Novak, W. R. P., El-Samad, H., & Baker, D. (2019). De novo design of bioactive protein switches. *Nature, 572*(7768), 205–210. https://doi.org/10.1038/s41586-019-1432-8.

73. Lee, M. J., Mantell, J., Hodgson, L., Alibhai, D., & Warren, M. J. (2018). Engineered synthetic scaffolds for organizing proteins within the bacterial cytoplasm. *Nature Chemical Biology, 14*(2), 142–147. https://doi.org/10.1038/nchembio.2523.

74. Lee, S. Y., & Kim, H. U. (2015). Systems strategies for developing industrial microbial strains. *Nature Biotechnology, 33*(10), 1061–1072. https://doi.org/10.1038/nbt.3365.

75. León-Buitimea, A., Balderas-Cisneros, F. D., Garza-Cárdenas, C. R., Garza-Cervantes, J. A., & Morones-Ramírez, J. R. (2022). Synthetic biology tools for engineering microbial cells to fight superbugs. *Frontiers in Bioengineering and Biotechnology, 10*, 869206. https://doi.org/10.3389/fbioe.2022.869206.

76. Li, C., Zhang, R., Wang, J., Wilson, L. M., & Yan, Y. (2020). Protein engineering for improving and diversifying natural product biosynthesis.

Trends in Biotechnology, 38(7), 729–744. https://doi.org/10.1016/j.tibtech.2020.01.002.

77. Li, Y.-P., Bu, Q.-T., Li, J.-F., Xie, H., Su, Y.-T., Du, Y.-L., et al. (2021). Genome-based rational engineering of *Actinoplanes deccanensis* for improving fidaxomicin production and genetic stability. *Bioresource Technology, 330*, 124982. https://doi.org/10.1016/j.biortech.2021.124982.
78. Liao, L., Cheng, Y., Liu, S., Zhou, J., An, S., & Lv, M. (2014). Production of novel antibiotics zeamines through optimizing *Dickeya zeae* fermentation conditions. *PLoS One, 9*, e116047. https://doi.org/10.1371/journal.pone.0116047.
79. Lin, T.-Y. Y., Lo, Y.-H. H., Tseng, P.-W. W., Chang, S.-F. F., Lin, Y.-T. T., & Chen, T.-S. S. (2012). A T3 and T7 recombinant phage acquires efficient adsorption and a broader host range. *PLoS ONE, 7*, 1–10. https://doi.org/10.1371/journal.pone.0030905.
80. Liu, G. S., Li, T., Zhou, W., Jiang, M., Tao, X. Y., Liu, M., Zhao, M., Ren, Y. H., Gao, B., Wang, F. Q., Wei, D. Z. (2020). The yeast peroxisome: A dynamic storage depot and subcellular factory for squalene overproduction. *Metabolic Engineering, 57*, 151–161. https://doi.org/10.1016/j.ymben.2019.12.007.
81. Lu, Z. M., Liu, N., Wang, L. J., Wu, L. H., Gong, J. S., Yu, Y. J., Li, G. Q., Shi, J. S., & Xu, Z. H. (2016). Elucidating and regulating the acetoin production role of microbial functional groups in multispecies acetic acid fermentation. *Applied and Environmental Microbiology, 82*(19), 01331–01316. https://doi.org/10.1128/AEM.01331-16.
82. Lubkowicz, D., Ho, C. L., Hwang, I. Y., Yew, W. S., Lee, Y. S., & Chang, M. W. (2018). Reprogramming probiotic *Lactobacillus reuteri* as a biosensor for *Staphylococcus aureus* derived AIP-I detection. *ACS Synthetic Biology, 7*, 1229–1237. https://doi.org/10.1021/acssynbio.7b00415.
83. Lv, X. Q., Wu, Y. K., Tian, R., Gu, Y., Liu, Y., Li, J., Du, G. C., Ledesma-Amaro, R., & Liu, L. (2020). Synthetic metabolic channel by functional membrane microdomains for compartmentalized flux control. *Metabolic Engineering, 59*, 106–118. https://doi.org/10.1016/j.ymben.2019.12.012.
84. Lv, Y. K., Qian, S., Du, G. C., Chen, J., Zhou, J. W., & Xu, P. (2019). Coupling feedback genetic circuits with growth phenotype for dynamic population control and intelligent bioproduction. *Metabolic Engineering, 54*, 109–116. https://doi.org/10.1016/j.ymben.2019.03.004.
85. Ma, T., Shi, B., Ye, Z., Li, X., Liu, M., Chen, Y., Xia, J., Nielsen, J., Deng, Z., & Liu, T. (2019). Lipid engineering combined with systematic metabolic engineering of *Saccharomyces cerevisiae* for high-yield

production of lycopene. *Metabolic Engineering, 52*, 134–142. https://doi.org/10.1016/j.ymben.2019.01.006.

86. Ma, Y.-R., Wang, K.-F., Wang, W.-J., Ding, Y., Shi, T.-Q., Huang, H., & Ji, X.-J. (2019). Advances in the metabolic engineering of *Yarrowia lipolytica* for the production of terpenoids. *Bioresource Technology, 281*, 449–456. https://doi.org/10.1016/j.biortech.2019.02.104.

87. Makitrynskyy, R., Tsypik, O., & Bechthold, A. (2021). Genetic engineering of *Streptomyces ghanaensis* ATCC14672 for improved production of moenomycins. *Microorganisms, 10*, 30. https://doi.org/10.3390/microorganisms10010030.

88. Massaiu, I., Pasotti, L., Sonnenschein, N., Rama, E., Cavaletti, M., Magni, P., Calvio, C., & Herrgard, M. J. (2019). Integration of enzymatic data in *Bacillus subtilis* genome-scale metabolic model improves phenotype predictions and enables in silico design of poly-gamma-glutamic acid production strains. *Microbial Cell Factories, 18*(1), 3. https://doi.org/10.1186/s12934-018-1064-2.

89. Masschelein, J., Clauwers, C., Awodi, U. R., Stalmans, K., Vermaelen, W., Lescrinier, E., et al. (2015). A combination of polyunsaturated fatty acid, nonribosomal peptide and polyketide biosynthetic machinery is used to assemble the zeamine antibiotics. *Chemical Science, 6*, 923–929. https://doi.org/10.1039/C4SC03202A.

90. McCarty, N. S., & Ledesma-Amaro, R. (2019). Synthetic biology tools to engineer microbial communities for biotechnology. *Trends in Biotechnology, 37*(2), 181–197. https://doi.org/10.1016/j.tibtech.2018.11.002.

91. Miano, A., Liao, M. J., & Hasty, J. (2020). Inducible cell-to-cell signaling for tunable dynamics in microbial communities. *Nature Communications, 11*(1), 1193. https://doi.org/10.1038/s41467-020-14988-1.

92. Miller, T. E., Beneyton, T., Schwander, T., Diehl, C., Girault, M., Mclean, R., Chotel, T., Claus, P., Cortina, N. S., & Baret, J. C. (2020). Light-powered CO_2 fixation in a ch□orop□ast mimic with natura□and synthetic parts. *Science*, 370(6522), 474–478. https://doi.org/10.1126/science.aaz6802.

93. Moosmann, D., Mokeev, V., Kulik, A., Osipenkov, N., Kocadinc, S., Ort-Winklbauer, R., et al. (2020). Genetic engineering approaches for the fermentative production of phenylglycines. *Applied Microbiology and Biotechnology, 104*(8), 3433–3444. https://doi.org/10.1007/s00253-020-10435-7.

94. Morris, M. K., Saez-Rodriguez, J., Sorger, P. K., & Lauffenburger, D. A. (2010). Logic-based models for the analysis of cell signaling networks.

Biochemistry, 49(15), 3216–3224. https://doi.org/10.1021/bi902202q.

95. Mundhada, H., Seoane, J. M., Schneider, K., Koza, A., Christensen, H. B., Klein, T., Phaneuf, P. V., Herrgard, M., Feist, A. M., & Nielsen, A. T. (2017). Increased production of L-serine in *Escherichia coli* through adaptive laboratory evolution. *Metabolic Engineering, 39*, 141–150. https://doi.org/10.1016/j.ymben.2016.11.011.

96. Mushnikov, N. V., Fomicheva, A., Gomelsky, M., & Bowman, G. R. (2019). Inducible asymmetric cell division and cell differentiation in a bacterium. *Nature Chemical Biology, 15*(9), 1–7. https://doi.org/10.1038/s41589-019-0320-3.

97. Myhrvold, C., Polka, J. K., & Silver, P. A. (2016). Synthetic lipid-containing scaffolds enhance production by colocalizing enzymes. *ACS Synthetic Biology, 5*(12), 1396–1403. https://doi.org/10.1021/acssynbio.6b00134.

98. Nattermann, M., Burgener, S., Pfister, P., Chou, A., Schulz, L., Lee, S. H., Paczia, N., Zarzycki, J., Gonzalez, R., & Erb, T. J. (2021). Engineering a highly efficient carboligase for synthetic one-carbon metabolism. *ACS Catalysis, 11*(9), 5396–5404. https://doi.org/10.1021/acscatal.0c05530.

99. Ng, A. H., Nguyen, T. H., Gomez-Schiavon, M., Dods, G., Langan, R. A., Boyken, S. E., Samson, J. A., Waldburger, L. M., Dueber, J. E., Baker, D., & El-Samad, H. (2019). Modular and tunable biological feedback control using a de novo protein switch. *Nature, 572*(7768), 265–269. https://doi.org/10.1038/s41586-019-1446-0.

100. Nguyen, P. Q., Botyanszki, Z., Tay, P. K., & Joshi, N. S. (2014). Programmable biofilm-based materials from engineered curli nanofibres. *Nature Communications, 5*(2), 486–495. https://doi.org/10.1038/ncomms4945.

101. Niels, K., & Daniel, S. (2010). Environments that induce synthetic microbial ecosystems. *PLoS Computational Biology, 5*, 15–24. https://doi.org/10.1371/journal.pcbi.1000554.

102. Opgenorth, P., Costello, Z., Okada, T., Goyal, G., Chen, Y., Gin, J., et al. (2019). Lessons from two design-build-test-learn cycles of dodecanol production in *Escherichia coli* aided by machine learning. *ACS Synthetic Biology, 8*(6), 1337–1351. https://doi.org/10.1021/acssynbio.9b00020.

103. Ottenheim, C., Nawrath, M., & Wu, J. C. (2018). Microbial mutagenesis by atmospheric and room-temperature plasma (ARTP): The latest development. *Bioresource and Bioprocessing, 5*(1), 12. https://doi.org/10.1186/s40643-018-0204-3.

104. Palazzotto, E., Tong, Y., Lee, S. Y., & Weber, T. (2019). Synthetic biology and metabolic engineering of actinomycetes for natural product

discovery. *Biotechnology Advances, 37*, 107366. https://doi.org/10.1016/j.biotechadv.2019.02.001.

105. Park, S. Y., Yang, D., Ha, S. H., & Lee, S. Y. (2020). Biosynthesis of dihydroquercetin in *Escherichia coli* from glycerol. *bioRxiv*. https://doi.org/10.1101/2020.11.27.401000.
106. Pei, R., & Lamas-Samanamud, G. R. (2014). Inhibition of biofilm formation by T7 bacteriophages producing quorum-quenching enzymes. *Applied and Environmental Microbiology, 80*(17), 5340–5348. https://doi.org/10.1128/AEM.01465-14.
107. Perrino, G., Hadjimitsis, A., Ledesma-Amaro, R., & Stan, G.-B. (2021). Control engineering and synthetic biology: Working in synergy for the analysis and control of microbial systems. *Current Opinion in Microbiology, 62*, 68–75. https://doi.org/10.1016/j.mib.2021.04.003.
108. Peters, J. M., Koo, B.-M., Patino, R., Heussler, G. E., Hearne, C. C., Qu, J., Inclan, Y. F., Hawkins, J. S., Lu, C. H. S., Silvis, M. R., Harden, M. M., Osadnik, H., Peters, J. E., Engel, J. N., Dutton, R. J., Grossman, A. D., Gross, C. A., & Rosenberg, O. S. (2019). Enabling genetic analysis of diverse bacteria with Mobile-CRISPRi. *Nature Microbiology, 4*(2), 244–250. https://doi.org/10.1038/s41564-018-0326-2.
109. Ramos, H. D., et al. (2017). Tat proteins as novel thylakoid membrane anchors organize a biosynthetic pathway in chloroplasts and increase product yield 4-fold. *Metabolic Engineering, 44*, 108–116. https://doi.org/10.1016/j.ymben.2017.10.005.
110. Reeve, B., Hargest, T., Gilbert, C., & Ellis, T. (2014). Predicting translation initiation rates for designing synthetic biology. *Frontiers in Bioengineering and Biotechnology, 2*, 1. https://doi.org/10.3389/fbioe.2014.00001.
111. Sakai, A., Mitsumori, A., Furukawa, M., Kinoshita, K., Anzai, Y., & Kato, F. (2012). Production of a hybrid 16-membered macrolide antibiotic by genetic engineering of *Micromonospora* sp. TPMA0041. *Journal of Industrial Microbiology and Biotechnology, 39*(11), 1693–1701. https://doi.org/10.1007/s10295-012-1177-6.
112. Sandberg, T. E., Salazar, M. J., Weng, L. L., Palsson, B. O., & Feist, A. M. (2019). The emergence of adaptive laboratory evolution as an efficient tool for biological discovery and industrial biotechnology. *Metabolic Engineering, 56*, 1–16. https://doi.org/10.1016/j.ymben.2019.09.003.
113. Schooley, R. T., Biswas, B., Gill, J. J., Hernandez-Morales, A., Lancaster, J., Lessor, L., et al. (2017). Development and use of personalized bacteriophage-based therapeutic cocktails to treat a patient with a disseminated resistant *Acinetobacter baumannii* infection. *Antimicrobial*

Agents and Chemotherapy, 61(10), e00954–17. https://doi.org/10.1128/AAC.00954-17.

114. Sharma, S., Chatterjee, S., Datta, S., Prasad, R., Dubey, D., Prasad, R. K., et al. (2017). Bacteriophages and its applications: An overview. *Folia Microbiologica, 62*(1), 17–55. https://doi.org/10.1007/s12223-017-0501-7.
115. Shomar, H., Gontier, S., Van Den Broek, N. J. F., Tejeda Mora, H., Noga, M. J., Hagedoorn, P.-L., et al. (2018). Metabolic engineering of a carbapenem antibiotic synthesis pathway in *Escherichia coli. Nature Chemical Biology, 14*(8), 794–800. https://doi.org/10.1038/s41589-018-0106-6.
116. Shong, J., Jimenez Diaz, M. R., & Collins, C. H. (2012). Towards synthetic microbial consortia for bioprocessing. *Current Opinion in Biotechnology, 23*(5), 798–802. https://doi.org/10.1016/j.copbio.2012.02.001.
117. Short, F. L., Blower, T. R., & Salmond, G. P. (2012). A promiscuous antitoxin of bacteriophage T4 ensures successful viral replication. *Molecular Microbiology, 83*(4), 665–668. https://doi.org/10.1111/j.1365-2958.2011.07962.x.
118. Siegel, J. B., Smith, A. L., Poust, S., Wargacki, A. J., Bar-Even, A., Louw, C., Shen, B. W., Eiben, C. B., Tran, H. M., Noor, E., Gallaher, J. L., Bale, J., Yoshikuni, Y., Gelb, M. H., Keasling, J. D., Stoddard, B. L., Lidstrom, M. E., & Baker, D. (2015). Computational protein design enables a novel one-carbon assimilation pathway. *Proceedings of the National Academy of Sciences of the United States of America, 112*(12), 3704–3709. https://doi.org/10.1073/pnas.1500545112.
119. Strecker, J., Ladha, A., Gardner, Z., Schmid-Burgk, J. L., Makarova, K. S., Koonin, E. V., & Zhang, F. (2019). RNA-guided DNA insertion with CRISPR-associated transposases. *Science, 365*(6448), 48−53. https://doi.org/10.1126/science.aax9181.
120. Taffs, R., Aston, J. E., Brileya, K., Jay, Z., Klatt, C. G., McGlynn, S., Mallette, N., Montross, S., Gerlach, R., Inskeep, W. P., Ward, D. M., & Carlson, R. P. (2009). In silico approaches to study mass and energy flows in microbial consortia: A syntrophic case study. *BMC Systems Biology, 3*, 114–123. https://doi.org/10.1186/1752-0509-3-114.
121. Tan, D., Wu, Q., Chen, J. C., & Chen, G. Q. (2014). Engineering Halomonas TD01 for the low-cost production of polyhydroxyalkanoates. *Metabolic Engineering, 26*, 34–47. https://doi.org/10.1016/j.ymben.2014.09.002.
122. Tsai, S. L., Goyal, G., & Chen, W. (2010). Surface display of a functional minicellulosome by intracellular complementation using a synthetic

yeast consortium and its application to cellulose hydrolysis and ethanol production. *Applied and Environmental Microbiology, 76*(22), 7514–7520. https://doi.org/10.1128/AEM.01105-10.

123. Ventola, C. L. (2015). The antibiotic resistance crisis: Part 1: Causes and threats. *Pharmacy and Therapeutics, 40*(4), 277.

124. Vo, P. L. H., Ronda, C., Klompe, S. E., Chen, E. E., Acree, C., Wang, H. H., & Sternberg, S. H. (2021). CRISPR RNA-guided integrases for high-efficiency, multiplexed bacterial genome engineering. *Nature Biotechnology, 39*(4), 480–489. https://doi.org/10.1038/s41587-020-00793-5.

125. Volke, D. C., Nikel, P. I. (2018). Getting bacteria in shape: Synthetic morphology approaches for the design of efficient microbial cell factories. *Advanced Biosystems, 2*(11), 1800111. https://doi.org/10.1002/adbi.201800111.

126. Volke, D. C., Turlin, J., Mol, V., Nikel, P. I. (2020). Physical decoupling of XylS/Pm regulatory elements and conditional proteolysis enable precise control of gene expression in *Pseudomonas putida. Microbial Biotechnology, 13*(1), 222–232. https://doi.org/10.1111/1751-7915.13470.

127. Wagner, J. M., Liu, L., Yuan, S. F., Venkataraman, M. V., Abate, A. R., & Alper, H. S. (2018). A comparative analysis of single cell and droplet-based FACS for improving production phenotypes: Riboflavin overproduction in *Yarrowia lipolytica. Metabolic Engineering, 47*, 346–356. https://doi.org/10.1016/j.ymben.2018.03.013.

128. Wang, B., Kitney, R. I., Joly, N., & Buck, M. (2011). Engineering modular and orthogonal genetic logic gates for robust digital-like synthetic biology. *Nature Communications, 2*, 508. https://doi.org/10.1038/ncomms1516.

129. Wang, G., Moller-Hansen, I., Babaei, M., D'Ambrosio, V., Christensen, H. B., Darbani, B., Jensen, M. K., & Borodina, I. (2021). Transportome-wide engineering of *Saccharomyces cerevisiae. Metabolic Engineering, 64*, 52–63. https://doi.org/10.1016/j.ymben.2021.08.005.

130. Wang, G., Zhao, Z., Ke, J., Engel, Y., Shi, Y. M., Robinson, D., Bingol, K., Zhang, Z., Bowen, B., Louie, K., Wang, B., Evans, R., Miyamoto, Y., Cheng, K., Kosina, S., De Raad, M., Silva, L., Luhrs, A., Lubbe, A., Hoyt, D. W., Francavilla, C., Otani, H., Deutsch, S., Washton, N. M., Rubin, E. M., Mouncey, N. J., Visel, A., Northen, T., Cheng, J. F., Bode, H. B., & Yoshikuni, Y. (2019). CRAGE enables rapid activation of biosynthetic gene clusters in undomesticated bacteria. *Nature Microbiology, 4*(12), 2498–2510. https://doi.org/10.1038/s41564-019-0568-7.

131. Wang, K., & Nicholaou, M. (2017). Suppression of antimicrobial resistance in MRSA using CRISPR-dCas9. *American Society for Clinical Laboratory Science, 30*, 207–213.

132. Wang, W.-F., Xiao, H., & Zhong, J.-J. (2021). Biosynthesis of a novel ganoderic acid by expressing CYP genes from *Ganoderma lucidum* in *Saccharomyces cerevisiae*. *Applied Microbiology and Biotechnology, 106*, 523–534. https://doi.org/10.1007/s00253-021-11699-w.

133. Wang, Y., Boghigian, B. A., & Pfeifer, B. A. (2007). Improving heterologous polyketide production in *Escherichia coli* by overexpression of an S-adenosylmethionine synthetase gene. *Applied Microbiology and Biotechnology, 77*(2), 367–373. https://doi.org/10.1007/s00253-007-1165-4.

134. Wang, Y., Ling, C., Chen, Y., Jiang, X., & Chen, G. Q. (2019). Microbial engineering for easy downstream processing. *Biotechnology Advances, 37*(6), 115–129. https://doi.org/10.1016/j.biotechadv.2019.02.007.

135. Wannier, T. M., Ciaccia, P. N., Ellington, A. D., Filsinger, G. T., Isaacs, F. J., Javanmardi, K., Jones, M. A., Kunjapur, A. M., Nyerges, A., Pal, C., Schubert, M. G., & Church, G. M. (2021). Recombineering and MAGE. *Nature Reviews Methods Primers, 1*(1), 7. https://doi.org/10.1038/s43586-021-00001-6.

136. Wannier, T. M., Kunjapur, A. M., Rice, D. P., Mcdonald, M. J., & Church, G. M. (2018). Adaptive evolution of genomically recoded *Escherichia coli*. *Proceedings of the National Academy of Sciences of the United States of America, 115*(12), 3090–3095. https://doi.org/10.1073/pnas.1719317115.

137. Wei, S. P., Qian, Z. G., Hu, C. F., Pan, F., & Xia, X. X. (2020). Formation and functionalization of membraneless compartments in *Escherichia coli*. *Nature Chemical Biology, 16, 1143–1148*. https://doi.org/10.1038/s41589-020-0579-9.

138. Wei, W., Sun, P., Li, Z., Song, K., Su, W., Wang, B., Liu, Y., & Zhao, J. (2018). A surface-display biohybrid approach to light-driven hydrogen production in air. *Science Advances, 4*(2), 263−272. https://doi.org/10.1126/sciadv.aao7263.

139. Wen, Z. Q., Wu, M. B., Lin, Y. J., Yang, L. R., Lin, J. P., & Cen, P. (2014). A novel strategy for sequential co-culture of *Clostridium thermocellum* and *Clostridium beijerinckii* to produce solvents from alkali extracted corn cobs. *Process Biochemistry, 49*(11), 1941−1949. https://doi.org/10.1016/j.procbio.2014.08.012.

140. Wheeldon, I., Minteer, S. D., Banta, S., Barton, S. C., Atanassov, P., & Sigman, M. (2016). Substrate channeling as an approach to cascade

reactions. *Nature Chemistry, 8*(4), 299–309. https://doi.org/10.1038/nchem.2447.

141. World Health Organization [WHO]. (2017). Who publishes list of bacteria for which new antibiotics are urgently needed. *Saudi Medical Journal.* Retrieved from: https://www.who.int/news-room/detail/27-02-2017-who-publishes-list-of-bacteriafor-which-new-antibiotics-are-urgently-needed.
142. Wu, J., Zhang, H.-B., Xu, J.-L., Cox, R. J., Simpson, T. J., & Zhang, L.-H. (2010). 13C labeling reveals multiple amination reactions in the biosynthesis of a novel polyketide polyamine antibiotic zeamine from *Dickeya zeae. Chemical Communications, 46*, 333–335. https://doi.org/10.1039/B918293D.
143. Wu, M. L., Tsai, C. Y., & Chen, T. H. (2006). Cell surface display of Chi92 on *Escherichia coli* using ice nucleation protein for improved catalytic and antifungal activity. *FEMS Microbiology Letters, 256*(1), 119–125. https://doi.org/10.1111/j.1574-6968.2006.00115.x.
144. Wu, M.-C., Struck, A.-W., Nunns, L., Sty□es, M. Q., Mick□efie□d, J., & Law, B. J. C. (2015). Engineered biosynthesis of enduracidin lipoglycopeptide antibiotics using the ramoplanin mannosyl transferase Ram29. *Microbiology, 161*, 1338–1347. https://doi.org/10.1099/mic.0.000088.
145. Wu, Y., Wang, C.-W., Wang, D., & Wei, N. (2021). A whole-cell biosensor for point-of-care detection of waterborne bacterial pathogens. *ACS Synthetic Biology, 10*, 333–344. https://doi.org/10.1021/acssynbio.0c00545.
146. Xiao, Y., Bowen, C. H., Liu, D., & Zhang, F. (2016). Exploiting nongenetic cell-to-cell variation for enhanced biosynthesis. *Nature Chemical Biology, 12*(5), 339–344. https://doi.org/10.1038/nchembio.2056.
147. Xin, F., Dong, W., Zhang, W., Ma, J., & Jiang, M. (2019). Biobutanol production from crystalline cellulose through consolidated bioprocessing. *Trends in Biotechnology, 37*(2), 167–180. https://doi.org/10.1016/j.tibtech.2018.07.010.
148. Xu, M., Wang, W., Waglechner, N., Culp, E. J., Guitor, A. K., & Wright, G. D. (2020). GPAHex—A synthetic biology platform for type IV–V glycopeptide antibiotic production and discovery. *Nature Communications, 11*(1), Article 5232. https://doi.org/10.1038/s41467-020-19058-0.
149. Xu, P., Gu, Q., Wang, W., Wong, L., Bower, A. G., Collins, C. H., & Koffas, M. A. G. (2013). Modular optimization of multi-gene pathways for fatty acids production in *E. coli. Nature Communications, 4*, Article 1409. https://doi.org/10.1038/ncomms2400.
150. Xue, C., Liu, F., Xu, M., Tang, I. C., Zhao, J., Bai, F., & Yang, S.-T. (2016). Butanol production in acetone-butanol-ethanol fermentation with in situ

product recovery by adsorption. *Bioresource Technology, 219*, 158–168. https://doi.org/10.1016/j.biortech.2016.07.041.

151. Yang, D., Jang, W. D., & Lee, S. Y. (2021). Production of carminic acid by metabolically engineered Escherichia coli. *Journal of the American Chemical Society, 143*(14), 5364–5377. https://doi.org/10.1021/jacs.0c12618.
152. Yang, D., Park, S. Y., & Lee, S. Y. (2021). Production of rainbow colorants by metabolically engineered *Escherichia coli. Advanced Science, 8*, Article 2100743. https://doi.org/10.1002/advs.202100743.
153. Yang, D., Park, S. Y., Park, Y. S., Eun, H., & Lee, S. Y. (2020). Metabolic engineering of Escherichia coli for natural product biosynthesis. *Trends in Biotechnology, 38*(7), 745–765. https://doi.org/10.1016/j.tibtech.2020.02.004.
154. Yang, H., Lu, X., Hu, J., Chen, Y., Shen, W., & Liu, L. (2018). Boosting secretion of extracellular protein by *Escherichia coli* via cell wall perturbation. *Applied and Environmental Microbiology, 84*(20), e01382-18. https://doi.org/10.1128/AEM.01382-18.
155. Ye, C., Ng, I.-S., Jing, K., & Lu, Y. (2014). Direct proteomic mapping of *Streptomyces roseosporus* NRRL 11379 with precursor and insights into daptomycin biosynthesis. *Journal of Bioscience and Bioengineering, 117*(5), 591–597. https://doi.org/10.1016/j.jbiosc.2013.10.013.
156. Yehl, K., Lemire, S., Yang, A. C., Ando, H., Mimee, M., Torres, M. D. T., et al. (2019). Engineering phage host-range and suppressing bacterial resistance through phage tai□fiber mutagenesis. *Cell, 179*(2), 459–469.e9. https://doi.org/10.1016/j.cell.2019.09.028.
157. Yim, G., Thaker, M. N., Koteva, K., & Wright, G. (2014). Glycopeptide antibiotic biosynthesis. *Journal of Antibiotics, 67*(1), 31–41. https://doi.org/10.1038/ja.2013.105.
158. Yim, G., Wang, W., Thaker, M. N., Tan, S., & Wright, G. D. (2016). How to make a glycopeptide: A synthetic biology approach to expand antibiotic chemical diversity. *ACS Infectious Diseases, 2*(9), 642–650. https://doi.org/10.1021/acsinfecdis.6b00057.
159. Yin, N., Zhu, G., Luo, Q., Liu, J., Chen, X., & Liu, L. (2020). Engineering of membrane phospholipid component enhances salt stress tolerance in *Saccharomyces cerevisiae. Biotechnology and Bioengineering, 117*(3), 710–720. https://doi.org/10.1002/bit.27241.
160. Yu, T., Zhou, Y. J., Huang, M., Liu, Q., Pereira, R., David, F., & Nielsen, J. (2018). Reprogramming yeast metabolism from alcoholic fermentation

to lipogenesis. *Cell, 174*(6), 1549–1558.e14. https://doi.org/10.1016/j.cell.2018.08.044.

161. Zhang, F. Z., Carothers, J. M., & Keasling, J. D. (2012). Design of a dynamic sensor-regulator system for production of chemicals and fuels derived from fatty acids. *Nature Biotechnology, 30*(4), 354–359. https://doi.org/10.1038/nbt.2149.
162. Zhang, F., Gao, D., Lin, J., Zhu, M., Zhuang, Z., Duan, Y., et al. (2020). Construction of inducible genetic switch for the global regulator WblA to sustain both overproduction of tiancimycins and on-demand sporulation in *Streptomyces* sp. CB03234. *ACS Synthetic Biology, 9*(6), 1460–1467. https://doi.org/10.1021/acssynbio.0c00190.
163. Zhang, J., Astorga, M. A., Gardner, J. M., Walker, M. E., Grbin, P. R., & Jiranek, V. (2018). Disruption of the cell wall integrity gene ECM33 results in improved fermentation by wine yeast. *Metabolic Engineering, 45*, 255-264. https://doi.org/10.1016/j.ymben.2017.12.007.
164. Zhang, X., Pan, Z., Fang, Q., Zheng, J., Hu, M., & Jiao, X. (2009). An auto-inducible *Escherichia coli* lysis system controlled by magnesium. *Journal of Microbiological Methods, 79*(2), 199–204. https://doi.org/10.1016/j.mimet.2009.09.005.
165. Zhao, E. M., Suek, N., Wilson, M. Z., Dine, E., Pannucci, N. L., Gitai, Z., Avalos, J. L., & Toettcher, J. E. (2019). Light-based control of metabolic flux through assembly of synthetic organelles. *Nature Chemical Biology, 15*(6), 589–597. https://doi.org/10.1038/s41589-019-0277-4.
166. Zhao, E. M., Zhang, Y., Mehl, J., Park, H., Lalwani, M. A., Toettcher, J. E., & Avalos, J. L. (2018). Optogenetic regulation of engineered cellular metabolism for microbial chemical production. *Nature, 555*(7698), 683–687. https://doi.org/10.1038/nature26141.
167. Zheng, Y., Cheng, F., Zheng, B., & Yu, H. (2020). Enhancing single-cell hyaluronic acid biosynthesis by microbial morphology engineering. *Synthetic and Systems Biotechnology, 5*(4), 316–323. https://doi.org/10.1016/j.synbio.2020.09.001.
168. Zhou, K., Qiao, K. J., Edgar, S., & Stephanopoulos, G. (2015). Distributing a metabolic pathway among a microbial consortium enhances production of natural products. *Nature Biotechnology, 33*(4), 377–383. https://doi.org/10.1038/nbt.3095.
169. Zhou, Y. J., Buijs, N. A., Zhu, Z., Gomez, D. O., Boonsombuti, A., Siewers, V., & Nielsen, J. (2016). Harnessing yeast peroxisomes for biosynthesis of fatty-acid-derived biofuels and chemicals with relieved side-pathway

competition. *Journal of the American Chemical Society, 138*(47), 15368–15377. https://doi.org/10.1021/jacs.6b09745.

170. Zhu, Q., Li, J., Ma, J., Luo, M., Wang, B., Huang, H., et al. (2012). Discovery and engineered overproduction of antimicrobial nucleoside antibiotic A201A from the deep-sea marine actinomycete *Marinactinospora thermotolerans* SCSIO 00652. *Antimicrobial Agents and Chemotherapy, 56*(1), 110–114. https://doi.org/10.1128/AAC.05276-11.

171. Zhu, Z. W., Hu, Y. T., Teixeira, P. G., Pereira, R., Chen, Y., Siewers, V., & Nielsen, J. (2020). Multidimensional engineering of *Saccharomyces cerevisiae* for efficient synthesis of medium-chain fatty acids. *Nature Catalysis, 3*(1), 64–74. https://doi.org/10.1038/s41929-019-0398-y.

INDEX

A

Abnormal growth 206
Actuators 71, 269
Acute dermatitis 142
Adoptive cell therapy (APC) 215
Agrochemicals 12
Albumin microparticles 210
Alcoholic extract 40, 41, 42, 43, 44, 45, 46, 47, 48, 49, 51, 52, 53, 54, 55, 56, 57, 58, 59, 60, 61, 62, 63
Aloe vera 47, 142
Alzheimer's disease 81
Amino acid producer 4
Amorphadiene synthase 80
Amylases 7
Amyloglucosidase 7
Anthocyanins 38, 66
Antibody target protein 207
Antifungal 38, 294
Antigen-antibody interaction 201
Anti-inflammatory 35, 38, 39, 41, 43, 44, 46, 47, 48, 49, 50, 55, 56, 58, 59, 63, 206, 210, 211, 212, 228
Anti-microbial 35
Antimicrobial proteins 191, 196
Antimicrobials 33, 280
Antiplasmodial 38
Anucleate cells 205, 206, 229
Aqueous leaves 47, 53
Area-under-curve (AUC) 171
Aromatic amino acid (AAA) 182
Artificial intelligence (AI) 87, 156
Autoimmune diseases 202, 207
Automated Recommendation Tool (ART) 179

B

Bacteria 4, 8, 17, 99, 205, 213, 230, 231, 268
Bacterial systems 213
B cell antigen 79, 90
Beer fermentation 7
Beverages fermentation 6
Bioactive proteins 205
Biochemical data 174
Biocontrol agent 10
Biodesign platform 87
Biological degradation 10
Biological pathways 78, 161
Biological signal 72
Biological systems 72, 247, 248,

254, 268
Bio-oscillators 71
Bioreactors 12, 140
Biosynthesis rate models 183
Biotransformation 8, 9, 10, 11
Blood-brain barrier (BBB) 203, 204, 211, 212
Blood circulation 209, 226
Bloodstream 204, 206, 207, 209, 230
Body system 200
Boolean logic circuit 82
Bread fermentation 4, 5
Bronchoalveolar 56, 57

C

Calvin-Benson-Bassham (CBB) 187
Cancer cells 78, 81, 82, 92, 205, 209, 218, 219, 229, 230, 237
Cancer immunotherapy 201, 215, 216, 222, 239, 241, 243
Cancer therapy 92, 199, 234, 238
Carbon monoxide 147
Cardiosphere-derived cells (CDCs) 203, 236
Carotenoids 64
Cell backpacks 203
Cell engineering 199, 200, 201, 221, 223, 240, 246, 276, 277, 278, 281
Cell factories 132, 144, 147, 151, 152, 153, 154, 245, 259, 270, 273, 281, 286, 292
Cell-free systems 70
Cell hyperplasia 49
Cell membrane 78, 201, 206, 209, 227, 228, 233, 239, 261, 276
Cell number 214
Cell surface 114, 201, 203, 206, 212, 228, 229, 254, 255
Cell therapy 199
Cellular immune response 50
Cellular level 203, 259, 262
Cellular membranes 200, 218, 222
Cellular sensors 71, 269
Celluloses 7
Cell-wall 8, 9, 10
Cheese fermentation 5, 6
Chemical antibodies 81
Chemical libraries 70
Chemica modification 201, 203
Chemical synthesis 80
Chemotherapy 80, 91, 142, 212, 213, 234, 240
Chimeric antigen receptors (CARs) 199
Chlorhexidine (CHX) 213
Chloroform fraction 62
Chlorogenic acids 143
Chlorophyll 65
Circulating tumor cells (CTCs) 209
Circulatory system 206, 211
Click chemistry 203
Clostridium perfringens 35
Cocoa fermentation 6
Coding sequences 70, 113
Combinatorial libraries 82
Compound Annual Growth Rate (CAGR) 2
Cucurbitaceae 51
Cystic fibrosis 142
Cytotoxic T lymphocyte (CTL) 47

D

Damaged cells 82
Daruharidra 47
Data-driven objective (DDO) 156
Data handling 87
Delayed type hypersensitivity (DTH) 41
Delivery of docetaxel (DTX) 202
Demonstrate biodegradability 204
Dendritic cells (DCs) 215, 216

Design-Build-Test-Learn (DBTL) 179, 268
Dexamethasone 206, 207, 232
Diabetes treatment 201, 223, 224
Dichloromethane 41, 50
Disease progression 200, 222
DNA biology 87
DNA production 49
DNA sequences 71
DNA technology 78
Docosahexaenoic acid (DHA) 142
Doxorubicin (DOX) 208
Doxycycline 72, 73, 82, 90
Drug delivery 200, 203, 204, 206, 208, 209, 212, 214, 218, 219, 220, 221, 226, 227, 228, 229, 230, 231, 232, 233, 234, 236, 238, 240, 242, 243
Drug discovery 77, 78, 79, 81, 88, 89, 92, 93, 268
Dynamic regulation system 85

E

Eco-friendly method 245
Eicosapentaenoic acid (EPA) 142
Electrical switches 72
Elevated interstitial pressures 213
Encephalomyelitis 202
Encephalomyocarditis virus (EMCV) 52
Endothelial cells 219, 227
Enhancing health 70
Entner-Doudoroff (ED) 113
Enzymatic biodegradation 10, 11
Enzyme function 170
Enzyme recognition sites (ERS) 207
Epithelial cell 48, 53
Ethionamide repressor (EthR) 78
Eukaryotic cells 70
Execution nanocarrier (NCB) 219

F

Fatty acid production 96, 110, 112, 153, 266
Fermented milks 4, 5
Fermented soybeans 4
Flavonoid 144
Fluorescein isothiocyanate (FITC) 202
Fluorescence-activated cell sorting (FACS) 202
Flux balance analysis 179
Food additive 4, 5, 6, 7, 78
Food allergy 42
Food and Drug Administration 199
Fungal species 20

G

Gene expression 62, 73, 75, 82, 85, 86, 88, 90, 91, 195, 196, 214, 215, 233, 265, 266, 270, 277, 278, 284, 292
Gene networks 78, 82, 163
Generally Recognized as Safe (GRAS) 97
Genetic circuits 71, 77, 214, 215, 253, 254, 278, 284, 286, 287
Genome editing 71, 79, 91
Genotype-to-phenotype 182
Ginsenosides 53, 143
Glaucoma 207
Global Market 132
Glucose oxidase 7
Glucose-signal amplifiers 224, 243
Glucose transporters (GLUTs) 223
Glucosinolates 66
Glutamic acid 4, 273, 288
Glycoengineering 203
Glycolipids 4
Glycolysis 179, 251
Glycoprotein accumulation 55

Glyoxylate cycle 100
Golundrina 38
G protein-coupled receptors (GPCRs) 78
Granulocyte-macrophage colony-stimulating factor (GM-CSF) 216
Green fluorescence protein (GFP) 214
Greenhouse gas 12, 99
Gycerol 3-phosphate dehydrogenases (GPDH) 113

H

Hemagglutination (HA) 50
Hematopoietic stem cells 206, 209
Hepatoma cells 53
Hepatoprotective nephroprotective 38
Herpes simplex virus (HSV) 57
High-throughput screening (HTS) 70, 75
HIV replication 43
Homoserine lactones 77, 253
Hormones 73
Human body 213, 226
Human genome 81, 160
Human microbiome 2
Hydroalcoholic 41, 43, 44, 45, 47, 48, 49, 52, 54, 55, 57, 58
Hydroxycamptothecin 213
Hydroxylation 10, 251
Hypolipidemic 38
Hypoxia 82

I

Immune responses 40, 200, 205
Immune system 40, 199, 204, 208, 209, 210, 215, 218, 227, 230, 270
Immunomodulators 215
Immunosuppressive functions 35
Induced pluripotent stem (iPS) 209, 230
Industrial production 7, 173
Inflammatory cytokines 56, 58, 60, 61, 210, 216
Irritable bowel syndrome 142

K

Kefir fermentation 6
Kimchi fermentation 5
K-nearest neighbors (KNN) 158
Knowledge-driven objective (KDO) 156

L

L-Asparaginase 207
Limonene production 180
Limonene synthase 180
Lipid markers 64
Listeria monocytogenes 35
Logical operation 71
Logic gates 71
Lymphocytes 41, 43, 46, 48, 52, 56, 83

M

Machine learning (ML) 87, 156
Macrophages 39, 41, 43, 45, 48, 51, 53, 56, 57, 58, 62, 63, 203, 204, 211, 227, 228, 229, 232, 237, 238, 244
Malic acid 4
Malolactic fermentation 5
Malonyl-CoA reductase (MCR) 99
Marburg virus 56
Mass spectrometry (MS) 156, 161
Matthews Corre□ation Coefficient (MCC) 162
Mean absolute error (MAE) 183
Meat extract 6, 7, 8

Meat fermentation 6
Medium-chain fatty acids (MCFAs) 108
Megakaryocytes (MKs) 209
Membrane engineering 201, 274
Mesenchymal stem cells (MSCs) 202, 203, 210, 211, 227, 229, 236
Mesenchymal Stem Cells (MSCs) 210
Metabolic disorder 83
Metabolic engineering 70, 71, 80, 83, 84, 86, 87, 88, 90, 91, 97, 98, 106, 108, 109, 112, 115, 116, 129, 131, 132, 143, 145, 150, 153, 154, 156, 173, 174, 179, 191, 195, 249, 265, 268, 272, 273, 277, 283, 284, 286, 287, 288, 289
Metabolic pathway 70, 81, 85, 156, 171, 173, 175, 176, 196, 246, 250, 251, 256, 266, 281, 296
Metabolite change 172
Metabolomics data 172
Methanol 41, 147
Microalgae 99
Microarray 156, 162
Microbes 2, 3, 8, 11, 15, 33, 35, 67, 70, 76, 80, 127, 129, 142, 143, 150, 151, 232, 271
Microbial cells 2, 76, 147, 248, 286
Microbial food market 33
Microbial protein 12
Microbiology 2
Microbiome market 2
Microenvironment dynamics 214
Mitotic index (MI) 41
Model system 79, 283
Molecular chaperones 81
Multicellular organisms 15
Multilayer Perceptrons (MLPs) 162
Multiple myeloma 210, 219, 235
Myeloma cells 79, 220
Myeloperoxidase 56

N

NAND-gates 71
Nanomaterials 200, 215, 216, 218, 220, 221, 240
Naringinases 7
Natural cells 200, 201, 220
Neural stem cells (NSCs) 202
Neutrophils 211, 212, 226, 227, 228, 230
Next-generation feedstocks (NGFs) 147
N-hydroxy succinimide (NHS) 204
Nitrogen conversion 12
Nonimmunogenic 220
Normoglycemic 202
Nucleic acid molecules 213

O

Ochratoxin 8, 9, 10, 11
Oral Suspension 143
Oscillatory behavior 214
Osteoarthritis (OA) 210
Osteoporosis 207
Oxygen concentration 214

P

Parkinson's disease 81
Phagocytic cells 45
Phagocytic function 50, 54, 60
Phagocytic index 49, 50, 52, 54
Pharmacokinetics 207, 226, 234, 239
Phenolic acids 64, 66
Photolithographic deposition 201
Phylum Rhodothermaeota 174
Phytestrogens 64
Plantar fasciitis 143

Plant growth 11
Plant natural products (PNPs) 152
Polyketides 80, 248, 276
Poly(lactic-co-glycolic) acid (PLGA) 203
Polyphenols 39, 64, 80, 143
Polysaccharide isolated 47
Polysaccharides, proteins 143
Porcine reproductive and respiratory syndrome virus (PRRSV) 42
Prebiotic fiber 65
Precision-recall (PR) 171
Principal Component Analysis of Proteomics (PCAP) 180
Pristinamycin-induced protein (PIP) 73
Probiotic culture 6
Production cells 106
Promoter regulation 85
Protein-protein interactions 82
Protein scaffolding 85
Protospacer-adjacent motif (PAM) 71
Psychomotor functions 143
Pyrethrin 40
Pyruvate decarboxylase 100, 135

Q

Quorum Sensing 77, 254

R

Receiver Operating Characteristic curves (ROC) 171
Red blood cells (RBCs) 200, 202, 206, 207, 208, 223, 224, 227, 229
Reductive tricarboxylic acid cycle (rTCA) 178
Release oxygen molecules 207
Reverse Engineering Gene Networks with Artificia Neura Networks 162
Ribosome binding site (RBS) 133, 187, 270
Riboswitches 71, 74, 88
Ribozymes 74, 75
RNA sequences 71

S

Salmonella pullorum 35
Secondary metabolites 70
Secreted alkaline phosphatase (SEAP) 214
Semliki forest virus (SFV) 52
Single cell protein 15
Skin homeostasis 143
Small regulatory RNAs (sRNAs) 71
Sourdough fermentation 5
Spatial organization 85
Specificity determining residues (SDRs) 170
Support vector machine (SVM) 158
Sustainable agriculture 11, 12, 68
Synthetic biology 70, 77, 78, 79, 83, 89, 90, 92, 109, 153, 195, 214, 232, 247, 248, 268, 286, 288, 289
Synthetic chemistry 70
Synthetic memory 82
Synthetic network 77
Synthetic pharmaceutical drugs 204

T

Tailor-made system 180
Target cells 202, 207
Targeted drug delivery 201, 222, 230, 235, 236, 240
Targeting tissues 204
Taxol synthesis 80
Terpenoids 38, 288
Tetracycline 72, 73, 78, 260, 265

Therapeutic delivery 205, 206, 212, 213, 232
Therapeutic molecules 205, 206, 207
Therapeutics 204, 206, 212, 213, 221, 229, 230, 235, 237, 238
Thiamine pyrophosphate 74
Thrombocytopenia 210
Thrombus 66, 210, 219, 220, 235
Thymelaeaceae 50
Time-series proteomics 172
Tissue engineering 199, 232
Tissue plasminogen activator (tPA) 219
Total oligomer flavonoids (TOF) 49
Triple-negative breast cancer (TNBC) 202
Triterpenoids 38
Tumorigenic transformation 206
Tumor microenvironment 213, 228, 234

U

Uric acid macrolides 73

V

Vinegar fermentation 4, 5
Von Willebrand factor 210

W

Waste management 15
Waste materials 15

X

Xylose isomerase (XI) 114

Z

Zearalenone 8, 9, 10, 11

FSC
www.fsc.org
MIX
Papier | Fördert gute Waldnutzung
FSC® C083411

Kolibri 360 GmbH
Ferdinand-Jühlke-Straße 7
99095 Erfurt, Deutschland
produktsicherheit@kolibri360.de